The Refracted Muse

The Refracted Muse

LITERATURE AND OPTICS IN EARLY MODERN SPAIN

Enrique García Santo-Tomás

Translated by Vincent Barletta

The University of Chicago Press CHICAGO AND LONDON

The University of Chicago Press, Chicago 60637
The University of Chicago Press, Ltd., London

Published 2017
Printed in the United States of America

26 25 24 23 22 21 20 19 18 17   1 2 3 4 5

ISBN-13: 978-0-226-37646-2 (cloth)
ISBN-13: 978-0-226-46573-9 (paper)
ISBN-13: 978-0-226-46587-6 (e-book)
DOI: 10.7208/chicago/9780226465876.001.0001

Library of Congress Cataloging-in-Publication Data
Names: García Santo-Tomás, Enrique, author. | Barletta, Vincent, translator.
Title: The refracted muse : literature and optics in early modern Spain / Enrique García Santo-Tomás ; translated by Vincent Barletta.
Other titles: Musa refractada. English
Description: Chicago : The University of Chicago Press, 2017. | Includes bibliographical references and index.
Identifiers: LCCN 2016058026 | ISBN 9780226376462 (cloth : alk. paper) | ISBN 9780226465739 (pbk. : alk. paper) | ISBN 9780226465876 (e-book)
Subjects: LCSH: Spanish fiction—Classical period, 1500–1700—History and criticism. | Literature and science—Spain—History—17th century. | Science—Spain—History—17th century.
Classification: LCC PQ6142 .G3713 2017 | DDC 863/.309—dc23
LC record available at https://lccn.loc.gov/2016058026

♾ This paper meets the requirements of ANSI/NISO Z39.48–1992 (Permanence of Paper).

Ninguna ciencia en cuanto ciencia engaña; el engaño está en quien no la sabe.
No science is deceptive in itself; the deception lies with the person
who doesn't understand it.

MIGUEL DE CERVANTES

Contents

Preface

The idea for this book originated in 2008 from a brief footnote published in a study previously unknown to me. It was titled *Boccalini in Spain: A Study of His Influence on Prose Fiction of the Seventeenth Century*, penned by Robert Haden Williams and published in 1946 through a small press in Menasha, Wisconsin. Its brevity belied an enormous amount of useful information regarding the intellectual exchanges between early modern Spain and Italy, especially the fruitful dialogue between science and satire in the seventeenth century. The footnote in question referred to the vitality of the motif of the *occhiali politici* (political lenses) in half a dozen Spanish satirists, and to learn more about this unexplored issue in Spain's cultural history piqued my interest. By analyzing how these lenses used to make political commentary came to life in the titles that were cited, I slowly began to widen the net of references, exploring the relationships between these authors, and from these first reflections I wrote an article that came out a year later in *PMLA* with the title "Fortunes of the *Occhiali Politici* in Early Modern Spain: Optics, Vision, Points of View."

I soon realized, however, that I was dealing with a phenomenon of a much broader scope, one involving several textual traditions. Boccalini was just one of the writers participating in this dialogue, and the motif of the *occhiali* was part of a much more complex system of citations, one that not only was built on the literary histories of these two nations but also touched on specific findings in disciplines like astronomy and optics. As it was a period of groundbreaking epistemological changes resulting from the so-called Scientific Revolution, I began to understand that to speak of the political lens was to speak of the matter and the properties of glass, and that its qualities inevitably took me to the field of optics, to the advances in optometry as well as to the elaboration of instruments of scientific exploration. Once I had stepped into this new territory, I felt myself immersed in the discoveries of astronomy, since many of these achievements were closely related. It was not just about expanding the field of inquiry for the sake of it: my primary sources demanded it, at times explicitly, at times in very subtle ways. When the great minds of the Spanish Baroque set out to write on the controversial *anteojo* (lens), they were choosing a term that, in its ambiguity, allowed them to reveal both capricious behaviors and delirious states—in the form of *antojo*, or "whim"—and to reflect on the power of corrective lenses, spyglasses, telescopes, and, by extension, on the act of seeing as a gesture of insolence and courage. No one embodied these traits better than the most daring and influential scientist of the time, Galileo Galilei (1564–1642).

The first steps in the elaboration of this book made me realize that my literary background was a limited tool when embarking on a project of this nature, but they also revealed to me something that I had been suspecting all along, that is, that the study of the history and philosophy of science in Spain was by no means detached from its fictional forms. Moreover, it was the existence of shared languages that fed and inspired one another. If the word *interdisciplinary* may sound a bit clichéd to some by now, the fact is that the rise of the novel in early modern Spain cannot be fully understood without taking into consideration the work carried out and published in specific areas of scientific exploration. And this dialogue is precisely one of the most promising areas of inquiry for the twenty-first-century scholar. *The Refracted Muse: Literature and Optics in Early Modern Spain* seeks to shed new light on this fruitful intersection, which remains, for the most part, unexplored.

My itinerary begins with a selection of novels published during the reign of Philip III and ends with the study of a handful of pieces that came to light at the very end of the seventeenth century. It doesn't aim to exhaust what is nothing but a rich catalog of narrative nuances and flavors, but rather seeks to identify a number of differential nodes in a very select group of testimonies, thus leaving the door open for future inquiry. Although the book's design is chronological, it tries to avoid a linear narrative, since, as we have been reminded recently, Baroque science was sometimes accidental in its progress, exposed as it was to continuous dead ends, false steps, and frequent rectifications, for either internal or external reasons.[1] This trajectory ran parallel to the mind-set of the Spaniards of the time, who were curious and open to innovation while feeling heavily constrained by the existing mechanisms of censorship. My analysis thus identifies a number of inherent tensions and contradictions within these texts, resulting from the doubts of those who authored them. If this was a time of technological achievements, it is also undeniable that these changes affected those who experienced them in their daily lives.

Many of the pieces I examine in this book must be read very cautiously, even with a reasonable amount of suspicion, in order to understand once and for all that the thesis of a "backward Spain" is reductive and inexact, given that its major writers were not as narrow minded and reactionary as they have been traditionally portrayed. There was, without a doubt, a curious attitude, but also a great deal of irony when tackling some of these (sometimes) thorny issues. Without this porosity with respect to the new on the part of the writers whom I study, this book would not exist. The muse of each of these Baroque voices is a refracted one who captures the light from abroad and projects it at new angles, achieving surprising results in genres and forms such as comedy, emblematics, sonnets, and the novel, with satire as its most successful mode of expression.

In the present book I have revisited a number of "classics" but also pieces that are fairly unknown to contemporary scholars. This includes a long poem that had never been studied before, Alonso Jerónimo de Salas Barbadillo's *Tratado poético de la esfera* (Poetic treatise on the sphere, 1609), which had been playfully embedded in a larger poetic composition of very little interest. I have also recovered the voice of a number of historians and philosophers of science reflecting on Baroque fiction, as well as that of literary critics dealing with the technicalities of scientific language. This not only has allowed me to reflect on disciplines like literature and aesthetics but also has led me to conduct extensive archival work of a philological nature in Spanish libraries. In addition, the perusal of several hundred entries of the *Nuevo diccionario histórico* (New historical dictionary) has once again put me in touch with the history of the Spanish language by surveying the diachronic changes of terms like *anteojo*—which has a wide range of meanings, from eyeglass to spyglass, and binoculars to spectacles—and *telescopio*, whose evolution across the centuries is a fascinating research topic.

The book is divided into eight chapters, preceded by an introduction; these are followed by a conclusion, a works cited, and an index. In the introduction, I establish the theoretical and methodological framework for the itinerary through the book. I also set the stage on which I will build the different networks between cultures and agents. From this historical perspective I pay close attention to the different social and scientific parameters that surrounded Galileo's work, his achievements, and his frustrations. I also include—albeit briefly, since this is not my primary interest—a summary of Galileo's tempestuous relationships with the civil and ecclesiastical authorities of his era, to provide a sense of what Spain did and did not absorb from all these polemics.

The first part of the book is titled "Writing on the firmament" and includes the single chapter "Observations." The chapter begins with the section "Galileo's telescope and the Spanish gaze," in which I examine the voyage of Galileo's telescope to the Spanish court and its arrival in Madrid's academic and scientific circles. I delve into the different diplomatic networks established between Tuscany, Rome, and Madrid, as well as into Galileo's own attempts to travel to Spain through contacts like the Spanish poet Bartolomé Leonardo de Argensola (1562–1631). I also include in this chapter a cursory survey of the history of the famous Academy of Mathematics (Academia de las Matemáticas)—later turned into the powerful Imperial College (Colegio Imperial)—paying particular attention to those figures who were pivotal in its development. Some of these scholars were on occasion honored by their students, as was the case of the famous playwright Lope de Vega (1562–1635), who wrote a number of pieces celebrating his education there. I also examine in this section the role of certain agents in the social and political arena—the Roget family in Catalonia, Venice, the Academy of Lincei (Accademia dei Lincei), and so on—that will later appear throughout the study. The second section of this chapter, "First symptoms: The 'new physics' and the treatises on optics," focuses on a number of important theoretical and practical considerations from the field of optics in Spain before zeroing in on what we understand today as the discipline of ophthalmology. I then analyze a number of passages from the most important and comprehensive treatise of its time, Benito Daza de Valdés's *Uso de los anteojos* (Use of eyeglasses, 1623). This is a text that incorporates a number of findings from Galileo's *Sidereus nuncius* (Starry messenger, 1610) without ever mentioning the name of the famous astronomer or his groundbreaking treatise.

The second part of the study is called "Galileo and his Spanish contemporaries" and includes three chapters with one section each. The first, "Foundations," concentrates on the years in which Spanish fiction begins absorbing new ideas, receiving new books, and understanding the uses of new measuring instruments as they arrived in the Iberian Peninsula. In "Science (and) fiction: Elements for a new mechanics," I examine a number of texts by Miguel de Cervantes (1547–1616), Luis de Góngora (1561–1627), Lope de Vega, Alonso Jerónimo de Salas Barbadillo (1561–1635), and Tirso de Molina (1579–1648). I highlight the different tensions that arise from a personal and from an institutional point of view when these writers—some of them members of the church and educated under a Ptolemaic vision of the cosmos—incorporate new ideas coming from the writings of Johannes Kepler, Tycho Brahe, and Nicolaus Copernicus, among others. In "Assimilations," I write on what I call the "Italian influence and the culture of knowledge" by choosing two pieces that are seminal to our understanding of why some optical tools reached such levels of controversy: the polemical anti-Spanish invective *Ragguagli di Parnaso* (Newsletter from Parnassus, 1612) by Trajano Boccalini (1556–1613)—a piece that was read by many Spaniards in its original language—and Tomaso Garzoni's *La piazza universale di tutte le professioni del mondo* (The universal plaza of all the world's professions, 1585) translated quite loosely by Cristóbal Suárez de Figueroa (1571–1644) as *Plaza universal de todas ciencias y artes* (The universal plaza of all sciences and arts, 1615). These two titles in particular help us understand why certain fictional scenarios—the city, the market, the optical shop, Mount Parnassus—allowed for the critique of the eyeglass as a symbol of vanity and, by extension, of a society in complete decay. The third chapter in this part, "Inscriptions," focuses on how this sharing of ideas facilitated the propagation of a new model of scientific endeavor in Spain that was influenced by the prestige of the Italian *accademia*. In "Visible intermittence: The voyage of the secret and the creation of the virtuoso," I argue that a figure like the famous collector Juan de Espina (ca. 1563–1642) is relevant to our understanding of how specific forms of scientific dissemination, and in particular those stemming from the *nuova scienza,* succeeded in Spain. Espina was a Spanish virtuoso, whose life and deeds intrigued and fascinated many of his neighbors, and whose network of contacts went all the way up to the monarchs. The mystery and appeal of his famous house, in which he stored a Galilean telescope, resulted in a number of literary tributes by the most illustrious writers of his time, including Alonso de Castillo Solórzano (1584–1647), Anastasio Pantaleón de Ribera (1580–1629), Juan de Piña (1566–1643), and Luis Vélez de Guevara (1579–1644).

"The science of satire" is the third part of the book, and it contains two very different chapters. The first, "Situations," examines a number of imaginary places—although in many cases one can identify specific locations—in the most important fictional accounts of the first thirty years of the seventeenth century. The chapter's two sections ("The city's refracted space" and "Watchtowers, visions, horizons") can be read as a long sequence in which to establish a number of parameters for the study of urban spaces in the Spanish Baroque. This approach helps me explain how a number of specific pieces captured the anxieties and concerns of the time. I analyze two satires by Rodrigo Fernández de Ribera (1579–1631) and Antonio Enríquez Gómez (1601–1661), which offer a fascinating glimpse of the city from an allegorical watchtower, to acquire a more precise and unbiased view of their surroundings. "Explorations" gives an account of a slightly different perspective, that of the aerial voyage, as it reveals the impact of Galileo's oeuvre on the writers who chose this narrative device. In "The social critique in the universe of glass," I write about the novel that most comprehensively contributed to this dialogue, Luis Vélez de Guevara's *El diablo Cojuelo* (The limping devil, 1641), which provides the first direct mention in Spanish of the famous astronomer. Finally, in "Dream/vigil: Moons, moles, and lunatics in the poetry of the Baroque" I traverse the literary skies with Juan Enríquez de Zúñiga (ca. 1580–1642) and Anastasio Pantaleón de Ribera in their particular conflation between the Ptolemaic and the Copernican, mixing the old and the new in a bitter denunciation of contemporary mores.

"The refracted muse," the fourth and last part of the book, is divided into two very different chapters. It covers the middle and final decades of the century, in which one can perceive a more obvious tension between old and new cosmogonies, at a time in which the use of glass in personal accessories triggers a number of satires that expose the weakness and "effeminacy" of the midcentury courtesan as symptomatic of the demise of the nation. In "Interventions," I reflect on the impact of the spyglass and the telescope in two political satires: "The political intervention I: The transatlantic prism" deals with the sophisticated view of the spyglass in the chapter "Los holandeses en Chile" (The Dutchmen in Chile) included in the satire *La Hora de todos y la Fortuna con seso* (The Hour of all and Fortune with sense, 1650) by Francisco de Quevedo (1580–1645)—a writer who may have met Galileo in Rome in 1616, and who portrays himself as a lynx in his treatise to Philip IV *El lince de Italia u zahorí español* (The lynx of Italy or the Spanish diviner, 1628). In "The political intervention II: The transalpine prism," I study an emblem, *empresa* 7 from Diego de Saavedra Fajardo's *Empresas políticas* (Political advice, 1640). With the motto *auget et minuit* (waxes and wanes) and a telescope as the *pictura*, or image, the Spanish moralist offers a fascinating meditation on the limits and abuses of absolutist power. Finally, in "Reverberations" I revisit a number of pieces published at the end of the century in which the understanding of new cosmology is still met with traces of doubt, and even resistance. In "Foreign muses, local verses," I open the discussion with the playwright Calderón de la Barca (1600–1681)—for some, a full-fledged Copernican[2]—and follow with two poets based in Northern Europe, Bernardino de Rebolledo (1597–1676) and Miguel de Barrios (*né* Daniel Leví, 1635–1701), whose work combines the modern tastes of academicism (from Copenhagen and Amsterdam, respectively) with the traditional flavor of Castilian poetics. These are, in Spain, the decadent years of Charles II (1665–1700), and the fatigue and disillusion so typical of the late Baroque can be felt in the creation of new scenarios like the bazaar, where one can purchase all kinds of eyeglasses that allow for a disenchanted examination of the present. In "Strained vision: The eyewear shop" I close this itinerary with two important moralists, the soldier and professor of mathematics Andrés Dávila y Heredia (fl. 1674) and the novelist Francisco Santos (1623–1698), who were very popular in their time, and who prove extremely useful when it comes to understanding how optics was used to comment on contemporary mores at the sunset of this kaleidoscopic century. We have reached, in fact, the so-called *periodo novator*, a period of renewal and innovation coinciding with the turn of the century that witnesses the first uses of a word that will become part of the private domain in the eighteenth century: *telescopio*. As I argue at the end of this section, the telescope as a familiar object to possess and share with others will be the topic of revealing meditations by the most important *ilustrados* of the time, from Benito Jerónimo Feijóo (1676–1764) to Martín Martínez (1684–1734), Diego de Torres Villarroel (1693–1770), and Gaspar Melchor de Jovellanos (1744–1811).

With the brief chapter "Conclusions," manifested in certain "lights, shadows, eclipses" of Spain's literary history, I end *The Refracted Muse*. My aim in these pages is to give some coherence to the previous chapters while offering new venues of study through a number of questions that remain open for discussion. These questions, I argue, not only are posed to the literary critic but also pertain to the domains of the history and philosophy of science, given that it is from these three territories that I began my analysis.

This book was written at the University of Michigan's Harlan Hatcher Graduate Library, and at Stanford University's Cecil H. Green Library, where I spent a sabbatical year. I am thankful to the staff at both institutions. A section of chapter 3 was previously published as "Visiting the virtuoso in early modern Spain: The case of Juan de Espina," in *Journal of Spanish Cultural Studies* 13, no. 2 (2012): 127–42; an abridged version of chapter 7 was published under the title "Saavedra Fajardo en la encrucijada de la ciencia," in *Crítica hispánica* 32, no. 2 (2010): 82–102. I thank both editors for allowing me to reproduce their contents here.

I have had the privilege to be invited to present parts of this book. I thank my generous colleagues Mercedes Alcalá-Galán and Steven Hutchinson (University of Wisconsin, Madison), Frederick A. de Armas (University of Chicago), María Mercedes Carrión (Emory University), María del Pilar Chouza Calo (Central Michigan University), Robert Davidson (University of Toronto), Barbara Fuchs (University of California, Los Angeles), Esther Gómez-Sierra (University of Manchester), Carlos Gutiérrez (University of Cincinnati), Rebecca Haidt (Ohio State University), Carmen Hsu (University of North Carolina, Chapel Hill), Donald S. Lopez Jr. (Michigan Society of Fellows, University of Michigan), Luce López Baralt (Universidad de Puerto Rico, Río Piedras), Joan Ramon Resina (Stanford University), Veronika Ryjik (Franklin & Marshall College), Germán Vega García-Luengos (Universidad de Valladolid), and Julio Vélez Sáinz (Universidad Complutense). I am also thankful to Jordi Aladro, Josiah Blackmore, Marina S. Brownlee, Frank P. Casa, Alison Cornish, Antonio Cortijo Ocaña, Edward H. Friedman, Luciano García Lorenzo, Guillermo Gómez Sánchez-Ferrer, Roland Greene, George Hoffmann, José Pardo Tomás, William Paulson, Randolph Pope and María Inés Lagos de Pope, Antonio Sánchez Jiménez, John D. Slater, and Ryan Szpiech for their sustained dialogue throughout the years. I would also like to thank the anonymous reviewers for their helpful comments and suggestions. Finally, I want to express my deepest gratitude to my editor Karen Merikangas Darling, as well as to all those involved in the production of the book.

I have been very fortunate to work with Vincent Barletta (Stanford University), who carried out the translation of this book with exemplary patience, rigor, and imagination.

I hope this book will be of interest not only to the literary critic but also to all those intrigued by the dialogue between the worlds of fiction and the realities of scientific practice. After all, that is how Galileo wanted it to be.

Palo Alto, California
September 2015

Introduction

Ignorato motu ignoratur natura.
If motion is not known, then neither is nature.
THOMAS AQUINAS

Parameters, panoramas

The present book explores the impact that advances in optics from the seventeenth-century Scientific Revolution had in Baroque Spain.[1] In more concrete terms, it focuses on the literary universe of this period, paying special attention to those texts and authors who incorporated references to the applications of glass in the field of astronomy during its gradual transition to a Copernican framework from a Ptolemaic one. Through this analytical focus, I work to connect two apparently distant forms of discourse—science and literature—demonstrating how, despite their misleading separation or apparent respective autonomy, there reverberated in the minds of key Spanish authors an anxiety about optics that became more pronounced as the century wore on. This book proposes, for example, that many of the allusions to Aristotelian cosmology were nothing but a traditional reaction to what already seemed inevitable, that is, the recognition of a new heliocentric vision of the cosmos. It also attends to the many shifts that took place during a long century that witnessed the gradual acceptance and integration into scientific texts of a much more correct and precise idea of the universe. Within this temporal framework—during which one understandably finds numerous readings situated halfway between the old and the new, for the most part shaped by ignorance, fear, or simply prudence—my analysis traces out the historical trajectory of two phenomena that intermittently unite and come apart throughout the seventeenth century: on the one hand, the reception in Spain of Galileo Galilei's scientific work, conditioned not only by religious, political, and even economic factors but also by the contingencies of all such semiclandestine acts of *translatio*; and on the other hand, the evolution of the *occhiali politici* literary motif from its original treatment in the Menippean satire of the day.[2] If the first of these issues already boasts an expansive and deeply interesting critical bibliography within the history of science, the second continues to be (and here I employ an image popular among seventeenth-century Castilian writers) in a state of *mantillas* (swaddling clothes)—that is, in its infancy.[3] A constant preoccupation with the faculty of vision played a central role in both cases, and this generated meditations, very typical of the period, on the ethical foundations of visual perspective and the correct interpretation of reality, once the veil of appearances had been lifted.[4] These two lines of inquiry—one philosophic and scientific, and the other social and literary—do not seem in principle to have much in common, and yet they emerge as inseparable once we take into account the discrete (but almost always resistant) threads that connected what ended up as a dual expectation: if Spain was Galileo's desired destination, given the enormous attraction that Spanish culture held for many Europeans,[5] then Spain's New World discoveries likewise sparked the interest of many thinkers of the period.[6] To write on the visual perception and interpretation of objects made no sense without reflecting on that which had begun to be questioned thanks to advances in optics: theory had no justification without practice—or, perhaps vice versa, the putting into practice of certain concepts thanks to narrative language lacked meaning without recognition of their empirical foundation. In fact, in a Spain supposedly characterized at the time by forms of belatedness and isolation that continue to be matters of academic debate and critical revision, the work of Galileo was known by a wide range of Spanish readers who incorporated it into many of their most important creations; and whether those readers praised or criticized Galileo, they almost always wrote with a hesitancy provoked by that which was radically new. This novelty caused early modern writers, in many cases, to write with extreme care and to engage in a kind of rhetorical and thematic contortionism. They did this to navigate successfully the minefield of an institutional censorship that, as evidenced in the front matter of many volumes, could be asphyxiating—an asphyxia, I would suggest, that often enough today has us taking texts to be much less audacious than they otherwise might have been. Nevertheless, this censorship did not prevent the flowering (as would continue with modern readings) of a certain preoccupation or curiosity caused by the sense that certain dogmas were beginning to stumble, or that some formerly rejected ideas were circulating more or less freely throughout Europe. Just such a translation constituted for Galileo an invitation—or even a push—to participate in Spanish cultural life, to *live* in Spain without actually living in it, as had occurred in other European countries where his work was also read, translated, and discussed.

The general consensus of modern criticism with respect to the course of inquiry adopted by the present book is that we must delimit three differentiated chronological phases: the first is that defined by a continuation of Renaissance science; this is followed by a second phase, coincident with the middle decades of seventeenth century, during which dispersed elements of the new physics were assumed without abandoning earlier ones; the third phase is associated with the arrival of the *novatores* around 1680 and involves an abandonment of the earlier approach to integrate the new scientific and philosophical currents (Navarro Brotons and Eamon 36). The present book is situated in the second phase, which corresponds more or less neatly with the reign of Philip IV (1621–1665), even if I do refer to texts produced immediately before and after this period when they enrich our understanding of the specific matter at hand. These are years that witnessed significant changes that should be seen not as isolated events but rather as part of a prolonged development in the science of optics over a broader time frame, and for this reason I make mention of texts and authors that correspond to the earlier reigns of Philip II and Philip III, as well as to that of Charles II, which saw many significant changes, even if the monarch was himself very weak. I do not intend to establish an ordered evolution of ideas and concepts, since the historical reality points to something quite different, that is, to a general back-and-forth, an oscillation between the ancient and the modern, between the familiar and the unknown, between the inherited and the new. This is a phenomenon, it is worth remembering, that manifested itself as much in the field of literature as in the intellectual evolution of its most notable authors. The tension between old and new, which finds expression in many of the most important texts of the time, is perhaps that which stands out most during this second, intermediate phase, which without a doubt becomes the most interesting aspect of this sequence of irreversible transformations. Over more than a century, in fact, a simple object such as the telescope was used as a metaphor by the most cosmopolitan authors to denounce some of Spain's worst defects. In Miguel de Cervantes's novella *La gitanilla* (The gypsy girl, 1613), for example, it is an allusion to jealousy: "jealous people always see things as through a telescope, which makes small things large, turns dwarves into giants, and suspicions truth" (*siempre miran los celosos con antojos de allende, que hacen las cosas pequeñas grandes, los enanos gigantes, y las sospechas verdades*). A less acclaimed intellectual such as Juan Eusebio Nieremberg (1595–1648) likewise makes reference to optics in his commentary on pride in *Obras y días: Manual de señores y príncipes* (Works and days: Manual for lords and princes, 1629): "and self-love always paints that which is ours with the best colors, and in such a way that they falsely make them larger than they really are . . . much as if one were to see something through a telescope that represents things larger" (*el amor propio siempre nos pinta nuestras cosas mejores, y a tal luz que se mienten mayores de lo que son,* [. . .] *al modo que si alguno viese alguna cosa por unos anteojos que supiese que representan las cosas mayores*). In his *Sueños morales, visiones y visitas con don Francisco de Quevedo* (Moral dreams, visions and visits with Francisco de Quevedo, 1727–1728), Diego de Torres Villarroel uses the telescope to lament the ignorance and impetuousness of the general public: "the common folk have always been an irrational judge of all things, . . . and with no head whatsoever they look through the spyglass of their apprehension, with no knowledge of ultimate differences and without the prolixity of examination" (*siempre el vulgo fue arbitrio irracional de todas las cosas,* [. . .] *y sin tener cabeza alguna mira por los anteojos de su aprehensión, sin conocer las últimas diferencias y sin la prolijidad del examen*).[7] At the center of the reception of scientific novelty, more than the theory or the complex language of numbers, is the measuring instrument: a silent actor of the study and the laboratory (but eloquent in its very presence), a material object often idolized and fetishized, a *mechanics of the new*.[8] The present book is thus about ideas and language; but it is also, above all, about the fascination that the object itself produces and the new way of looking that, as the three examples here reveal, an interplay of lenses offers us.

The history of science in the Iberian Peninsula would never have existed as a field of inquiry had there not been an intense dialogue among the three cultures (Christian, Jewish, and Muslim) that defined its medieval period. Serious analyses of optics took place primarily in the Islamic East after the fall of the Roman Empire, and until the late Middle Ages there emerged no important European contribution to the field. The Muslim astrologer Albumazar (787–886) was enormously influential in the West, and his popularity in the Iberian Peninsula stemmed principally from the famous treatise *Introductorium in astronomiam* (Introduction to astronomy, 849–850), which was perhaps "the most widely read of astronomical treatises in Spain" during the period (Halstead, "Attitude of Tirso" 425);

with respect to optics, the most important text was Alhazen's well-known *Kitāb al-manāẓir* (Book of optics, ca. 1030).[9] Having benefited from a technological and scientific tradition the most notable achievements of which can be attributed to the Muslim presence in the Iberian Peninsula and the legacy of Alfonso X, the Wise (r. 1252–1284), Spain experienced sustained development throughout the fifteenth and sixteenth centuries in fields such as mathematics (applied also to navigation with great success), engineering, mining, medicine, and geography. These flourished in urban centers as noninstitutionalized practices and related texts found their way, albeit in small numbers, into Renaissance libraries. Astronomy was one of the most cultivated fields during this period, in spite of continuous pressure from the church, which became all the more acute during the third and fourth decade of the sixteenth century with the existence in the Iberian Peninsula of harshly persecuted religious practices. In the wake of the late sixteenth-century work of Tycho Brahe and his disciple Johannes Kepler in the Czech observatory of Bénakty nad Jizerou, the exploration of new fields of vision, principally under the rubric of either science or fiction, was followed very closely by the Inquisition and by the Sacred Congregation of the Index (Sagrada Congregación del Índice, dedicated to the revision and censorship of books) at the turn of the seventeenth century.[10] Unforgettable judgments and condemnations by both institutions would weigh heavily on the collective consciousness, such as the cases of Miguel Servet in Geneva (1553), Giordano Bruno in Rome (1600), and, years later, the famous witch trial and five-year imprisonment of Katherine Kepler for circulating the famous *Somnium* (Dream, 1634) signed by her son. It is not surprising that in a country progressively closing its borders under the isolationist measures of Philip II there should be a reduction in the appearance of foreign titles, especially after 1559, when the inquisitorial indices of prohibited and expurgated books grew exponentially.[11] Juan Pimentel ("La monarquía" 55) has gone so far as to claim that Spanish science was at this time "a failure dotted with periodic achievements by those who bothered to throw a stone or two against an impossibly heavy and obsolete edifice; an account of how and with what difficulty advancement reached the Peninsula, generally late, from outside and from above" (*un fracaso salpicado de logros puntuales, un recuento de quienes se molestaron en lanzar alguna piedra contra un edificio demasiado pesado, demasiado obsoleto. Un repaso de cómo y con cuántas dificultades las luces llegaban a la península, generalmente tarde, desde fuera y desde arriba*).

It would be a mistake, however, simply to dismiss all branches of Spanish science during the period. As early as 1550, Cristóbal de Villalón, for example, called for a universal and modern erudition that might stand in contrast to "the decadence of the scientific formation of the period, making it clear that the real state enters into conflict with the ideal state of utopian thought" (*a la decadencia de la formación científica de la época, aclarando que el Estado real entra en conflicto con el Estado ideal del pensamiento utópico*) (Schmelzer 202). In seventeenth-century Spain, astronomy was, after medicine, the scientific discipline that boasted the largest number of publications (López Piñero, *Ciencia y técnica* 122).[12] Its development had followed during the previous century two very different lines of investigation: cosmography, understood as the theoretical knowledge of the universe, and "prognostics," which connected astronomy to the calendar and with a subjective interpretation of individual destiny. Prognostic astronomy, also denominated by some as "judiciary astrology," was finally condemned by Pope Sixtus V in his 1585 bull *Coeli et terrae* (Of heaven and earth, published in Spain in 1612), and it was commonly criticized as antiscientific and capricious (*antojadiza*) by the writers of the time. Literature of the period in fact frequently equated astrology with other marginal or heterodox figures of the urban landscape, such as witches, prostitutes, and gypsies. Such a reading of the cosmos was defined as "a false and superstitious art" (*arte falaz y supersticiosa*) insofar as it turned prognostics into a determinist interpretation, denying the doctrine of free will and ignoring "the coincidental nature of many natural events" (*la azarosidad de muchos sucesos naturales*) (López Piñero, et al., 209). Cardinal Gaspar Quiroga, who became inquisitor general in 1573, was particular severe with judiciary astrology, which he condemned in his 1583 *Index* of prohibited books.[13] Despite all the distaste that it provoked, this form of astronomical interpretation ended up being as important as its more serious counterpart, in that it became a symbol for all pseudoscientific practices attacked as much by serious scholars as by less learned enemies of superstition and opportunism. It was, beyond this, one of the most common targets for satirists during the seventeenth century. Through such attacks, these writers achieved the double objective of constructing a mordant personal joke while keeping in place, on another level, a didactic message that warned of the consequences of such tricks. It was therefore easier and more profitable to speak of "false science" than of serious science, which was much more complex and difficult to understand and, perhaps because of this, much more resistant to scorn. We should not be surprised, therefore, that nearly all of the texts analyzed in the present book have their origin in cities that also were the most important and dynamic centers of the Iberian Peninsula for the cultivation of astronomy; that is, Seville's House of Trade (Casa de Contratación), the University of Valencia, and Madrid's Academy of Mathematics (Academia de las Matemáticas), which had managed to unite the most illustrious cosmographers of the last third of the sixteenth century. These three institutions, together with certain literary academies and universities such as Salamanca or specific loci of scientific exploration in cities like Barcelona, were instrumental for the development of science in the Iberian Peninsula. Pasqual Mas i Usó has emphasized "the participation of women (who appear in the Valencian academies only as spectators), the dressing up of participants as shepherds in the Arcadian Academy in Rome (1690, a practice repeated in the Valencian Academy in 1705), and the work in the physical sciences by the Roman Academy of Lincei (an academy to which Galileo belonged and that would influence the Valencian Academy at the end of the seventeenth century, etc.)" (*la participación de mujeres (que sólo aparecen en las academias valencianas como espectadoras), o el disfrazarse de pastores como en la Accademia degli Arcadi (1690) en Roma (que se repite en la Academia Valenciana de 1705), o el tratar de ciencias físicas como en la romana de los Lincei (1603) a la que perteneció Galileo (que influye en las academias valencianas de fines del* [*siglo*] *XVII), etc.*) (52). This atmosphere of curiosity and exploration made Spain, together with England, the only country in Europe to promote—albeit with a series of limitations that we will soon consider—the theses of Copernicus.[14] A series of parallel events in the sphere of technology and science had also taken place and must be taken into account; for example, it was during the reign of Philip II that the reform of the old Julian calendar (named in honor of Julius Caesar, who had promoted it) took place—a crucial event for the correct computation of time and one tightly linked to astronomy, with very obvious social repercussions.[15] There was no opposition to heliocentrism from any institution at the time, and as such, this was used by Pope Gregory XIII to reform the calendar in 1582. The new Gregorian calendar was therefore based on the work of Copernicus, and throughout Europe the church actively promoted it. Spain was by this time already characterized by a ferment of progress that had taken root decades before: under Cardinal Gaspar Quiroga, Copernicus became recommended reading in the 1561 statutes of the University of Salamanca. This was perhaps because of the presence of Juan Aguilera, who held the chair in astronomy at that institution from 1551 to 1560, and it is extraordinary in itself when we take into account the fact that other universities, such as those of Zurich (1553), the Sorbonne (1576), and Tübingen (1582), had prohibited the teaching of heliocentrism. In the so-called Covarrubias Reform (1559), mathematics and astrology at Salamanca were divided into three years of study; the first year dealt with astrology; the second with Euclid, Ptolemy, or Copernicus *ad vota audientium,* that is, by student (not professorial) demand; and geography was finally introduced in the third year. In the university statutes of 1595, Copernicus's *On the Revolutions of the Heavenly Spheres* (1543) and Erasmus Reinhold's *Prutenic Tables* (1551) became required reading. The *Prutenic Tables* were based on Copernicus's published work, and they quickly began to rival their model in importance; they were read during the second year along with Ptolemy's second-century *Almagest* and the *Tablas alfonsíes* (Alphonsine tables, first printed in 1483).[16] It is in this faculty of astrology, the first in Spain and the most important faculty at the University of Salamanca, that *Don Quixote*'s ill-fated shepherd Grisóstomo had supposedly studied, as we are told after his suicide: "he knew the science of the stars, and that which happens in the heavens, the sun and the moon" (*sabía la ciencia de las estrellas, y de lo que pasan allá en el cielo, el sol y la luna*).[17] Spanish theologians should receive credit for having created an environment of such tolerance and comprehension while at the same time in Paris the celebrated Petrus Ramus was expelled from the Collège de France for similar reasons. In the end, he was murdered by his own colleagues—or at least at their instigation—during the St. Bartholomew's Day massacre in 1572.[18] While it is not known whether these Copernican texts were actually used as didactic material in the classrooms of the University of Salamanca, it remains clear that there was established there, at the very least, a relatively open and flexible plan of study that favored an approach to that which, in a very schematic way, could be considered "modern science":

> In the Faculty of Astrology, over eight months during the first year will be read the Sphere and Theory of Planets and some Tables; or as a substitution, the Astrolabe. During the second year, the six books of Euclid and Arithmetic, up to square and cubic roots, and the *Almagest* of Ptolemy or the Epitome of Monte Regio, or Geber or Copernicus, according to the desire of the students; as a substitution, the Sphere. During the third year will be read Cosmography or Geography; an introduction to judiciary practices and perspectives, or an instrument, according to the will of the students: as a substitution, whatever seems best to the professor, subject to the approval of the rector.

> *En la Cátedra de Astrología, el primer año se lea en los ocho Esfera y Teóricas de planetas y unas tablas; en la sustitución, Astrolabio. El segundo año, seis libros de Euclides y Aritmética, hasta las raíces cuadradas y cúbicas y el* Almagesto *de Ptolomeo o su epítome de Monte Regio, o Geber o Copérnico, al voto de los oyentes; en la sustitución, la Esfera. El tercero año Cosmografía o Geo-*

> *grafía; un introductorio de judiciaria y perspectiva, o un instrumento, al voto de los oyentes; en la sustitución, lo que paresciere al catedrático comunicado con el Rector.* (López Piñero, et al., 86)

The Copernican vision, as is already known, enjoyed in Spain a series of fundamental supports in texts such as Fray Diego de Zúñiga's *In Job commentaria* (Commentary on Job, 1585); Andrés García de Céspedes's *Theoria de los planetas según la doctrina de Copernico* (Theory of the planets according to Copernican doctrine, 1606); the *Doctrina general repartida por capítulos de los eclipses de sol y luna* (General doctrine divided into chapters on the eclipses of the sun and moon), authored by the Mexican mathematician and astronomer Diego Rodríguez (1596–1668); as well as the work of Pablo de Alea.[19] Zúñiga, in particular, argued that the Holy Scriptures were not opposed to the movement of the earth. In glossing Job 9:6 ("He shakes the earth from its place and makes its pillars tremble") and Ecclesiastes 1:4 ("Generations come and generations go, but the earth remains forever"), he defended two theses: that the movement of the earth and the heliocentric system of Copernicus did not contradict the Bible, and that the Copernican system was superior to the traditional one from an astronomical point of view.[20] Domingo Natal Álvarez reminds us that Galileo himself, who had studied in the Roman College with the Spanish Jesuits Benito Perera and Francisco de Toledo, cited Zúñiga more than once in his writings. Only after the formal condemnation of Copernicus, whose work was included—together with that of Zúñiga—in the 1616 *Index* of the Sacred Congregation, did religious authorities in the Iberian Peninsula begin to deal with the situation. Even so, with regard to the teaching of astronomy, the 1625 constitutions of the University of Salamanca continued reproducing point for point the statutes of 1595, repeating: "during the second quadrennium [students] read Nicholas Copernicus" (*el Segundo cuadrenio léase a Nicolás Copérnico*). This attitude nevertheless was eclipsed little by little by a much more cautious posture adopted in the face of new advances that were then taking place in Europe. López Piñero (*Ciencia y técnica* 145) has pointed out various factors that contributed to this situation. In the first place, he suggests that the isolation of the Counter-Reformation, which increased progressively up to the last third of the seventeenth century, was not so much the consequence of repressive measures but rather

> the manifestation of a process that affected Spanish society as a whole . . . the growing incapacity for integration of minority groups, adversities of structures and economics, the regressive change of mentality of politi-

cians in charge, the social weight of religious fanaticism and the retreat of secularization.

> *la manifestación de un proceso que afectó a la sociedad española en su conjunto* [...] *la creciente incapacidad de integración de las minorías, las adversidades de las estructuras y de la coyuntura económica, el cambio regresivo de la mentalidad de los grupos políticos dirigentes, la vigencia social del fanatismo religioso y retroceso de la secularización.*

López Piñero also mentions, in more concrete terms, the expulsion of the Jews, which he considers intimately connected to "the ruin of the earliest Peninsular bourgeoisie and the dissolution of its group mentality, a fact that we must keep very much in mind if we are to understand why in Spain there were not the same conditions that led to the scientific revolutions that occurred in the rest of Europe" (*la ruina de la primera burguesía peninsular y la disolución de su mentalidad de grupo, hay que tenerlo muy en cuenta para entender por qué en España no se dieron las condiciones que en el resto de la Europa occidental condujeron a la Revolución científica*) (*Ciencia y técnica* 77). Together with this, the trials of three public figures—Juan Piquer, who had studied at the University of Valencia and was a disciple in Naples of the controversial Giambattista della Porta; the Valencian priest Juan Ramírez; and Luis Rosicler, an embroiderer trained in the Academy of Mathematics in Madrid (and brother-in-law to Lope de Vega)—did little to help the national mood. Some of these figures are of significant relevance to the transmission of the image of the scientist as a literary archetype, and I return to them soon.

As the preceding examples demonstrate, it is necessary to revise somewhat the popular notion of Spain as a country relegated to backwardness, to the caboose of the train. Within this complex process of reception, the isolationist inertia of Philip II was not the only factor that undercut the intellectual influence of Galileo in Spain; also significant were the many other obstacles that after 1612 managed to marginalize the Florentine astronomer's work and transform him into a paradigm for the conflictive relations between science and the Roman Curia.[21] But not everything was so simple in a country that offered much more than closure and censorship. There were also points of innovative and rigorous scientific activity, and the very existence of curiosity in the fields of astronomy and cosmography for figures such as Copernicus or Galileo demonstrates that the supposed turn-of-the-century clampdown was neither complete nor uniformly felt. According to María Portuondo:

> Judged solely by their reliance on empirical methods and use of mathematics as a tool for achieving utilitarian results, Spanish cosmographers fall squarely within the tradition of Italian and English mathematical practitioners considered by some to be the first exponents of a nascent Scientific Revolution. Perhaps their closest counterparts in continental Europe were mathematical practitioners, along the lines of Niccolò Tartaglia (1500–1557), Simon Stevin (1548–1620), John Dee (1527–1608), and of course Galileo. (21)

Take, for example, a late sixteenth-century humanist such as Pedro Simón Abril, a supremely pragmatic figure, who in his mathematical treatise *Apuntamientos de cómo se deben reformar las doctrinas y la manera de enseñarlas* (Notes on how to reform doctrines and how to teach them, 1589) praised the transformative character of Copernicus, who "changes the situation" (*trueca la suerte*), and for whom calculations "come out right" (*sálenle bien*) (qtd. in Pardo Tomás, *Un lugar* 49); or the case of Diego Pérez de Mesa, who in *Comentarios de sphera* (Commentary on spheres, 1596) asserts that the earth does, perhaps move. Víctor Navarro Brotons and William Eamon thus are quite right when they assert:

> Vestiges of the Black Legend continue to perpetuate the stereotype of sixteenth-century Spain as fanatical and Inquisitorial, and as an enemy of progress and innovation. Yet it is not without significance that it was the Inquisition in Rome, not Spain, that prosecuted Europe's leading Copernican, Galileo, while the major Spanish defender of the Copernican doctrine, Diego de Zúñiga, was allowed to publish his opinions freely, without real threat of prosecution. (32)[22]

We already know that the intellectual and political history of Spain and Italy is characterized by a very volatile landscape, extremely sensitive to any threat to both religious and lay doctrines of power. If Rome was at this time a boiling pot of polemics between different sectors of the church (including, as we will see, the influential role of the Jesuits), Spain for its part found itself stuck at a crossroads that was by any standards paradoxical: it allowed the reformist winds of the scientific revolution to pass it by, even as it tried, through a select number of institutions, to connect itself to Europe. As a result, and as has occurred on not a few occasions throughout history, knowledge of Galileo's work was assimilated in a more or less Manichaean fashion when not incomplete, distorted, or simply erroneous: Galileo himself was generally seen as the tireless observer of the lunar surface, the discoverer of constellations, and/or the anticlerical rebel who gave the popes so many headaches, and yet his own treatises reveal a scientist at times hesitant, not completely free of Aristotelian and Ptolemaic influence (especially in his early writing), and sometimes even contradictory.[23] He was always, in any case, constantly evolving—and always polemical.

I deal with Galileo in more detail later in this book; however, there exists a series of vital patterns that merit emphasis here. Galileo was born a year after the closure of the Council of Trent (1563), a council through which the church would become increasingly bureaucratic, largely to support papal designs with respect to the Catholic offensive against Protestantism. The institution of the *Index of Prohibited Books* (1559) under Pope Paul IV carried with it serious consequences over the medium and long term for Spanish libraries. Paul IV's successor, Pius V, would in fact become an instrument of active repression and censorship. As we will see, those Spaniards residing in Italy were not untouched, as their own texts demonstrate, by the turbulent scenario of achievements and polemics that were then taking place. The most important advancements of the moment situated figures such as Cristóbal Suárez de Figueroa, Francisco de Quevedo, and the Murcian diplomat Diego de Saavedra Fajardo at a veritable intersection of Ptolemaic and Copernican thought, placed the scientific endeavor at the center of political intrigues, and converted tools of study into metonymies of power. As Paolo Rossi has reminded us (74), Galileo himself was part of this discourse during the first phase of his career: although his *Treatise on the Spheres; or, Cosmography* (1597) was essentially a geocentric letter dedicated to his students, in a letter to Kepler written in the same year, he confessed that he had converted to Copernicanism even if he had not dared to publish his discoveries for fear of what had occurred to his master. Upon observing these "new heavens," assisted by recent advances in optics, Johannes Kepler published his seminal *Paralipomena* (1604), and shortly afterward, Galileo began his observations of the Pleiades, Orion, and the Milky Way, also identifying the three—later converted into four—moons of Jupiter. He named these new stars (Io, Europa, Ganymede, and Callisto) *Mediceas* in honor of the family of Grand Duke Cosimo II de' Medici, through whose support he was able to compose what would later become the fifty-eight-page pamphlet *Sidereus nuncius*, published in Venice on March 13, 1610, and considered soon afterward to be a radical break with what was then dominant.

With the option of being translated as *Message of the Stars* or *Messenger of the Stars*, *Sidereus nuncius* ended up being the most controversial study of its time. Galileo himself was crowned as messenger, transforming him into a much more mystical figure than was then customary for a scientist, even if he was quite adept at transmitting an air of mystery or secrecy to each one of his discoveries, as Mario Biagioli has more recently reminded us. But in 1633, after seventeen years of defending the theories of Copernicus, Galileo was brought to trial in Rome. His *Dialogue concerning the Two Chief World Systems* (1632) had put the Aristotelian geocentric view expressed by Pope Urban VIII in the mouth of the fool, Simplicius, who was often caught in his own errors. Although Galileo indicated in the preface that the character was named after a famous Aristotelian philosopher, his portrayal of this "simpleton" made the text appear to be an attack on Aristotelian geocentrism and a defense of Copernican theory. Although Galileo did not act out of malice and certainly was blindsided by the reaction to his book, the pope did not take the insult lightly. Up to that point, it bears mentioning, Urban VIII considered himself Galileo's friend, and even while still a cardinal he had praised Galileo's advances with the telescope in the poem *Adulatio perniciosa* (Pernicious adulation, 1620)—written four years after the famous condemnation of 1616 by the Holy Office. As is well known, Galileo was accused in 1633 of holding opinions opposed to the Holy Scriptures and condemned as a heretic. This trial was, as Beltrán Marí has indicated, a "thicket of irregularities" (*enjambre de irregularidades*) in which "neatness and thoroughness were conspicuous through their absence" (*la pulcritud y exigencia brillaron por su ausencia*) (*Talento y poder* 579, 581). The conservative faction of the church, made up of Jesuits, Dominicans, and some cardinals associated with the Inquisition, held the reins of the trial, even if it was Urban VIII himself who was already making use of the Inquisition as a powerful instrument for his own decisions.

Galileo's trial was followed with great interest in all of Europe: he was eventually forced to abjure under suspicion of heresy, and his sentence was commuted to reclusion and confinement in the church of the Santissima Trinità dei Monti. The famous scientist ended his days under house arrest, afflicted by the loss of close relatives and limited by a progressive blindness that nevertheless could not completely put a stop to his feverish activity. Even in absentia he continued to be the most famous person of his time for his great talent and enormous audacity in calling attention to—and for some, even openly attacking (Shea, *Galileo in Rome*)—the extensive power of the church. The many portraits of him executed by Ottavio Leoni, Santi Di Tito, Domenico Cresti da Passignano (Passagnani), Giusto Sustermans, Jacopo Tintoretto, and Francesco Villamena—as well as his influence upon contemporaries such as John Milton (1608–1674); Margaret Cavendish (1623–1673); Giambattista Marino (1569–1625);

René Descartes (1596–1650), who became so interested in dioptric lenses; and the optician Baruch Spinoza (1632–1677)—confirm his enormous magnetism for the philosophers of his time as well as for writers of fiction.[24] All of these biographical and political reverberations also were felt in the literary field of his Spanish contemporaries.

The present book brings to light, through the identification and analysis of certain astronomical instruments in a selection of important Spanish works, the ways in which writers, especially satirists, in early modern Spain made strategic rhetorical use of "corrective lenses" (*anteojos de mejor vista*), that is, the telescope and related optical devices. It connects in this way with what was perhaps the most universally known facet of Galileo's career: as the creator of the telescope (which in reality he did not invent but rather perfected), an instrument that provided access to new realities. Several generations of writers interested in analyzing their present through the inherited canon of Menippean satire (in prose and verse) transformed these new realities into narrative material, albeit reshaped them according to national tastes, with a genuinely unique style.[25] Nearly a century of poetic, theatrical, and narrative production observed the sun, the moon, the planets, the stars, and the constellations in a new light thanks in large measure to travel and/or advances in optics, and this fact contributed to the development of Baroque reflections on power and perspective. In a local setting filled with misery, disorder and optical illusions became urgent matters, insofar as certain moral defects such as vanity and hypocrisy came to define a social fabric immersed in the most excruciating decadence. For the practitioners of satire, more than any other genre, the object of scorn was not as important as how one presented the work; and the resources offered by "corrected" vision came to be the seed of thematic development in a great number of texts.

The trajectory of scientific and literary innovation was linked to more than one scientific discipline. The seemingly infinite expanse of the sea, for example, also occasionally inspired scholars to explore the celestial sphere. It should come as no surprise, then, that during this period of successive technological advances, of constant improvements over the status quo ante, the telescope was at times understood as a spying tool, and on occasions the latter was scarcely differentiated from the former. This was the case, as we will see, with the famous Dutch optic tube that a clearly irritated Francisco de Quevedo denounced as an instrument of colonial power. From another perspective, the existing ambiguity between the words *antojo* (craving) and *anteojo* (lens) in Spanish would permit a playful flexibility when referring to an object such as a telescope. If the experts, such as Benito Daza de Valdés, were able to define each optic tool with great precision, the masses tended to receive these trinkets in a more spontaneous fashion, one less connected to any method and therefore much more improvised and free. What united all of these groups was a curiosity to reach new distances, to explore what was prohibited, and to seek out new horizons. We find evidence of this in the numerous Spanish texts from the period in which what was truly crucial was the act of looking itself, not so much the physiognomy of the object observed. In this book, I focus on some of the sixteenth century's more perspicacious and playful reflections on the faculty of vision mediated by new technological advances: the telescope as a means of detecting truth, as a long-distance lens that allows one to focus on distant details from the moral watchtower of writing, that provides clarity of vision, and that supports precise word choice.[26] Throughout the present book, I respond to questions that even today—and perhaps more than ever—continue to concern us: To what degree does technology liberate the individual? How does one quantify the limits of knowledge? What is the role of art in the incessant search for what the beyond holds? How does one train the eye, and how does one take that training and transform it into something that might be taught and shared? What is the exact distance, as Carlo Ginzburg has asked, that we require to see things as they are?

To present possible answers to these questions, it is useful to remember once again that the texts under analysis articulate something more than a simple critique. They invite the reader to participate in what Christine Buci-Glucksmann has described as "the madness of vision" (*la folie du voir*); and I understand such madness, in this case in particular, as "the overloading of the visual apparatus with a surplus of images in a plurality of spatial planes" (Jay 48), a madness that is at once positive, kaleidoscopic, and fecund in its manifestations. This "plurality of spatial planes" announces a series of related concerns with respect to the use of lenses for scientific ends, given that the telescope can also be interpreted as an indicator of existing tensions in Spain between creative liberty and legal restrictions. As a result, we find ourselves faced with pieces that reveal fascinating oscillations between freedom and authority, which point to evidence of what Eduard Diksterhuis refers to as "the mechanization of the world picture," that is, the mental transformation that permitted modern science to flourish in Europe through discoveries such as the quantitative law of refraction, Newton's study of white light and colors, and new lenses that could see objects from new perspectives—as George Berkeley would announce in his fundamental study *An Essay towards a New Theory of Vision* (1709). In this sense, as has been demonstrated by an entire critical genealogy that proceeds from Marshall McLuhan to Bruno Latour, objects such as the telescope and other catoptric devices advanced systems of knowledge as "extensions" or "exteriorizations" of the human body—even as prosthetics, as has been suggested[27]—in which the object emancipates the subject, creating new modes of perceptions and structures of experience, "but that, due to the increased concealment of their mechanisms, progressively withdraw from the control and access of their users" (Kramer 77).

It is worth remembering that modern science from the beginning was also conceived as a philosophical endeavor. Donald P. Verene has written that the connection between light, the eye, and thought is also the connection between the mind's eye and truth, and certainly to that truth that was of such concern to figures such as Saavedra Fajardo: "Modern optics," argues Verene, "is the analogue for the modern conception of the intellect as a source of 'reflective' knowledge." This is so because

> from the reflection of Narcissus to the reflections seen outside Plato's cave to the analogue of the Sun and the Good, light is the medium of knowledge, our primary access to the objects of the world. The eye, being the organ of sight, is the primary sense of knowing. The ancient notion of the inner and the outer eye is tied to the phenomenon of light. The mind, the *mens*, is most like the eye. Like the divine *mens*, it can see ideas. The divine *mens* is omniscient, and it is all-seeing. The human *mens* depends upon the object as conveyed by light. Light transports the image. The power of light is to reflect and refract what is there. (77)[28]

The pieces selected here therefore reveal the energy expended by the conflict between tradition and innovation. They also echo, at times timidly, the ideas originating from official centers such as academies and universities, which, thanks to new advances in optics, assimilated and promulgated what Karsten Harries has called "the laws of the earth" (224). In their discursive journey from the earth (the case, for example, of Fernández de Ribera) to the heavens (as occurs in Vélez de Guevara) and then back to the earth (in someone such as Dávila y Heredia), the texts that I examine reveal the telescope to be a democratic instrument of progress and fantasy that takes from our planet its tenacious centrality "while liberating man from the narrowness of a finite world" (Battistini, "Telescope" 22–23). In examining the use of *occhiali politici* in these novels in light of the technical advances of the period, we can identify certain previously invisible connections between scientific achievement and court culture, even as this rereading through fiction identifies two very definite forms of anxiety that obtained during the period: the adoption of the telescope as a mark of social distinction in a society suffering from the same violence that these works denounce, and the tensions between astronomy and religion, which derive from the use of lenses to examine the "new heavens."

The present book allows me to enter into a type of investigation that gives equal exposure to science and literature, something that has been previously carried out in important pieces by scholars such as Michel Serres (*Feux*) and Peter Gallison (*Image and Logic*). I also try to demonstrate, following the argumentative line taken by Peter Dear, and continued more recently by critics such as Frédérique Aït Touati, Eileen Reeves, and Howard Marchitello, that the separation of science and literature—so entrenched in various languages—is a creation that is at once artificial and fragile, since the circulation of ideas between one sphere and the other has been and continues to be constant. A schism such as this, in the case of the Iberian Peninsula, serves only to make more difficult our understanding of Baroque culture and the development of these two fields of study in a historical moment in which both domains were still taking form: "uniting literary and scientific texts," as Aït Touati has argued, "does not imply an attempt to reduce their heterogeneity, still less to deny their essential differences in semiotic and epistemological terms" (4). One must rather highlight, as she puts it, "common ways of thinking and similar writing strategies, to demonstrate the appropriation of poetic ideas, and to identify themes that cut through different texts," in order to offer "an outline not of boundaries between disciplines, but of specific strategies in literary and in scientific writing and in common poetic tools" (4).[29] In parallel fashion, the present study presents numerous fictional works containing aspects of scientific exploration, scientific texts that participate in poetic creativity and hybrid creations, halfway between one and the other discursive field. Examples of these abound: was not the *Dialogue concerning the Two Chief World Systems* a deliberate attempt by Galileo to situate science at least partially within the realm of fiction, that is, to bring the two fields closer, to remain faithful to a certain didacticism while attending to his desire for dissemination? There can be no doubt that Galileo's text should be read as a scientific treatise, but it can also be enjoyed as a piece of entertainment in the form of an academic dialogue, a dialogue that through its didactic zeal connects with the earlier dialogues of the Spanish humanists, and brothers, Juan and Alfonso de Valdés (to give but two examples), unquestionable pillars of any course on early modern Spanish literature.

In the end, we find ourselves faced with forms of expression that frequently go beyond their own perceived boundaries. In light of this, it can seem somewhat odd that at a moment when the greatest portion of scholarly work on the early modern period has opted for an interdisciplinary approach, so little attention has been given to the impact of the Scientific Revolution on Spanish authors of the period, many of whom did not back down from the impoverished and difficult landscape that surrounded them. In a recent article, Agustín González-Cano has suggested that "we are still without a definitive study of the use of the telescope in literary works of the Spanish Golden Age, a period of extreme importance for our cultural history and yet a period still somewhat marginalized with respect to the development of scientific theories regarding lenses and vision" (*No disponemos aún de un estudio definitivo sobre el empleo de los anteojos en las obras literarias del Siglo de Oro español, periodo de máxima importancia en nuestra historia cultural y época aún fronteriza en lo que concierne al desarrollo de las teorías científicas sobre las lentes y la visión*) ("Un poema" 35). The present book seeks to fill this critical lacuna thanks largely to recent scholarly achievements in the history of Spanish science and in literary studies, not to mention the excellent theoretical work that has been published (mostly) through US and Canadian university presses. From the monumental work of José María López Piñero to the exacting efforts of scholars within concrete disciplines or specific centers of investigation, readers today enjoy a wide array of options to understand better the scientific activity that took place in sixteenth- and seventeenth-century Spain and its connections to the fiction of the period. A sampling of some of the best-known Spanish works of these two centuries—*Don Quixote, La vida es sueño* (Life is a dream), *El diablo Cojuelo* (The limping devil), *El criticón* (The critic)—reveals the great interest that Miguel de Cervantes, Pedro Calderón de la Barca, Luis Vélez de Guevara, and Baltasar Gracián had in what was then taking place in Europe. Glass, for example, had been converted into a fertile metaphorical element from which to construct a theory of the good ruler. Were not the mirrors of princes of earlier centuries a "third dimension" from which to observe and reflect on exemplary conduct?[30] It should not seem strange, therefore, that in the first years of the seventeenth century one should witness a growing fascination with lenses as instruments of power and progress, or that the second half of the century should give us scholars situated halfway between fictional creation and scientific experimentation, between the pleasure of the text and the didacticism of science: Juan Bautista Corachán (1661–1741), Juan Caramuel y Lobkowitz (1606–1682), Andrés Dávila y Heredia (fl. 1674), and many others.

One should also not forget that science and war, as Dávila y Heredia (a military engineer all too familiar with the situation in Europe) reminds us, were conjoined during the seventeenth century. Alicia Cámara has recently spoken of this *saeclum bellicum*, or violent century, arguing that it was "universally accepted that mathematics were the common base of professions such as architecture, the military, and engineering. . . . Mathematics was necessary in times of peace and war, as much to construct palaces as to organize squadrons, to elaborate plans and descriptions of territories, and to create mechanical devices" (*era universalmente aceptado que las matemáticas eran la base común a profesiones como la de arquitecto, militar, ingeniero* [. . .] *La paz y la guerra necesitaban de las matemáticas, tanto para levantar un palacio como para organizar los escuadrones, para realizar planos y descripciones de los territorios, para crear artificios mecánicos*) (68). If faith could move mountains, it is also the case that the impetus and motivation for many major episodes of early modern imperial history was unmistakably belief in the reliability of the empirical: the spatial ordering of surveyors, the mechanics of weapons that helped to open new fronts on the battlefield, the clock as the emblem of political discipline, the astrolabe that guided overseas explorers, the magnetic compass and the pair of compasses as emblems of moral rectitude (as Boccalini himself wrote in the first section of his *Ragguagli*), the art of fortifications that so amazed the writers and mercenaries of the period, and so on.[31] It should thus not come as a surprise that when Galileo presented his *cannocchiale* to Leonardo Donato, the doge of Venice, in a letter dated August 24, 1609, he did so as an instrument of war, as a defensive weapon with which to see enemy ships from a distance. As Mauricio Jalón has put it: "The 1600s were a mechanistic century" (*el Seiscientos fue un siglo maquinístico*) (155).[32]

Satire is a form of literary discourse that lends itself more or less perfectly to the question of scientific innovation as a poetic problem, insofar as it decisively captures the fears and suspicions of ordinary people with respect to important breakthroughs in technology. The existence in the Iberian Peninsula of a solid satirical literary tradition, represented most fully in its political dimension by the burlesque creation of the *arbitrista* (political schemer) facilitated the transformation of the "machinery" of which Jalón speaks into a fertile narrative motif.[33] It is practically obligatory to mention, for example, the passage in *El buscón* (The swindler) in which Quevedo makes fun of the "republic" and its "fourteen-year plan" to lay siege to the Flemish seaside city of Ostend with enormous sponges "to lower the sea level at that point by twelve fathoms" (*para hundir la mar por aquella parte doce estados*) (106–7). The passage is openly comical; however, behind this example of impossible mechanics also pulses, as in many of the texts under analysis in the present book, a very palpable anxiety in the face of so much stupidity and opportunism. This anxiety would in some cases manifest itself concretely through invocations of ingenuity, vision, and creative madness, all of which had been previously rendered canonical through the Italianate figure of the virtuoso thanks to inventors such as Leonardo Fioravanti, Giambattista della Porta, and Galileo himself. These invocations found their way to the Iberian Peninsula in variants characterized above all by great literary fecundity. Let us take, for example, the fictitious re-creations—and the verbal or metaphorical conceits—of the Italian-born engineer Juanelo Turriano (in *El buscón*) or those of the musicologist and collector Juan de Espina (in *El diablo Cojuelo*), which are invoked, at times in burlesque fashion, as representations of the impossible dreamer or the mad scientist, but almost always are examples of figures moving against the current.[34] It is for this reason that many of the disciplines most in fashion during the early modern period, such as alchemy or geometry, served early modern Spanish thinkers as means to develop new equations imbued with an inevitable ethical and moral component. What was ultimately at stake for them was the search for a science that is useful, positive, and even corrective—a science, in other words, that from its precision might calibrate and adjust the detours and absurdities of the mind. Consider Boccalini's sophisticated rereading of the pair of compasses in the early section of *Ragguagli* devoted to his famous store:

> Also sold in that store are compasses, not made of silver, brass, or steel, but of the purest interest of fine reputation that can be found in all the minerals of honor; and they are remarkable in measuring one's actions. As experience has made all aware, compasses made of lesser materials of capriciousness, and of interest alone, prove themselves unjust to those who in their affairs wish to draw parallel lines. These compasses are exceedingly good for those who know how to use them exactly, to take the correct measure of the latitude of those ditches that some, out of respect for their reputation, are forced to jump over entirely so as not to run the risk of falling in the middle of them and shamefully burying themselves in the disgusting mud of imprudence.
>
> *También se venden en aquella tienda algunos compases, no ya labrados de plata, latón o acero, sino de puro interés, de la más fina reputación que se halla en todos los minerales de la honra, y son admirables para medir con ellos las*

propias acciones; pues la experiencia ha hecho conocer a todos, que los compases labrados de la materia vil del propio parecer, y del interés sólo, salen poco justos a los que en sus negocios desean tirar las líneas paralelas: demás, que semejantes compases a los que exactamente poseen el arte de saber bien usar de ellos, salen excelentes para poder tomar las medidas de la latitud de aquellos fosos, que algunos por respeto de su reputación, les es forzoso saltar indemnes sin correr peligro de caer en medio de ellos, y vergonzosamente sepultarse en el asqueroso lodo de la imprudencia. (3r)

Boccalini's analogy may seem somewhat cryptic, and there can be little doubt that he more or less forces the compass into the service of the critique that he wishes at all costs to put forward. For him, when utilized by a capable and learned hand, the compass traces out the contours of virtue, delimiting an exceptionally small space, the miniscule island of truth. In this way, the points of the pair of compasses become the accurate measurement of reality, the correct interpretation of the Baroque surroundings that so oppress human subjects through their visual apparatus. Boccalini transforms the pair of compasses, a simple yet grandiose mechanism for drawing circles and lines, into a possible path toward rectitude by linking the term *medida* (*measurement*) to the quality of measure (*mesura*), which he presents as the opposite of imprudence. In this sense, a lack of discernment can cause one to lose his or her way (and here, once again, we see the importance of vision), to misunderstand the true raison d'être of each thing. As Saavedra Fajardo argues, one must possess proper judgment and dexterity, a kind of sprezzatura and mental agility; that is, he or she must be able to trace out clean, clear, and forceful lines, and with a steady hand. Boccalini, still speaking of the pair of compasses, makes this connection explicit: "politicians sell a great number of compasses used by surveyors; these are very necessary to determine in all places those with whom one should enter into serious business and share important secrets" (*Venden también los mismos políticos gran número de brújulas usadas de los agrimensores, que son muy necesarias, para bien cuadrar por todas partes aquellos con quienes algunos deben tratar negocios graves, y conferir secretos de importancia*) (3v–3r).[35] Put another way, humans live their lives disoriented, and as a possible redemptive force, science can help them to find a "true north" through the correct measure of things, through the faithful interpretation of reality. In this way, a personal, intimate, collected object corresponding to the bourgeois and aristocratic classes, an artifact that measures and orients, can also become a new kind of idolatry. The technical, the precise, the intricate, that which guards secret functions, is thus cared for and col-

FIGURE 1. Peter Paul Rubens and Jan Brueghel, *Allegory of Sight* (1617)

lected and made use of in intimate spaces, adding one more mechanism to the collection, a marvel with which to take delight; and this Baroque *Wunderkammer* is, evidently, one filled with countless possibilities.[36]

The pair of compasses and its kind may be marvels, but one must also "domesticate" or assimilate them in both individual and broader cultural terms. It is precisely this varied adaptation and application of the new mechanics that numerous satirical Spanish texts take up. And this is true not only of literary fiction; the history of portraiture, for example, presents us with some marvelous examples: in Peter Paul Rubens and Jan Brueghel's painting *Allegory of Sight* (1617), one finds one of the earliest known images of a telescope, resting on the ground in the lower corner of the canvas. The painting captures the numerous objects littering the study of the Spanish princess Isabella Clara Eugenia (daughter of Philip II) and her husband and cousin the Archduke Albert of Austria, and the telescope they had received it as a gift from the Marquis of Spinola (fig. 1).

Also telling is Jusepe de Ribera's painting *Vision* (1613) (fig. 2). Ribera's canvas belongs to a series of paintings on the five senses, and in this allegory of vision, a laborer looks directly at the spectator. Ribera bathes his subject's weathered face in sunlight and places a telescope in his strong hands. In the lower part of the composition rests a pair of eyeglasses, themselves an equally crucial element within the Baroque "cabinet" of objects of optical measurement.

FIGURE 2. Jusepe de Ribera, *La vista* (1613–1616)

In his study "El *Quijote* espectral," Fernando R. de la Flor has written:

> It is easy to recognize, through the totality of scholarly work devoted to early modern Spanish culture, that we are still unable to establish the boundaries and the depth of the vast field of Baroque visual perception; the ways of seeing and perceiving, and the corresponding analogous processes of reconstructing reality in terms of image, that which we could term the constitution of the high modern vision of the world, the scopic

regime that coincides with the modern age. Much of this is unknown to us, as are (and particularly so) the complex processes of constructing the *imago mentis*, of figuration, of vision, of the phantasmagoric perception of unreal realities so abundant in Hispanic culture, which found in this field of perceptive derangement and numinous vision an allegory for the eccentric and distorted position of man in the world. To all of this we could also add a certain lack of attention in our historiography (including the most recent sort) with respect the technical process by which the "Baroque optic" emerged.

> *Es fácil reconocer, por el conjunto de los estudiosos que ahora mismo operan en el campo de la cultura siglodorista, el que no estamos todavía en condiciones de establecer los límites y la profundidad de lo que es el vasto campo de la percepción visual barroca; los modos del mirar y el percibir, y los correspondientes procesos analógicos de reconstrucción de la realidad en términos de imagen, lo que podríamos denominar la constitución de la mirada altomoderna sobre el mundo, el régimen escópico que corresponde a la Edad Moderna. Todo esto nos es, en buena medida, desconocido, como así mismo lo son también, y particularmente éstos, los complejos procesos de construcción de la* imago mentis, *de la figuración, de la visión, de la percepción fantasmática de realidades irreales, en los que, por cierto, sobreabunda la cultura hispánica, que encontró en este campo del desarreglo perceptivo y la visión numinosa una alegoría de la posición excéntrica y distorsionada del hombre en el mundo. A todo ello, añadiríamos también una cierta desatención de nuestra historiografía, incluso de la más reciente, por lo que se refiere al propio proceso técnico en que está embarcada la "óptica barroca."* (n.p.)

As I have already suggested, it makes little sense that there should exist such a sparse bibliography with respect to the intersections of fiction and optics in early modern Spain. This is even more the case when we take into account that the figure of the astronomer was an enormously attractive one during this period and unquestionably relevant. This fascination with the astronomer's person and work has endured over time, as evidenced by numerous studies and the attention that both received upon the fourth centenary of the invention of the telescope, in 2009. The file on Galileo in particular provides a faithful account of what has been, over time, an unquestionable fascination with his figure: jealously guarded in the Vatican Secret Archives until the nineteenth century, this file fell at that time into French hands in 1811, in the wake of Napoleon's sack of Rome, and was returned to its place of origin in 1845 only with the promise that it be published.

Maurice A. Finocchiaro and Thomas F. Mayer have examined not only the famous 1632–1633 inquisitorial condemnation, based mainly on events that had occurred in 1613 with the edict promulgated by Cardinal Robert Bellarmine, but also Pope John Paul II's controversial "rehabilitation" of Galileo from 1979 to 1992.[37] Both Finocchiaro's and Mayer's work masterfully presents this fascinating process, which brings together the critical reception of the work and the person of Galileo over four centuries, the incomprehension that his legacy generated, and some of the most important fictions revolving around his mythic figure.[38] They reveal, in the end, a very healthy resistance to certain ideas that were formerly received as solid truth. And, as Carmen Mataix has reminded us, not only did Galileo "overcome the Aristotelian conception of the universe and, above all, incorporate a new form of understanding nature that inaugurated what came to be called the New Science" (*con Galileo* [. . .] *se superó la concepción aristotélica del universo y, sobre todo, se incorporó una nueva forma de entender la naturaleza que inauguró lo que se llamó la Nueva Ciencia*) (131); he also employed rational and empirical arguments to defend his heliocentric theory, "a new reality and a parallel anthology of the world" (*una nueva realidad y una ontología paralela al mundo*) (138). If the cosmos was the personal creation of the great "Artisan," according to the felicitous coinage of Galileo's contemporary Johannes Kepler, then for Galileo nature was a book with its own language that had to be read correctly, that needed to be deciphered: his famous premise, expressed in *Il saggiatore* (The assayer, 1623), that nature is "written in the language of mathematics" (*scritto in lingua matematica*) coincided with these two facets, the technological and the humanistic, presenting him as a *mathematicus*, or a philosopher deeply versed as much in the domain of mathematics as in astronomy and astrology. This is what appears, for example, on the title page of the first edition of *Sidereus nuncius*: *philosophis* precedes *astronomis*. With this a new, much more modern scientific spirit became crystalized. As Peter Dear puts it: "The mathematical astronomer merely described and modeled the motions of the celestial bodies; it was the job of the natural philosopher to explain *why* they moved" (*Revolutionizing* 42). Astrology (particularly judiciary astrology) never much pleased the Spaniards (Pimentel 60), despite it having been the most important academic discipline during Galileo's years at the University of Padua. It was certainly a branch of knowledge that granted Galileo large benefits: his *Sidereus nuncius*, it bears repeating, was dedicated to his sovereign, and patron, the Grand Duke Cosimo II de' Medici, and in it he proposes that the new moons of Jupiter be named in honor of him, given the great interest of the entire Medici family in that material (Kollerstrom 422; Battistini, "Telescope" 10). If up to that time the stars were given the names of divinities, the audacity of the Medicis supposed a clear change of paradigm, and with it the luck of Galileo, who replaced his humble university professor's salary with an annual stipend that was much more generous.

The 550 printed copies of *Sidereus nuncius* sold out in just a few days. Galileo's first Aristotelian phase was relegated to the past, and a new century opened up—one both full of possibilities and sensitive to the achievements and limitations of this new science. According to Pamela Smith:

> Noble interest in the objects of science, such as preserved natural specimens, objects for the burgeoning *Kunstkammern,* territorial and new world maps, and instruments such as telescopes affirmed the potential of natural knowledge to celebrate reputation and establish credit—both of the ruler and the natural philosopher—to produce commercially valuable and aesthetically pleasing objects, and to open up unknown worlds. (350)

Galileo became a central figure through the interweaving of cosmology, art, and technology with commercial politics, even while he struggled for social legitimation through theology and religion in an Italy that, as I have indicated, never ceased looking to Spain. The history of these relations is at once necessary to understand and fascinating in its development. But I wish to begin this voyage a bit earlier, in the final decades of the sixteenth century, through an analysis of earlier Spanish writers' worldview, determined still by the legacy of a Ptolemy whose theories would very soon come into question.

I

Writing on the firmament

1
Observations

Yo aplaudo a los hombres sabios y prudentes que nos han traído el telescopio.
I applaud the sage and prudent men who have brought us the telescope.

ÁNGEL GANIVET

Galileo's telescope and the gaze of Spain

The history of the circulation of the telescope in seventeenth-century Europe is nothing less than a crucible of different events, at times unfolding almost simultaneously. At the time that Galileo began his studies at the University of Pisa, in 1589, Copernicus's *On the Revolutions of the Heavenly Spheres* was still an important reference, even if, despite the appearance of a second edition in Basel (1566), it did not seem quite as new as it once had. In fact, the most important scientist at the time was likely Tycho Brahe, who had determined that the planets orbit the sun, and that the sun in turn orbits the earth. This idea deeply concerned the Jesuits, as demonstrated by Giovanni Battista Riccioli's *Almagestum novum* (New almagest, 1651), which refined some of the points in Brahe's text and became for many the single most influential work until the publication of Isaac Newton's *Philosophiae naturalis principia mathematica* (Mathematical principles of natural philosophy, 1687) (Shea, "Galileo the Copernican" 41–60). However, the social life of the telescope would also be determined by Kepler's aforementioned *Ad Vitellionem paralipomena* (Paralipomena to Witelo, 1604), whose theories about the image in the ocular retina demarcated the line between the eye as an instrument of perception and the exterior world, ruled by laws of physics and geometry, thus paving the way for future studies on distance and perspective, as well as for Cartesian rationalism. This was also a highly significant book in the history of ophthalmology, as it dealt with the functions of the eye, the crucial role of the retina, the process of refraction, and the first scientifically correct explanation of myopia.

We know that by the summer of 1609 the English astronomer Thomas Harriot was in London observing and drawing a new map of the moon, while almost simultaneously, Galileo was investigating—as he had since the 1590s—the possibility that the earth turns on its axis.[1] This idea was considered controversial, if not outright heretical, given that this was a decade of ironclad ideological control defined by a series of famous condemnations, from the arrests of Giambattista della Porta, Cesare Cremonini, and Tommaso Campanella to the burning at the stake of Francisco Pucci and Giordano Bruno. However, the manipulation of concave and convex lenses beginning in the 1590s in Italy, arriving in the Low Countries around 1604, and spreading throughout all of Europe by 1609—first as an instrument of navigation and later as an astronomical tool—radically changed the prevailing understanding of the cosmos such that, in just a few months, there were a variety of new proposals regarding the characteristics of the solar system.[2]

There is also what might be called a prehistory of the long-range lens available to us. Today we have access to ancient testimonies about the existence of instruments in the Middle East that could have been used as spyglasses. We also know, for example, that the Church of St. Nicholas in Treviso was home to the first known representation of a person with glasses: Cardinal Hugh de Saint-Cher, in Provence, painted by Tommaso da Modena in 1352.

The first glasses were made for farsightedness and were convex. Concave lenses for myopia would appear a century later. A real revolution in book reading began with the invention of the printing press in 1436, and the demand for glasses rose along with it. At this point, lens production ceased to be a monastic art, and the first dedicated workshops cropped up in places like Nuremberg, Haarlem, and Venice. It was in Nuremberg that the first guild for master optical lens makers was founded in 1438. The literary history of Europe is, in fact, sprinkled with anecdotes about farsightedness and its sufferers; Petrarch, for example, wrote in 1364 that he needed to use glasses due to his age, and the French poet François Villon donated his reading glasses to the poor in 1461 (fig. 3).

Some of the first real telescopes of which we know today appeared in Holland, with the first patent application filed in 1608 by Hans Lippershey and Zacharias Janssen in Middleburg; Jacob Metius (or Jacob Adriaanszoon) soon joined them from Alkmaar. Galileo made a series of improvements on his own design in the following months, working parallel to the Jesuit Niccolò Zucchi, who would develop the reflecting telescope in

FIGURE 3. Tommaso da Modena, *Hugues de Saint Cher* (1352)

1616 using a curved mirror instead of a lens as an objective. It was Girolamo Sirtori, however, one of Galileo's students, who in 1618 wrote what is now considered the first treatise on telescopes (*Telescopium sive Ars perficiendi novum illud Galilaei visorium instrumentum ad Sydera*; Telescope, or a performance of the art and means to Galileo's new vision of the stars, in three volumes), in which he maintained that the first telescopes came not from the aforementioned Dutch astronomers but from Catalan lens makers. In his text, Sirtori recounts his European travels in 1609 and 1610 and his meeting with Joan Roget, a Catalan lens maker who died between

1617 and 1624. He comments that upon arriving in Girona he met with Roget, who showed him "the plating or the iron supports of a telescope completely eaten by rust" (*la armadura o los hierros de un telescopio tomados de orín*), and "the forms of the instrument outlined in a book" (*las formas de un instrumento delieneadas en un libro*), on which Sirtori took copious notes.

The terms that appear in both Castilian, "long-range glasses" (*anteojos de larga vista*), and Catalan, "long telescope embellished with brass" (*ullera larga guarnida de llautó*), "long-range telescope," and "tin telescope for viewing the moon" (*ullera de llauna per mirar de lluny*), are no doubt meaningful, and they speak to an established lens-making tradition in Catalonia that coincides with the terminology used in several inventories written in Barcelona between 1593 and 1613. In the same way, we know that Joan Benimelis, the Majorcan doctor, historian, and mathematician who died in 1616, had owned a "tube, for looking at the moon, and another tube" (*trompa, de mirar de lluni i altre trompa*) since the beginning of the century.[3] The Sevillian poet Juan de Salinas (1559–1643) would sketch out this geographical trajectory in an interesting "riddle" in which he remarks in humorous terms on the innovations coming from this region:

News from Barcelona

(Riddle)
Two brothers arrived
Onshore in a ship
They came from foreign lands
To give Spain a sight
Of illustrious, ingenious, appearance,
and of very clear ancestry.
As their noble blazon
They carry two moons on their arms;
Of that splendid family
They are those that attend and protect
The great Lord on his throne
From traitorous traps.
With rigorous examinations,
They received their degrees in Italy,
And in every Department
They make obvious the most obscure things.
Oh, Great Queen of Sheba—

If you were to come to our age,
What a test you would give them,
And in what varied subjects!
With the great company
Of a splendid and lucid fleet
(which was a sight to be seen) they made
Their entry into Barcelona.
They have been well received
By princes and monarchs,
And the people by way of them
Achieve a thousand impossibilities.
GLASSES.

Nuevas de Barcelona

(Enigma)
Dos hermanos arribaron
en una nave a la playa,
que de tierras extranjeras
vienen a dar vista a España.
De ilustre ingenioso aspecto,
de clarísima prosapia,
que por blasón de nobleza
traen dos lunas en las armas;
de esta espléndida familia
son los que asisten y guardan
al gran Señor en su trono
de alevosas asechanzas.
Con examen riguroso
le dio sus grados Italia,
y en todas las Facultades
lo más oscuro declara.
¡Oh, tú, gran Reina Sabea,
si nuestra edad alcanzaras,
qué pruebas hicieras de ellos,
y en qué materias tan varias!
Con gran acompañamiento
de una muy lucida escuadra
(que eran para ver) hicieron
en Barcelona su entrada.

Han sido bien recibidos
de Príncipes y Monarcas,
y el pueblo por medio de ellos
mil imposibles alcanza.
LOS ANTOJOS[4]

This lens was an essential instrument in the development of science of the time—for example, in his *Sidereus nuncius* Galileo already had made mention of the French scientist Jacques Badovère's important role in the adoption of the telescope (Lewis 91–112; Baumgartner). Since July 1609, Galileo had been in Venice attempting to convince the wealthy patrons who controlled the University of Padua to increase his salary, and it was there that he learned that Maurice of Nassau had received as a gift a device that allowed faraway objects to be seen with uncommon detail. By the time he returned to Venice, on August 21, Galileo already had an instrument that allowed for eight times magnification. Even so, some of his contemporaries, like his enemy Cesare Cremonini, refused to look through it. However, Galileo would continue to perfect his rudimentary instrument—with which he was able to achieve up to thirty times magnification—and he would continue to observe the heavens. His telescope would be a refractive model, relying on a system of lenses to refract the light rays and make them converge on a focal plane, using a convex lens in the objective and a concave lens in the eyepiece. He did not know that he would not be the first to adopt such tools, just as he did not know that his embryonic telescope was already far superior to existing ones.[5] But it did present him with a new challenge, as Pimentel has noted:

> In 1610 Galileo's observation of sunspots alone could not overthrow the theory of the incorruptibility of the celestial spheres. He had to demonstrate how to look through his telescope, he had to discipline his sight, and he had to relocate the origin point of authority on the natural world. Instruments had no credibility; all credibility belonged to witnesses, depending on who they were, along with the Bible and whoever held the monopoly on its interpretation.

> *En 1610 la observación de Galileo de manchas solares no podía derribar por sí sola la incorruptibilidad de las esferas celestes. Había que enseñar a mirar por su telescopio, había que disciplinar la vista, había que reubicar el lugar de donde manaba el flujo de la autoridad sobre el mundo natural. Los instrumentos no tenían credibilidad; los testigos, depende quién; la Biblia y quien detentaba el monopolio de su interpretación, la tenían toda.* (56)

These months, then, constituted a crucial period as much in the life of Galileo as in the larger sphere of European science. Because of its novel character, this "new mechanism" made a very convenient form of social and economic capital, particularly as a gift that could be used to strengthen alliances and forge new friendships. The limitations of this early optical device were rapidly resolved in an empirical sense by Kepler, who designed another model of the refracting telescope more suited to astronomical use. By the end of 1610, to give just two examples, observations made with the telescopes of the Roman College (Collegio Romano) and in the Convent of St. Anthony in Lisbon. As such, it is not surprising that telescopes began to be commissioned very soon after by many of the great statesmen of the moment, like Maximilian I, Duke and Elector of Bavaria, as well as Ernest of Bavaria, archbishop and Prince-Elector of Cologne, and even Cardinal Francesco Maria del Monte. Galileo quickly sensed the possibility of a handsome payoff from this fascination with the heavens, whether it stemmed from suspicion or amusement. With the permission of Cosimo II de' Medici, Grand Duke of Tuscany, he attempted to send five telescopes to the respective heads of state of Spain, France, Poland, Austria, and the papal state of Urbino, taking advantage of the frenzy—well documented by William Shea (*Galileo in Rome*)—that the telescope had caused among European nobility and royalty. From Rudolph II of Prague to Cardinal Scipione Borghese, nephew of the pope; from Cardinal Alessandro Peretti of Montalto to Queen Marie de' Medici; from Cardinal Odoardo Farnese to the French court, which would request that a new planet be named in honor of King Henry IV—everyone wanted to enjoy the new realities promised by such an appealing invention (fig. 4).

Galileo presented the telescope to the Venetian Senate on August 21, 1609. He mounted his optical device on the bell tower of the Piazza San Marco, to the delight of those in attendance. The islands of Murano, situated at a distance of about a mile and a half, appeared to be only about nine hundred feet away. He willed the rights to the telescope to the Republic of Venice, which was very interested in its possible military applications. In a letter to Leonardo Donato, doge of Venice, dated August 24, 1609, Galileo explained the uses of his *cannocchiale* to spot enemy ships "from a distance of two hours before they are visible with the naked eye," and he promised Donato that he would keep it a secret. With the telescope he analyzed the moon and its phases in detail, realizing that it was not the perfect sphere described by Aristotelian theory. This new vision of the lunar surface not only broke with astronomical belief but also troubled the theological waters; as Frederick A. de Armas reminds us from the per-

FIGURE 4. Jan van Eyck, *The Madonna with Canon Van der Paele* (1436).

spective of literary history: "the moon could not contain true spots, seas or craters, since its light and purity stood for both the perfection of the planetary spheres and the immaculate conception of the Virgin" ("Maculate" 60). Galileo affirmed that the transition between shadow and light on the surface of the moon was irregular, which proved the existence of a mountainous surface rather than a perfectly spherical and smooth surface like the one depicted by Aristotle. He also discovered the nature of the Milky Way, where he was able to count stars in the nebula of Orion to find that certain objects taken to be stars were in fact clusters of smaller stars. As his observations continued, he discovered the phases of Venus—for him definitive proof of Copernicus's heliocentric hypothesis. However, the most notable discovery of all was that of the existence of four small stars near the planet Jupiter, which, after several days of observation, were taken to be four satellites orbiting the great planet: Io, Europa, Ganymede, and Callisto. For Galileo this was proof that Jupiter and its satellites formed a small model of the solar system. As Beltrán Mari (*Talento y poder*) reminds us:

> The epicycles are real, as demonstrated by the movement of the Medicean planets around Jupiter, and of Mercury and Venus around the Sun, that is to say, the movements around a center that is not the Earth. The same thing happens with the orbital eccentricities, which are proven because the movements of Mars, Jupiter, and Saturn include the Earth's orbit but do not orbit around the Earth.
>
> *Los epiciclos son reales en cuanto designan los movimientos de los planetas mediceos en torno a Júpiter, de Mercurio y Venus en torno al Sol, es decir, los movimientos en torno a un centro distinto de la Tierra. Lo mismo sucede con las excéntricas, que son reales en cuanto los movimientos de Marte, Júpiter y Saturno comprenden a la Tierra pero no orbitan con centro en ésta.* (225)

Galileo then attempted to demonstrate that Aristotle's perfect circular orbits did not exist and that the heavenly bodies did not orbit the earth. He would name these new stars *Mediceas,* in honor of the family of the ruling prince, Cosimo II. Shortly after, he undertook the writing of the text that would become *Sidereus nuncius,* published in Venice on March 13, 1610. If his fame spread in Italian political life to the point at which he became legendary among his contemporaries, his achievements also spurred a wide range of international tributes, like Scot Thomas Seggeth's "Laudatory in Nine Epigrams," published in the appendix of Kepler's *Narratio de Observatis a sequatuor Iovis sattelitibus erronibus* (Account of my observations of Jupiter's satellites), or the verses in an introduction to *Il saggiatore* undertaken by the German Johann Faber (Shea, "Galileo the Copernican" 49). There was also, as María Bayarri has noted in "Universos poéticos" and "Galileu Galilei" (56–57), an interesting correspondence between Galileo and the writer Margherita Sarrocchi (1560–1618), as well as the writing of a number of poetic compositions by Lucrezia Marinella (1571–1653), another friend of his. Additionally, there are fascinating references to Galileo's work in other disciplines, such as the frescoes in the Salone della Meridiana painted by Anton Domenico Gabbiani in 1692–1693 and designed by Vincenzo Viviani, who was not only Galileo's student and his secretary in his last years but also the official mathematician of the Grand Duke of Tuscany after Evangelista Torricelli (Frangerberg).[6] But these were, as we know, turbulent years of active anticlerical agitation in Venice, especially against the Roman Curia and against the Jesuits, entrenched as the vanguard of the Counter-Reformation. Theological polemics between Paolo Sarpi (Venice) and Cardinal Bellarmine (Rome) were also taking place during these months. The Society of Jesus would do everything possible to discredit the University of Padua as a stronghold of heretics, although without necessarily mentioning Galileo himself. The communication between the two camps reached extraordinary levels of intensity, a true "rite of dissimulation" (*rito de disimulación*) in the words of Antonio Beltrán Marí (*Talento y poder* 264). Although Galileo has been thought of at times as a martyr of science, he always avoided that condition. As scholars such as Arturo Fernández Luzón have argued, his condemnation was much more a product of Pope Urban VIII's personality than of the challenge that his works presented.

In 1611 Galileo was invited to the Roman College by Cardinal Maffeo Barberini—the future Pope Urban VIII—so that he could present his findings. As an alternative to the cultural politics of the Jesuits, the Academy of Lincei (Accademia dei Lincei), founded by Federico Cesi in 1603, received Galileo with enthusiasm and admitted him as its sixth member, giving a notable impulse to the spread of his work and the popularization of his persona.[7] However, Galileo's perceived intransigence in rejecting the idea of a possible accord between the heliocentric theories of Copernicus and Ptolemy's geocentric hypothesis precipitated Cardinal Bellarmine's decision to order the Inquisition to carry out a "discreet" investigation of his theories. A substantial portion of modern criticism has pointed out that the Inquisition was very patient, even benevolent, with Galileo, whose inflexibility was nothing less than a direct attack on ideas that had dominated Christian thought for centuries. Aristotle, for example, was still untouchable because he had been adopted by great Christian theologians like St. Albert the Great and St. Thomas Aquinas. In light of this fact, uncertainty was everywhere from this moment on, and suspicion was Galileo's constant companion. As Ofer Gal and Raz Chen-Morris have argued: "Instruments embedded sophisticated mathematical knowledge and fine artisanal skill. They could undoubtedly aid faulty vision, but they were still suspect" (79). The Jesuit Odo Van Maelcote demonstrated this by offering a summary of the discoveries of Galileo in his *Nuncius sidereus Collegii Romani* when he was invited to the Roman College, making use of sarcasm to avoid endorsing the terms *valleys* and *mountains* in reference to the orography of the moon, whose surface he did not believe was rough. Likewise, he gave a caricatured description of the phases of Venus, the argument for which had not entirely convinced him. He was following in the footsteps of Christoph Clavius, the all-powerful professor of mathematics at the Roman College. "While the Jesuits simply accepted the mountainous nature of the moon," Beltrán Marí writes:

Clavius did not deny that the moon was apparently mountainous but he put another possible solution forward: the irregular appearance of the moon's surface could be the visual effect of the unequal density of its different parts. This is a theory that had been developed more broadly and in greater detail by Colombe in his text *Di Ludovico delle Colombe contro il moto della Terra,* which was violently dismissed by Galileo.

> *Clavio no negó que presentara un aspecto montañoso pero adujo otra posible solución: el aspecto irregular de la superficie lunar podía ser un efecto visual debido a la desigual densidad de sus distintas partes. Se trataba de una teoría que había desarrollado más amplia y detalladamente Colombe en su escrito* Di Ludovico delle Colombe contro il moto della Terra, *que recibió un durísimo varapalo de Galileo.* (113)[8]

This move toward humor—or perhaps indignation disguised as humor—was significant, inasmuch as it would be one of the most common strategies of Baroque narrative: humor as a counterpoint to what was, after all, evident. Juan Pimentel and José Ramón Marcaida López have established an interesting parallel between the description of the scientific object as sketched by Galileo and certain still lifes of the period: "Compared with many depictions of fruits, particularly apples and oranges, these studies of shaded spherical volumes display remarkable similarities" (105). And in the realm of literature, as we will see later, the turn to humor would become apparent through figures like the satirist Luis Vélez de Guevara. In his masterwork *El diablo Cojuelo* (The limping devil), written a few years later, he would construct one of the most interesting celestial visions of his time, similar to Van Maelcote's in its sarcasm and in the imagery used to maintain a prudent distance from certain thorny issues.

Galileo, meanwhile, would keep writing pieces of unsurpassed importance. *Il saggiatore,* for instance, focused on natural philosophy, more specifically on the structure of matter and the mathematical character of the "book of nature." By that time Paul V and Cardinal Bellarmine were already dead, and Gregory XV, the new pope, proved himself relatively benevolent toward the famous scientist. Perhaps the most notable point, however, was the radical change undertaken at the intersection of ontology and methodology. Galileo insisted that mathematics was not solely a useful instrument for the study of physics but also absolutely necessary when attempting to know nature fully; this led him to write the famous phrase "nature is written in mathematical characters." In any case, by that time the controversy had reached the highly philosophical register that

Galileo himself had hoped for: "The kind of knowledge that mathematical practices tended to promote was not simply utilitarian," Peter Dear has written, "its elevation to philosophical importance by such figures as Galileo implies a reevaluation of mathematical characteristics as being important to true understanding of nature" (*Revolutionizing* 79). This understanding of nature was propagated in Spain through various institutions until the eighteenth century was well under way, all while the eyeglasses and the telescope became ever more familiar parts of the urban landscape. As we will see, however, this growing comprehension was never exempt from controversy. Galileo's science was received in varying ways in different fields of knowledge and in different geographical areas of the Iberian Peninsula. The results would be symptomatic of a lively interest, sustained across decades of intellectual exploration that extend to the present day. It would not be an interest, however, that would necessarily translate into an adoption of the new without reservations.

We have seen how the scientific panorama of the Iberian Peninsula included a series of institutions and figures that, beginning with the reign of Philip II, facilitated the advance of astronomy in Europe and negotiated a variety of obstacles and limitations. Starting from the artisanal lens-making initiative of the Roget family in Catalonia, passing through various figures associated with Andalusian intellectual circles and culminating in the institutional tasks of the Academy of Mathematics (1583) and the Imperial College of Madrid (1629), the turn of the century brought with it the necessary conditions for the reception and assimilation of the technical instrument that we know today as the telescope. However, we have also seen that the seventeenth century kicked off in a disheartening way with the 1600 execution of the philosopher, mathematician, and astrologer Giordano Bruno, and that its first three decades witnessed Galileo's fruitless attempts to travel to Madrid in order to continue his philosophical-scientific activity away from Rome's stifling yoke. As Víctor Navarro Brotons has pointed out, though, the conditions for scientific activity in Spain during this century were already very different from those of the preceding one, and the ideological contexts in which this activity took place were particularly complex, taking into account "the difficult conditions in which those Spanish astronomers who risked taking the path of cosmological discussion worked during this time" (*las difíciles condiciones en las que trabajaban los astrónomos españoles en esta época, que se aventuraban por los caminos de las discusiones cosmológicas*)

("Galileo y España" 816; see also Cardenal Iracheta 15–24; López Piñero, "Galileo" 51–58). The following pages focus on this conjunction of varying and often contradictory forces, so decisive for the development of the sciences and the arts in the Spain of the time.

The beginning of the seventeenth century offers a series of curious paradoxes. For example, it is ironic that the polemics between Cardinal Cesare Monti—at the time the papal nuncio in Spain—and the Grand Inquisitor favored the dissemination of Galileo's work, in however limited a form (Pardo Tomás, *Un lugar* 186–90). It is already well known that one of the events that marked the development of scientific and philosophical thought in seventeenth-century Spain was the 1616 condemnation of Copernicus's heliocentric theory, which severely curtailed debate. Additionally, despite the fact that the works of Galileo never appeared in the Spanish indexes of banned books, even in the face of pressure from Rome, the abjuration of the work by Cardinal Bellarmine, appearing in the same year, did have considerable influence in Spain. We already know that Copernicus's *De revolutionibus* was not included in the catalog of censored authors in the Spanish inquisitorial indexes, but that does not mean that the Inquisition was indifferent to the condemnation of the heliocentric theory by the Congregation of the Index. In the *Novus Index* of 1632, prepared by the Jesuit Juan de Pineda, Copernicus did not appear (because of an error, it is thought); however, numerous authors who defended heliocentric theory did appear, beginning with Rheticus, whose works were already included in the 1584 *Index*, as well as Kepler and other Copernicans. What is more, in the censorship of Kepler we can see an order for the expurgation of the repeated defenses of heliocentrism in the twenty or so works that are listed. Additionally, although Copernicus was not named in the body of the catalog, he did figure in the *index universalis* at the beginning of the work, even if incorrectly identified with another of the authors referenced. In any case, his work never lost its relevance, accepted as it was not only by the Protestants but also by the reformist factions of the Catholic Church, who ended up losing the battle. By the time Galileo made his first public declarations in favor of Copernicanism in 1610 and 1612, after his initial discoveries with the telescope, "the major holes of the Copernican system had already been filled" (*había colmado ya la mayor laguna del sistema copernicano*) (Beltrán Marí, introduction to *Diálogo*, xxxviii).

Although precise documentation exists regarding the arrival of the Roman decree in Spain, the Spanish Inquisition did not feel obligated to adopt the repressive measures taken by Rome. While it is true that the decree prohibited the *Dialogue concerning the Two Chief Systems* in addition to twenty-five other pieces, this prohibition of Galileo's work, originating in Italy, never took place; when the *Novissimus Index* was published in 1640, the name of Galileo did not appear anywhere. Monti created a conflict of interests when he sent the decree to all dioceses for publication without consulting the Council of the Inquisition, such that the council denounced the statement to the king, emphasizing that the measures taken by the papal nuncio were an attack on the prerogatives of the inquisitor general and, as such, on the Crown itself. The work of Galileo found itself wrapped up in a conflict that had little to do with its content but that in the end favored its dissemination in Spain, a dissemination that was somewhat limited given the condemnation of heliocentric theory and the prohibition or censorship *donec prodeat expurgatio* that weighed on the works of many European astronomers, especially Protestants.

The intellectual ties between Galileo and Spain were cemented in the different aspects of the activities that were developing in the realms of cosmography, natural history, natural philosophy, and engineering.[9] The most significant aspect of the communication established by the different agents who attempted without success to negotiate Galileo's move to Madrid concerned the controversial issue of the determination of geographical longitudes, as well as the application of the telescope. The first was based on the measuring of latitudes via astronomical methods, calculating the height of the sun or of the North Star, using the tables of solar declination—in the so-called method of the Sun—or the tabulated corrections (or indexes), assuming that the North Star was not exactly at the pole, giving rise to the so-called method of the polestar.

The problem of determining geographical longitude was of the greatest concern to navigators. The issue had reached maximum interest when the Treaty of Tordesillas (1494) fixed the meridian situated at 370 leagues west of the islands of Cape Verde as the dividing line between Spanish and Portuguese conquests. The Catalan cosmographer Jaume Ferrer de Blanes was charged with tracing this ideal line, and he had to be inventive to do so. The astronomical processes for the determination of geographical longitude, like those based on lunar eclipses, on the latitude of the moon, on the eclipsing of stars by the moon, and on lunar distance, were very difficult to determine once on board a ship. Facing the difficulties of these processes, the recognition of magnetic declination and its variation from one place to another gave rise to the idea that there was a simple relationship between this variation and geographical longitude. This in turn catalyzed the invention of instruments to measure the declination and thus solve the problem. In the mid-1590s, Philip II established a prize for whoever resolved this problem, offering, among other things, a prize of six thousand ducats and a lifetime pension. Among those who attempted to win the prize were not only scholars of Galileo's stature but also all manner of conmen and tricksters. One of the best-known disputes took place between Juan Arias de Loyola, former chronicler of the Indies and a professor for some time in the Academy of Mathematics, and Luís da Fonseca Coutinho, a Portuguese scientist. The proposals of these two authors were apparently similar and were based on magnetic declination, but Fonseca's, praised by prestigious cosmographers like Juan Bautista Labaña, generated more interest than Arias's, even though in the end both were dismissed. Around 1612, when Fonseca had already withdrawn of his own will from the competition, Arias found a powerful friend in the Count of Lemos, and his plan was finally able to be heard. In July of the same year the king issued a royal charter by which he would give the prize to Arias if his proposal gave the promised result.

The influence and reception of Galileo was made manifest, then, in two fundamental ways: one built by diplomatic and political contacts during his life and a second established indirectly by the translation and dissemination of some of his major works. With respect to the first, Jesús Sánchez Navarro has studied the negotiation with Spain between 1612 and 1618, "undertaken indirectly and through diplomatic means (Picchena, d'Elci, Argensola), only relatively involved with Galileo" (*llevada a cabo indirectamente y a través de medios diplomáticos (Picchena, d'Elci, Argensola) sólo relativamente comprometidos con Galileo*).[10] In 1612 he asked for permission to make his telescope commercially available, planning to bring about one hundred of them to Spain; as compensation he asked for a cross of St. James and a salary of four hundred escudos, with the added hope of being named to the Order of Santiago later on. Conceded all of these things by Philip III, Cosimo II de' Medici then demanded the free passage every year of two ships to the West Indies as the price of his permission for Galileo to leave Florence. That summer certain negotiations took place between the Tuscan and Spanish governments. At that time, Galileo clarified his intentions and the nature of his proposal, showing his willingness to move to Spain in order to personally instruct cosmographers and navigators on the matter. Additionally, he offered to direct the making of at least one hundred telescopes (indispensable for the observation of Jupiter's satellites), to publish yearly chronicles of the daily movements of those "stars," to edit a text about "all of this portion of the new astronomy" (*toda esta parte de la nueva astronomía*), and to instruct cosmographers and navigators in the use of the *celatone* (headgear), a type of "bracket fitted to the head of the observer and designed to facilitate observations from a moving ship" (*soporte ajustado a la cabeza del observador y diseñado para facilitar las observaciones desde una nave en movimiento*) (Navarro Brotons, "Galileo y España" 811). Galileo edited a short text to this effect on the problem of determining longitudes, where he emphasized that of all known methods the best was the lunar eclipse method, which nonetheless suffered serious limitations. He made references to his proposal, affirming that he had discovered in the heavens "things completely unknown in past centuries, which are equivalent to more than one thousand lunar eclipses every year, observable with great precision and, what is more important, reduced to calculations and to exact and exquisite tables. And all of this will be dedicated to the King's great majesty" (*cosas totalmente desconocidas en los siglos pasados, que equivalen a más de mil eclipses lunares cada año, observables con muchísima precisión y lo que es más importante, reducidas al cálculo y a tablas justísimas y exquisitas. Y todo este asunto será consagrado a la gran Majestad del Rey*) (qtd. in Navarro Brotons, "Galileo y España" 811; see also Galileo, *Le opere* 419–23). Thus he offered his solution to the problem of longitudes based on his observation of the occultation of the satellites of Jupiter. The king wanted Galileo's *Proposto por la longitudine* to be considered, but it was Count Orso d'Elci himself, the Tuscan ambassador in Madrid, who presented some objections, perhaps influenced by conversations with members of the House of Trade.

After this first attempt failed, Galileo presented himself again in 1616, by means of the ambassador from Florence, still hoping to resolve the problem of longitude and teach the use of his telescope, but this time with much more modest pretensions.[11] Finding himself in Rome, summoned by the ecclesiastical authorities for his first trial, he resumed negotiations with Spain personally and, following d'Elci's advice, he wrote two letters about the matter, together with a "general explanation of his invention," to the Duke of Lerma and the Count of Lemos. Bartolomé Leonardo of Argensola wrote to Galileo to tell him that Lemos, well versed in the topic of navigation, was interested in his proposal. In fact, the count sent Argensola to Rome to become acquainted with Galileo, and it seems that they struck up a friendship; from the Tuscan's correspondence it is known, in fact, that he was always grateful to his Spanish friend for the gesture, despite that he did not achieve his goal.[12] Otis H. Green took this even farther, writing:

> In his poems, Bartolomé Leonardo shows an uncommon interest in astronomy, and especially in Archimedes. These discussions with Galileo must have been a revelation for him. The fact that Galileo esteemed him as much as his letters suggest leads us to believe that they not only understood each other but that they felt themselves to be intellectual colleagues.

En sus poesías Bartolomé Leonardo demuestra un interés poco común en la astronomía, y sobre todo en Arquímedes. Esos coloquios con Galileo debieron ser una revelación para él. El hecho de que Galileo le estimaba tanto como da a entender en sus cartas nos hace creer no sólo que los dos simpatizaron sino que llegaron a sentirse compañeros intelectuales. ("Bartolomé" 383)

In a letter to d'Elci dated November 13, 1616, Galileo was already talking about his telescope, or "glass," with which he had been able to magnify normal vision by forty or fifty times; he also mentioned another instrument, binoculars, that helped him to follow the movement of objects without losing sight of them. In the end, though, he was never able to travel to Spain; the king sent Galileo's proposal to the president of the Council of the Indies to be evaluated by experts, since other proposals by Arias de Loyola and the swindler Lorenzo Ferrer Maldonado, "a drifter between marginal subculture and academic science" (*aventurero a caballo entre la subcultura marginada y la ciencia académica*) (López Piñero, *Ciencia y técnica* 117), had already proved useless. In a letter from Philip III to the Duke of Osuna dated January 28, 1620, the monarch communicated Galileo's offer to "demonstrate the method by which to measure longitude and to facilitate and secure navigation on the sea and [he] offered another invention for the Mediterranean galleys with which the ships of the enemy could be seen from ten times farther away than with ordinary vision" (*dar el modo para poder graduar la longitud y facilitar y asegurar la navegación del océano y* [. . .] *ofrecía también otra invención para las galeras del Mediterráneo con que se descubrían los bejeles del enemigo diez veces más lejos que con la vista ordinaria*).[13] The proposal was renewed by the scientist on two more occasions in the midst of unfortunate episodes; in 1620, encouraged by the Tuscan ambassador in Madrid, Giuliano de' Medici—a defender of Galileo's interests who had advocated for him with the royal favorite and minister Baltasar de Zúñiga—convinced the king himself to insist that the Duke of Osuna reopen the case. Galileo was invited to Naples to meet with the viceroy, Cardinal Gaspar de Borja, but nothing came of it. He made another attempt around 1629, and the wind seemed to be in his favor—the Duke of Medina Sidonia, for example, was very interested in him—and Galileo was informed by Gianfrancesco Buonamici, secretary of the Tuscan ambassador in Madrid, that Philip IV was interested in buying a number of telescopes. He agreed to sell one, even though he had never done so, nor did he plan to do so again, but he did not prepare the parcel in time for the emissary of the Grand Duke of Madrid, Esau del Borgo, to take it with him on his trip to Spain. It seems that he was able to send another later on, but when the monarch received the telescope it broke accidentally—or perhaps even arrived broken. By that time, the Duke of Medina Sidonia had pointed out certain objections to the issue of the moons of Jupiter, especially to the notion that they could be observed from a ship in motion, and once again, it all came to nothing.

The institutional structures by which Galileo's advances were transmitted were just as complex. As Fernando Bouza (*Imagen*) has rightly signaled, science was an essential part of the image of the monarchy as constructed by Philip II once Madrid was inaugurated as the seat of the court in 1561. As a result, the academies took on an essential role, competing with universities like Salamanca in an attempt to occupy the empty space in Madrid, which lacked a comparable center.[14] Mauricio Jalón has pointed out their importance:

> The academies of the seventeenth century were a type of "clearing house"; they developed criticism of new works, acted as moderators in intellectual controversies, and disseminated modern knowledge. It is precisely the lack of these institutions in Spain until the final years of the century that serve as an index of both the country's backwardness and the difficulties faced in any attempt to achieve progress.
>
> *Las academias del siglo XVII fueron una especie de "centros de compensación"; desarrollaban una crítica de los nuevos trabajos, ejercían de moderadores en las polémicas intelectuales y difundían el saber moderno. Justamente la carencia en España de estas asociaciones, hasta finales de la centuria, a la vez será índice de su atraso e impedimento para salir de él.* (70)

Initially, the standout figures were academics like Rodrigo Zamorano, a member of the House of Trade from 1575 until his retirement in 1613, and the author of the *Compendio de la arte de navegar* (Compendium of the art of navigation, 1581), whose tables of the declination of the sun were corrected to match the results of Copernicus, among others. In 1583 the first lectures were given in the academy—at the time called the Royal Mathematical Academy of Madrid—closely following those given in the Department of Astrology and Mathematics of the University of Salamanca in 1561. The building had been constructed on a site near the Royal Alcázar and the Gate of Balnadú, which had formed part of the old convent of St. Catherine of Siena. During its first eight years of existence, the academy was dependent on the palace for its administration, and the control of its activities was given to the head chamberlain Juan de Herrera, a close friend of the monarch (Piñeiro, "Los cosmógrafos" 128–29). The principle interest was in the pragmatic application of the academy's teachings, especially in disciplines like mathematical finance, cosmography, astrology, navigation, and fortification. Víctor Navarro Brotons and William Eamon give a summary of the academy's trajectory in these years, reminding us of the aforementioned role of Jerónimo Muñoz—who so interested Tycho Brahe:

> It would be difficult to detect any significant differences between the manner in which the new cosmology was taught in the Spanish universities and that of their European counterparts. Nor did Spain lack critics of Aristotle. Jerónimo Muñoz, a distinguished mathematician, astronomer, and geographer who taught at Valencia and Salamanca, was a fierce critic of Aristotelian cosmology. Muñoz became widely known for his precise observations and theoretical conclusions regarding the supernova of 1572, and in his *Libro del nuevo cometa* refuted the Aristotelian dogma of the incorruptibility of the heavens. Vestiges of the Black Legend continue to perpetuate the stereotype of sixteenth-century Spain as fanatical and Inquisitorial, and as an enemy of progress and innovation. Yet it is not without significance that it was the Inquisition in Rome, not Spain, that prosecuted Europe's leading Copernican, Galileo, while the major Spanish defender of the Copernican doctrine, Diego de Zúñiga, was allowed to publish his opinions freely, without real threat of prosecution. (32)

The academy and the activities of the faculty were well known in Europe, as demonstrated by the widespread correspondence of some Spanish Jesuits with Cristoph Clavius, the aforementioned professor of the Roman College. However, the separation of disciplines meant that the Jesuits did not entangle themselves with delicate topics like natural philosophy or astronomy. Undoubtedly, there was some attempt to give this institution an autochthonous flavor in order to reenergize a somewhat stagnant scientific field; this would be the case, for example, of the lectures of the Portuguese mathematician Juan Bautista Labaña (1555–1624), which were in Castilian, not Latin, as were the textbooks used.[15] This was an issue that, as José Manuel Floristán Imízcoz has argued, greatly delayed the beginning of the lectures. "It is important to emphasize this radical vernacularization of an institution dedicated to teaching," according to José Pardo Tomás, "because it is a feature that allows for a very clear delineation of court cosmography from that of the university, which was produced, taught, and communicated in Latin" (*es importante subrayar esta radical vernacularización de una institución dedicada a la enseñanza porque es un rasgo que permite distinguir muy nítidamente la cosmografía cortesana de la universitaria, que se producía, se enseñaba y se comunicaba en latín*) (*Un lugar* 69).[16] In this way, the court cosmography addressed itself to a primarily Castilian audience that could go on to careers in technical work at the service of the monarchy. The work plan of the academy professors included the translation of necessary scientific texts to Castilian; thus, in the first years of the academy's functioning, Euclid's *Optics*, along with *Catoptrics* (attributed to him), Theodosius's *Spherics*, Archimedes's *Equiponderantes*, and Apollonius's *Conics*, were all apparently translated under Pedro Ambrosio de Ondériz. Mauricio Jalón has written of these translations:

> For almost the entire first half of the seventeenth century, Italian, after Latin, was the language of science. Galileo himself, by publishing lengthy works in his mother tongue, promoted the discussion of opinions and the internal and external exchange of news. However, there was also a notable rise in *italophobia* as a reaction to the values of the Renaissance. At the same time, Latin was a feature of scientific modernity, precisely for the purpose of exchange, such that Bacon and Descartes, Newton and Leibniz wrote under this *European sign*, even if they also prepared vernacular editions of the same texts or edited versions depending on whom they wished to address.

> *Durante casi toda la primera mitad del siglo XVII, el italiano era, detrás del latín, el idioma de las ciencias. El propio Galileo, al publicar grandes escritos en su lengua materna, fomentó la discusión de opiniones y el intercambio interno y externo de noticias, aunque paralelamente se advierte una progresiva* italofobia *como reacción frente a los valores del Renacimiento. Al mismo tiempo, el latín fue un rasgo de modernidad científica, justamente por motivos de intercambio, de modo que tanto Bacon como Descartes, Newton o Leibniz escriben en esa especie de* signo europeo, *si bien redactan en vulgar el mismo texto u otro retocado según a quién deseen dirigirse.* (163)

During the final years of the sixteenth century, the academy went through various transformations and changes in direction. Beginning in 1591 it became part of the Council of the Indies; Juan Arias de Loyola took the place of Labaña, also receiving the title of chief chronicler of the Indies, one of the two positions into which the title of chief chronicler and cosmographer of the Indies had been divided. Four years later Philip II forced Arias to retire and named the Milanese engineer Giuliano Ferrofino as chair of mathematics; he also worked as the academy's only professor. With the appointment of Juan López de Velasco as secretary of the Council of Trade in 1591, the office of chief cosmographer fell to the previously mentioned Ondériz, by that time a professor of cosmography at the mathematics academy. He had to maintain his faculty and translation activities in the academy and carry out the task of emending and correcting maps and nautical instruments following the instructions of Juan de Herrera. However, Ondériz occupied the post for only five years, until he died suddenly in the first days of 1596. In 1599 the governor of Murcia, Ginés Rocamora y Torrano, published a piece in the Madrid press of Juan de Herrera titled *Sphera del universo* (On the sphere of the universe), in which he summed up the scientific explanations given during his stay at the royal court. In the introduction to the text, Rocamora gives effusive praise to applied mathematics, mentioning the golden age in which the academy found itself, and including a Spanish translation of Johannes de Sacrobosco's *Sphaera mundi* (On the sphere of the world) in the last part of the work (Fernández de Navarrete 585–90). Mariano Esteban Piñeiro ("Las academias") sums up the events:

> During the 150 years that followed, the succeeding chairs—appointed by the king, nominated by the Director of the College, and chosen by the preceding report from the Council of the Indies, which continued to defray the cost of the College's salaries and operating costs—remained obligated to teach the same subjects and the same "plan of study" offered by García de Céspedes, although an analysis of their activities shows that the readings were very current, such that "Herrera's department" was the principal center for the diffusion of European scientific advances in Spain during the seventeenth and eighteenth centuries. This is obvious in light of the fact that the chair was occupied by Jesuits, Spaniards, and foreigners of the scientific caliber of, for example, Claudio Ricardo, Carlos de la Faille, Jacobo Kresa, Pedro de Ulloa, Alexandro Berneto, Nicasio Gramatici, Manuel de Campos, Carlos de la Reguera, Pedro de Fresneda, Juan Wendlingen, Cristiano Riegen, and Tomás de la Cerda.

> *los sucesivos Catedráticos, durante los siguientes ciento cincuenta años, seguían obligados por sus nombramientos—otorgados por el monarca, a propuesta del Director del Colegio y previo Informe del Consejo de Indias, institución que continuó sufragando sus salarios y los gastos de funcionamiento de la cátedra—a impartir las mismas materias y el mismo "plan de estudios" ofrecido por García de Céspedes, aunque el análisis de sus actividades revela una gran actualización de las lecturas, de forma que la "cátedra herreriana" fue el principal centro difusor de los progresos científicos europeos en España durante los siglos XVII y XVIII. Lo que resulta evidente si se comprueba que ocuparon la cátedra jesuitas, españoles y extranjeros, del nivel científico de, por ejemplo,*

Claudio Ricardo, Carlos de la Faille, Jacobo Kresa, Pedro de Ulloa, Alexandro Berneto, Nicasio Gramatici, Manuel de Campos, Carlos de la Reguera, Pedro de Fresneda, Juan Wendlingen, Cristiano Riegen y Tomás de la Cerda. (11)

When the Court moved to Valladolid in 1601, the academy ceased all activity until 1607. With the return of the Court to Madrid, the Council of the Indies decided to renew the academy's activities, offering the chairmanship to Dr. Andrés García de Céspedes, the chief cosmographer of the Indies at the time. He had spent seven years in the Portuguese Court, learning from prestigious cosmographers and acquiring an education that would lead him to be considered one of the finest scientists in Europe (Piñeiro, "Los cosmógrafos" 126–27, 132–33). Until his retirement in 1611, García de Céspedes developed the *cursus*, or course of study, of mathematics, which did not differ from what had been taught in the Department of Mathematics and Astrology at the University of Salamanca in 1561. In the academy of the Court, Copernicus's work was not explicitly listed on the syllabus, while at Salamanca it was, regardless of whether it was actually studied.[17]

In 1611, Philip III named Juan Cedillo Díaz, dean of the cathedral of Pastrana and personal chaplain of the influential Marqués de Moya, as the new chief cosmographer of the Indies and chair of mathematics and cosmography at the Academy of Mathematics in Madrid. He remained in the chairmanship until his death in 1625. On the study of geometry in the academy, the Galician scholar Cristóbal Suárez de Figueroa would later write in discourse 23 ("On Surveyors, Measurers, or Architects, and Weighers") of his *Plaza universal de todas las ciencias y artes* (The universal plaza of all sciences and arts, 1615):

This department is so virtuous and of such genius very few can attend. Knowing its importance, it is read by order of his Majesty publicly in Madrid. Today the chair and its salary of eight thousand ducats are held by Juan Cedillo Díaz, extremely well versed in mathematics. He succeeded the distinguished Andrés García de Céspedes, a great investigator of the science, on which he composed not a few volumes, even if he only printed two: one on geometric instruments, and one on navigation.

Por ser esta facultad tan virtuosa y de tanto ingenio la siguen pocos. Conociendo su importancia se lee por orden de su Majestad públicamente en Madrid. Tiene hoy su cátedra con salario de ochocientos ducados el Doctor Juan Cedillo Díaz, versadísimo en Matemáticas. Sucedió al insigne Andrés García de Céspedes, grande inquiridor de esta ciencia, sobre que compuso no

> *pocos volúmenes; si bien imprimió solos dos; uno de instrumentos Geométricos, y otro de navegación.* (209)

It is believed that Cedillo studied in Salamanca, where Jerónimo Muñoz and his disciples were teaching, which would explain many of his scientific and technical ideas. Cedillo undertook the difficult task of translating a variety of scientific texts, as required by his appointment as chair of the academy, and he also wrote several original treatises on engineering, cartography, mathematics, and astronomy, in accordance with his faculty duties (Jalón 156–57). He also defended heliocentrism and brought together the previous theories of Tycho Brahe and Giordano Bruno on the immateriality of the celestial spheres; furthermore, he located the planets as orbiting in circles because of the "intelligence"—that is, the spiritual substance that governed the heavens—they provided from a distance. This theory followed the ideas of Kepler, Scaligero and Gilbert, an idea that had not yet appeared in the work of Copernicus. He also continued the plan initiated by Ondériz, translating between 1616 and 1625 a number of important works: Pedro Núñez's *Del arte de navegar* (Art of navigation, 1562), the first six books of Euclid's *Elements*, Niccolò Tartaglia's *Nova scientia* (New science, 1537), and Giovanni Antonio Magini's *Novae coelestium orbium theoricae congruentes cum observationibus N. Copernici* (New theories of the celestial orbs congruent with the observations of Copernicus, 1589).[18] The manuscripts of Cedillo's translations—four volumes on parchment—are conserved in Madrid's Biblioteca Nacional.[19] Cedillo's incomplete translation of Copernicus, the first in the Spanish language, is preserved through the thirty-fifth chapter of the third book, but the translation of Osiander's text, Schönberg's letter, and the preface addressed to the pope are missing. Cedillo titled the work *Ydea astronomica de la fabrica del mundo y movimiento de los cuerpos celestiales* (Astronomical idea of the fabric of the world and movement of celestial bodies), a translation, it happens, in which his name did not appear anywhere. However, beginning with the second book, when Copernicus speaks of "us" in referring to his observations, Cedillo does name him, perhaps showing his uncertainty regarding the presentation of the work (Navarro Brotons, "Galileo y España" 815). One of the drafts of the translation even includes a text that does not come from Copernicus and that seems to be the introduction that Cedillo himself prepared, attempting to better communicate Copernicus's concerns:

> I already knew well when I decided to bring the fruit of my studies to light that many learned men would reproach me for being one of those who

> seems to bring greater novelties to the world than anyone has brought before: because although it is true that our elders have examined them somewhat vaguely, all together as one, I suspect no one has examined them such that they come out to the end result I will give them here.
>
> *Bien sabia yo quando determinava sacar a luz los trabajos de mis estudios que me avían de reprehender muchas partes de los hombres doctos por ser yo unos de los que parecen traer mayores novedades al mundo que ninguno hasta nuestros tiempos a traydo: porque aunque es verdad que las cosas que digo las an tratado difusamente nuestros mayores, todas juntas en un cuerpo y que salgan a un fin como aquí las declararé sospecho que no las a tratado ninguno.* (180r–181r)

It has also been suggested that some authorities had reprimanded Cedillo, who then abandoned the translation without publishing it and prepared the new introduction identifying himself with Copernicus. However, in this same introduction, Cedillo presents some cosmological ideas that do not completely coincide with Copernicus's vision. Although he does place the sun at the center of the cosmos, he argues that the planets move through cosmic air like fish through water, following his teacher Jerónimo Muñoz. He also maintains that the epicycles and eccentrics of the planets are not spheres, but circles, moved by intelligences (the earlier-mentioned spiritual substances) situated at the center of the planet. Curiously, he affirms that the retrograde motions, directions, and appearances of planets are explained by eliminating the spheres. However, in Copernicus's system there is no retrograde motion, because this is explained as a mere appearance resulting from the observation of the planets from a moving platform. As such, this mention of retrograde motion could be interpreted to mean that, before becoming inclined toward Copernicus's theory, Cedillo had already abandoned the idea of the spheres, as Muñoz, Pérez de Mesa, and other authors had before him, although these others did not defend the idea of the Earth's movement.

At the time of Cedillo's death in 1625, edicts were once again issued in Spanish universities in search of another high-level mathematician; but, just as in 1604, no one with sufficient prestige was found. The Council of the Indies, wishing to avoid the closure of the Royal Academy, opted for a provisional solution: until a suitable candidate was found, the lessons would be carried out by members of Madrid's Society of Jesus in the usual academy annexes near the Gate of Balnadú. The instructors would be chosen by the director of the recently created Imperial College of San Isidro, opened with the support of the Count-Duke of Olivares. This is how the following three academic years (1625–1628) were held, and the teachings most likely had a very different makeup than before. No diplomas or certifications were awarded during this time, and in 1628 the royal superintendent of general studies, Juan de Villela, pressured the king so that the Jesuits could give the courses in cosmography, mathematics, and architecture in the buildings of the Imperial College itself. The king issued a royal decree to this effect on September 10 of that same year. García de Céspedes's courses were recognized, but whether they were actually taught is not known, nor is the extent to which Copernicus's work was imparted in the following years.

Nuria Valverde and Mariano Esteban Piñeiro have pointed out that the Imperial College became the principal circulator of scientific knowledge in Spain at this time. On the site of Calle de la Colegiata, Calle Toledo, and Calle de los Estudios, the Jesuits had established the college where Lope de Vega, Calderón de la Barca, and Francisco de Quevedo would study one day (1572–1602). Their interests lay primarily in mathematical physics, and they worked to remain in communication with the rest of Europe. Astronomy was still subject to philosophical doctrine in its theoretical aspects, meaning that its renewal had to deal with thorny questions of cosmology and natural philosophy. This renewal had to overcome the traditional division between science and technology; to do so, the practical application of scientific theory had to attain a level of social prestige that eclipsed the toxic notion that manual labor was dishonorable.

Driven by a spirit of competition with the three great universities of Alcalá, Valladolid, and Salamanca, the Imperial College became tremendously powerful in just a few years. Around 1625, as a result of the negotiations between the Spanish government and the superior general of the Jesuit order Mutio Vitelleschi, a royal institute was founded with the principal objective of educating the children of nobles. The curriculum of studies included "lower studies" of Latin grammar and "higher studies," where the later comprised seventeen departments, including two of mathematics; one dedicated to the spheres, astrology, astronomy, astrolabes, perspective, and prognostications; and another to geometry, geography, hydrography, and clocks. Additionally, the Jesuits took on the Court chair of mathematics that had belonged to Cedillo Díaz, which included the title of chief cosmographer of the Indies. This meant that, during the controversial 1630s—and, in reality, for a large part of the century—the Imperial College was the center for the reception and discussion of Galileo's astronomical work.[20] Its members attempted to stay in contact with what was happening in Europe, always trying to bring the continent's greatest scientists as visiting professors. During the 1627–1628 academic year, the

Swiss German Juan Bautista Cysat lectured on mathematics, although in 1629, for unknown reasons, he was no longer in Madrid. Vitelleschi attempted to attract without success the Belgian Grégoire de Saint-Vincent, although he filled the position of professor of mathematics with one of his finest students, the Belgian Jean-Charles della Faille, who took the post in 1629. The *Tratado de la teoría de los planetas, de la fábrica y uso del astrolabio, de la fábrica y uso del anteojo de larga vista, de la refracción y sus propiedades* (Treatise on the theory of the planets, on the making and use of the astrolabe, on the making and use of the long-range telescope, on refraction and its properties) is attributed to him. That same year Claude Richard, a scholar from Burgundy, was also named professor of mathematics. Along with della Faille and Richard, in the first decades of the royal institute's existence at the Imperial College, the Polish Alexius Sylvius Polonus, the Scotsman Hugh Semphill (Hugo Sempilius), the Italian Francisco Antonio Camassa, the French Jean François Petrey, and the Czech Jakub Kresa, as well as the Castilian mathematician José Martínez and the Basque Jesuit Francisco Isasi, who taught military arts, all lived and taught at the institution. These authors published few works, but they left behind an important body of manuscripts related to their classes and scientific works that enable contemporary scholars to reconstruct their teachings and activities. In his *De mathematicis disciplinis* (On the mathematical disciplines), published in Antwerp in 1635, Hugo Sempilius tried to offer a general vision of the varying mathematical disciplines that interested him, namely geometry, arithmetic, optics, statics, music, cosmography, geography, hydrography, meteors, astronomy, astrology, and the calendar. Additionally, though, Sempilius dedicated a lengthy chapter to the controversial issue of whether mathematics was a true science, closely following and endorsing the theories of Giuseppe Biancani. He also wrote extensively on the usefulness of mathematics, emphasizing his need to discuss questions of natural philosophy, including those related to movement. The defense of mathematics undertaken by Sempilius was directed, without a doubt, to support and promote his teaching in the Imperial College as being at the same level as the other disciplines and, in general, to call the attention of the governing elites to its importance and utility. More important for our purposes, though, were the many references to Galileo's contribution to astronomy; in the section on optics the Scotsman described the Galilean telescope, emphasizing that its creator was the first to apply the instrument to astronomical observation. In the chapters on astronomy, the author highlighted Galileo's observations of the satellites of Jupiter and the appearance of Saturn, his controversial surface of the moon, and the observation of sunspots. Regarding the lunar relief,

Sempilius insisted that no interpretation was possible other than the one offered by Galileo in *Sidereus nuncius*, but even so, he took precautions, observing that the goal of mathematics was only that which related to the quantity, figure, movement, position, and number of the heavens. By 1620, it should be noted, *Sidereus* was already being taught by professors at the House of Trade in Seville.

Outside of the work undertaken by Sempilius, the astronomical discoveries of Galileo were incorporated during these decades into texts like Juan Eusebio Nieremberg's *Curiosa y oculta filosofía* (Curious and hidden philosophy, 1630) and *Fábrica y uso del anteojo de larga vista* (Craft and use of the long-range telescope) by the aforementioned Jean-Charles della Faille, who described the telescope in what survives of a manuscript preserved at Madrid's Academy of History. The former focused his efforts on the "renewed philosophy of the heavens" (*filosofía renovada de los cielos*) in which he declared himself against Copernican theory. However, he considered the Ptolemaic system to be obsolete and showed himself to be partial to Tycho Brahe. He also described Galileo's astronomical discoveries, denying the solidity of the "celestial spheres" and citing various observations of comet trajectories, novae, and planetary movements. Finally, he defended the corruptibility of the heavens and that the stars were in motion by themselves, such that the stars were of the same nature as the Earth. In his treatise *Natural History*—in which he included a number of materials from the work of Francisco Hernández—Nieremberg also addressed astronomical and cosmological issues in similar terms to the ones he had employed in *Curiosa filosofía* (Curious philosophy). These contributions complemented the research on mechanics and the science of movement that was produced in the following decades, although the diffusion and discussion of these works was much slower. Traces of Galileo's mechanics cropped up in works of military architecture, in treatises on astronomy and geography, in attempts to create mathematical compendiums that included the so-called mixed mathematics, and in many manuscripts on varying subjects. This is indicative of how his work was received and assimilated in a country like Spain, which had to wait for the arrival of a new century to adopt definitively what it had been cultivating for nearly one hundred years.

First symptoms: The "new physics" and the treatises on optics

The preceding pages have identified some of the reference points of the technical-scientific angle that defined both the diplomatic relationship between Italy and Spain and the public standing of each of the parties involved during these first four decades of the seventeenth century. Many of these references were reconstructed from the correspondence between scientists—often Jesuits—and from the exchange of books and instruments, as well as from the rumors and secrets that circulated throughout Europe. A figure like Galileo demonstrates how delicate the negotiations between courts, patrons, and governments could be when a man who was no longer a scholar in the traditional sense, but rather the embodiment of an entirely new vision of the cosmos during a period of upheaval, attempted to move to a new country. While the genius of Galileo celebrated, in a certain sense, advances in technology thanks to the reputable glass provided by the local industry, Habsburg Spain slowly became more open to his ideas with the writing of scientific treatises and with translations that testified to the impact of the new in the Iberian Peninsula. From manuals on tobacco consumption and the correct preparation of chocolate to treatises on cordials, alcoholic beverages, and new tropical fruits, the new products were surrounded with a halo of mystery, defined as much by curiosity about them as by the caution with which they were praised (García Santo-Tomás, *Espacio urbano*).

This attitude affected many branches of knowledge related to the use of concave lenses, particularly in Castile, which was generally much more conservative than the Crown of Aragon, home to a flourishing lens-making industry that had existed for decades. Soon we will pause to examine the debates generated by the use of glasses in the social sphere, which took on a fascinating form in artistic expression. Glasses were linked to reflections that hinged on ideas of *vanitas* and *gravitas*, as much for the literary character as for the subject painted on canvas. *Anteojos, gafas, quevedos, espej(uel)os*—the terminology is as varied as it is meaningful, and it speaks to the morphology of the object as well as its journeys beyond local frontiers. Humor and sarcasm, brought on by the strangeness or the ostentatiousness the visual effect of eyeglasses, also found their way into a range of responses. For example, who can forget the mention by Lope de Vega of Cervantes's spectacles, which looked like "poorly fried eggs" (*huevos estrellados*), as he wrote to the Duke of Sessa after borrowing them to read a sonnet at an academic gathering.[21] Glass was a gateway for reflection on and appreciation of access to new realities, even if only as a filter and even if those realities were complicated and made less perceptible by its imperfections.

Cervantes and Lope de Vega were not the only writers of the time in need of a good pair of eyeglasses—one of the most common terms for a type of them, *quevedos*, comes from the satirist Francisco de Quevedo, who was notoriously nearsighted (fig. 5). What is more interesting, how-

FIGURE 5. Portrait of Francisco de Quevedo, attributed to Juan van der Hamen

ever, is the treacherous waters that the ever-changing field of optometry had to navigate during these decades of turmoil, undoubtedly troubled by doubts and fears surrounding the advantages and inconveniences of using eyeglasses. Given that the average citizen of the time was not able to draw a clear distinction between the benefits and risks of glasses, they became highly controversial in the delicate balance between the natural and social orders. We already know that convex lenses had been used to combat presbyopia, and that later on the use of concave lenses to treat myopia also became common. Additionally, knowledge of the camera obscura, the laws of refraction, and the double pinhole occluder for testing visual acuity brought about both progress and mistrust in physical optics, as they advanced a more complete systematization of physiological optics, which in turn would be fundamental for the nascent field of medical optics. As lenses developed, the design of the frames transitioned from the articulated monocle—which disappeared over time in the sixteenth century—to a two-piece frame with a round bridge, supporting itself on both sides of the nose. These were the aforementioned *quevedos,* which eventually came to be attached to the ears with cords to create an effect between comical and bizarre, when one was not displaying a certain sense of vanity (Clark 10). Such was the case, as we will see later, of the two celestial travelers in Vélez de Guevara's *El diablo Cojuelo,* who tied their lenses to their ears with guitar strings. The literary use of optometry was thus pointing to two parallel concerns: the adoption of glasses as a sign of social prestige in a society that suffered from the same moral blindness that many novels denounced and the growing tension between religion and astronomy, which resulted from the use of magnifying lenses to observe the heavens.

This complex intersection would become more apparent in light of the discoveries that would later change the scientific panorama of Europe. If *Sidereus nuncius* was distributed initially in Spain in the community of Sevillian cosmographers and some intellectuals associated with that city's powerful university, its contents did not go unnoticed in the realm of optometry. Optometry as a discipline had already made interesting scientific contributions in the previous century; early studies like *Libro del ejercicio corporal y de sus provechos, por el cual cada uno podrá entender qué ejercicio le sea necesario para conservar su salud* (Book of corporeal exercise and of its benefits, by means of which one may understand what exercise is necessary to conserve his health, 1533) by Cristóbal Méndez, had recommended a type of "ocular gymnastics" to improve vision (González-Cano, "Eye Gymnastics"). However, it would be the manual of a notary of the Inquisition and member of the House of Trade, the Cordovan Benito Daza de Valdés, who would most accurately and completely reflect the existing debates in the field of optics, offering contemporary readers a complete look at the ways in which the use of eyeglasses was perceived at the time.

Referring to the manuscript of a treatise on perspective by Antonio Moreno, a cosmographer and professor at Seville's House of Trade, Daza made his *Uso de los antojos para todo género de vistas: En que se enseña a conocer los grados que a cada uno le faltan de su vista, y los que tienen cualesquier antojos* (Use of eyeglasses for all kinds of eyesight: In which it is taught everyone's prescription, and those of their lenses, 1623) into the most ambitious study of optics in Spain. With the goal of reaching the widest possible audience, the author skillfully integrated the theoretical with the fictional in a superb demonstration of didacticism and erudition. It is thought that, in his work as a notary, Daza could have had extensive contact with Dominican monks, who in Spain held a monopoly on the Holy Office. These monks played an essential role in the dissemination of glasses in Europe. What interests me here is that this is the first known text that incorporates some of Galileo's theses in the field of optics, interspersing passages from *Sidereus* without citing its author. It is also an extremely valuable testimony that contains pivotal information about the lens-making and scientific centers of Spain, and in particular the use of glasses in the society of the time both by those who needed them and by those who used them as tokens of distinction.

The main text, preceded by a short prologue, is composed of three books. The first, titled "De la naturaleza y propiedades de los ojos" (On the nature and properties of eyes), is divided into eleven chapters, in great detail describing the visual functions and examining the defects that require optical correction. The second book, "De los remedios de la vista por medio de los antojos" (On the remedies for sight by means of eyeglasses), is divided into ten chapters, analyzing the optical properties of concave and convex lenses, and the proper use of gradation to avoid visual inaccuracy. The last book is made up of four "Dialogues," in which a master (craftsman) and a doctor are the central figures. Daza makes use of this dialectical tool to emphasize the need to combine technical skill with a theoretical expertise. In the first pages of his book, the Cordovan scholar identifies and treats the most common symptoms (myopia, hyperopia, cataracts) that afflict his contemporaries. At the same time, he offers interesting solutions to everyday problems: how to find the correct prescription, how to get the appropriate frames, how to clean the lenses or prevent them from "clouding," and so on. At the beginning the different properties of the eye are revealed, including a series of anatomical considerations taken from Galen's *De usu partium* (On the use of the parts), from Juan Fragoso's *Chirurgia universal* (Universal surgery) as well as the contributions to optometry from Matteo Realdo Colombo's *Re anatomica* (On things anatomical). Daza praises the powers of vision, occasionally using a rhetoric that borders on the mystical. He also devotes a portion of the work to analyzing the different "kinds of sight" (*vistas*) and to a lecture on who does and does not need eyeglasses.

From a historical point of view, the second book is the most interesting, in that it praises the invention of something as valuable as eyeglasses. However, as the author of the prologue to the 1923 edition points out:

The author *either does not know or does not want to say anything* about the inventor of glasses. The latter seems more likely for a man who demonstrates his erudition in the history of other topics, and it might be that reasons primarily of a religious or dogmatic nature, which must be kept in mind for a notary of the Inquisition, made him feel obligated to say nothing. In this he resembles the brothers Giordano da Rivalto and Alexandro della Spina three centuries before, who hypothesized that Bacon had been the inventor, but whose name, suspected of heresy, they dared not cite.

> *el autor* o no sabe o no quiere decir nada *sobre el inventor de los antojos. Más bien parece ser lo segundo en un hombre que demuestra ser tan erudito en las historias de otros asuntos, y acaso razones, de índole religiosa y dogmática sobre todo, muy de tener en cuenta en un notario de la Inquisición, hiciesen que él se creyese obligado a callar, como tres siglos antes los hermanos Giordano da Rivalto y Alexandro della Spina, en la hipótesis de que hubiera sido Bacon el inventor, cuyo nombre, sospechoso de herético, no se atrevían a citar.* (50)

The explanation that Daza gives about different types of lenses in the first paragraph turns out to be very telling (Reeves, *Painting* 209–10). He analyzes rock crystal, as both mirror crystal and glass crystal. The latter is a type of glass that is "very fine, made in Muran, a pleasant place near Venice, where they craft such excellent glasses that they almost compete with the best ones made of rock [crystal]" (*finísimo que se hace en Muran, lugar ameno junto a Venecia, de que se labran los anteojos tan excelentes que casi compiten con los mejores de roca*) (119–20). At a time in which Venetian glass was symbolically associated with treason, espionage, and aggression, Daza's praise is no doubt striking. The rest of the second book discusses the differences between convex, concave, and "conservative" lenses, before closing with advice on how to order glasses for weak or strained vision.

The third book is made up of a series of dialogues between those who search for remedies, in this case a master craftsman, whom today we would call an optometrist, and a doctor. In the first dialogue, a fictional character named Claudio goes to consult the master craftsman for a vision exam. The exchange of impressions is very useful not only because it helps us understand the state of scientific research in Seville during the first third of the seventeenth century, but also because it sheds light on many practices and customs that are very familiar today. This is especially true regarding the relationship between doctor and patient; the doctor offers excellent advice to Claudio, who suffers from tired vision, and his nearsighted friend Marcelo, also a sufferer of "strained vision" (169). Daza also describes other optical instruments like the camera obscura of Giambattista della Porta—whom he cites without any embellishment—and mirrors. The two concerns to which I have already alluded, unmerited social status and pseudoscientific practice, become the most pressing topics of the debate when it touches upon the phenomenon of the *visorio*, or telescope, in the fourth dialogue, and upon what these characters have seen in the moon (252–56). There is not a single mention of Galileo in these pages despite the fact that Daza includes entire passages from *Sidereus nuncius*. The book, which in principle could have been a simple treatise on glasses, also becomes an exercise in diplomacy in which the author makes himself into a sort of social commentator, displaying the internal conflict between what he wants to say and what he is able to say.

The fourth dialogue brings together the doctor, the master craftsman, Julián, Alberto, and Leonardo. Leonardo and Alberto travel together to Seville and to what is presumably the Giralda tower, a landmark that would become one of the most popular settings for the Menippean satire of the seventeenth century in the depiction of Seville as a sinful Babylon. The ramp of this tower is the site for the preparation of the lens, slowly adjusted and focused until it reaches the correct setting. Once he has reached the top, the doctor looks down at the horizon with a glass; he is able to see for a distance of six leagues, discerning the *chapitel* (spire) of the tower of the church of San Felipe in Carmona and saying, "and it seems that I could count the little birds that were flying there" (*y me parece que podía contar los pajarillos que por allí andan*). Leonardo comments that "I have seen with some of those telescopes that are about a yard long . . . buildings three or four leagues away" (*yo he visto con algunos de estos visorios de a vara* [. . .] *los edificios a tres y cuatro leguas*), while Alberto remains skeptical at all of this "enchantment." The telescope, then, does not entirely lose its magical aura, although it should not be assumed that the enchantment is entirely negative. Leonardo sees the towers of Guadaíra, while Julián admits to having surveyed the whole horizon with his gaze. The craftsman then insists that "it is a great thing to see with a telescope what the eye can not reach, and it is even better because one can see more easily and with greater clarity" (*gran cosa es ver con un visorio lo que la vista no alcanza, y más siendo bueno, porque se ve con más descanso y claridad*). The inquiry is carried out very prudently, with the horizon of the city as the objective (the source of the "great thing" the craftsman mentions), and not the landscape of the heavens.

However, the dialogue soon moves on to matters that are both weightier and decidedly more controversial. When Leonardo asks the craftsman about the size of his telescopes, he responds:

> Leaving aside the small ones of four or five fingers, which are more useful and agreeable for a walk or to find people in the plaza, with one about a yard long it seems to me that you can see anything. And last night I tested all of these by looking at the moon, and even though the longer ones showed the craters and rough spots of the moon more clearly, with the yard-long one I could see almost the same thing with more ease. But since the final goal of the instrument is to see as far as one can, I do not notice the difficulty and cumbersome nature of the longer ones, if one knows how to use them.
>
> *Dejando aparte los pequeños de cuatro a cinco dedos, que son más prestos y agradables para de camino o para reconocer la gente de una plaza, con uno de a vara me parece a mí que basta para ver cualquier cosa. Y anoche hice la prueba en la Luna con todos éstos, y aunque los más largos mostraban más aquellas concavidades y asperezas de la Luna, con este de a vara veía casi lo mismo y más descansadamente. Pero como el fin de este instrumento es para ver cuan lejos se pueda, no reparo en la penalidad y embarazo que tienen los largos, como se sepa mirar con ellos.* (253)

Alberto confesses to having seen the same craters:

> The other night I looked at the Moon with a telescope about three hands long, and even though it was not one of the best, on it I saw the craters that you mentioned, and they are even more manifest when it is waxing or waning, which makes it seem that they are on the front part of the moon, and not in the circumference, for when it is full, the edges can be seen to be smooth and perfect.
>
> *La otra noche vide la Luna con un visorio de tres cuartas de largo, y aunque no era de los muy aventajados, descubrí en ella aquesas concavidades que decís, y manifiéstanse más cuando va creciendo o menguando, por donde parece que están en esta parte frontera de la Luna, y no en la circunferencia, pues cuando está toda llena, la vemos alrededor lisa y muy perfecta.* (254–55)

This is, obviously, one of the central theses of *Sidereus nuncius*, presented in this dialogue as an inarguable truth and without the slightest irony. Even if it is executed very cautiously, Daza's stance is tremendously bold, given that he is defending what he already sees as an irrefutable fact: the ruggedness of the lunar surface detected by Galileo only a few years earlier.

Also on loan from Galileo are the words of the doctor when he weighs in on the issue:

> It seems to me that those [craters] that appear in the moon as eyes and a mouth are highs and lows, although up to today, before telescopes [*visorios*] appeared, we had understood them to be simply caused by the moon being more dense in some places than others; but seen with the telescope, as much when it is waxing as when it wanes again, we find certain branches and points of light in the darkness of the part that is not illuminated. A disciple of the craftsman, having seen them, came to say that the moon had little hairs. But these adornments are not always visible, except on the day when the waxing or waning of the moon comes close to their part. However, ordinarily we see that edge as very rough and spongy and spotty, with a few bright highlights on the highest parts, such that a good painter would know better than I how truly high or low they are. But leaving this aside, it surprises me even more that these telescopes do not make the stars seem larger, but rather smaller, although more lively and brilliant. From this we can gain a greater understanding of the immense distance of the stars, for even when the telescope brings them closer, as other objects have shown that it does, they remain just as small when seen with a telescope as they seem to be without it.

> *Yo tengo para mí que aquellos que parecen en la Luna como ojos y boca son altos y bajos, aunque hasta ahora que salieron los visorios habemos entendido que se causaban solamente por ser la Luna más densa por unas partes que por otras; pero mirada con el visorio, así cuando va creciendo como cuando vuelve a recogerse, hallamos que salen a lo oscuro de la menguante ciertos ramillos o partes luminosas, las cuales habiendo visto un discípulo del señor maestro, vino a decir que la Luna tenía melenas. Mas estos plumajillos no todas veces se manifiestan, sino en tal día que llega a aquella parte la creciente o menguante de la Luna. Pero de ordinario le vemos aquel canto muy áspero y como esponjoso y avirolado, con algunos retoques de mayor luz en las partes que son más altas, por donde un buen pintor conocerá mejor que yo cómo aquéllos son verdaderamente altos y bajos. Pero dejado ahora que lo sean o no, me admiro más de que estos visorios no agranden las estrellas, sino que antes las hagan menores, aunque más vivas y resplandecientes. Por donde venimos en mayor conocimiento de su inmensa distancia, pues con acercarlas tanto a nosotros, como vemos por otras cosas, con todo eso se quedan tan pequeñas miradas con los visorios, como parecen sin ellos.* (256)

These "touches of greater light on the parts that are higher" are the "highlights" already explained by the Tuscan astronomer in a summary of his observations of the moon with a telescope, the *visorio* that the doctor refers to. Here, these lunar mountain ranges are something evident that cannot be refuted; they are seen with the eyes that observe. Equally fascinating are the words of the master craftsman concerning the technical advances of the time, because they indicate the strong and steady will of the text, intent upon revealing what for many people was still a heresy:

> It would be lengthy to explain, if we had to refer to everything that has been discovered in the field of telescopes, but speaking of what I have seen and the experiences that I have had with them, I can tell you that this instrument with two lenses does not make the stars large and visible, no matter how long it is and no matter the degree of the concave lens that you put on your eye; the great enlarging effect can be seen only on the body of the Moon, which is much closer, and other faraway things on earth.

> *Largo sería de contar, si hubiéramos de referir las cosas que se han añadido en materia de visorios; pero hablando de lo que yo he visto y de las experiencias que he hecho con ellos, sé deciros que este instrumento de dos lunas no alcanza a mostrar grandes las estrellas, por largo que sea y por muchos grados que tenga la cóncava que se aplica a los ojos; sólo en el cuerpo de la Luna, que está más cerca, y en otras cosas de acá de la tierra se echa de ver lo mucho que engrandecen.* (257)

After the theoretical explications of the craftsman on the makeup and types of telescopes, Alberto cannot help but confess, "This time, I am departing as a great master of telescope-making, and Julián is leaving behind the incredulous stance he has toward its wonders, for this afternoon we have seen things that seemed impossible" (*de esta vez salgo gran maestro de hacer visorios, y el señor Julián, de la incredulidad que tiene de sus maravillas, pues hemos visto aquesta tarde cosas que parecían imposibles*) (260). Additionally, he includes an affirmation that is even more surprising coming from an inquisitorial notary: that "this science is curious and worthy of study because it holds so many secrets, and I would learn it happily were I able to do so" (*curiosa es, por cierto, aquesta ciencia y digna de saberse, por tantos secretos como tiene, y de buena gana la aprendiera yo, si fuera para ello*) (262). "Worthy of study" is extremely high praise for a discipline that should be taught as a science. At the last minute, the book becomes a defense of post-Copernican work, a defense without quotes, without erudition, and without any sources, but crystal clear in its intent and accessible to every new reader. However, Daza reminds readers that this new science should remain in the hands of experts, with the doctor's final warning against the dangers of sight, which can confuse "sometimes in such a way, that one believes things that are not, making the air appear as a terrifying figure, burning fires, battles of armed men, three suns, openings of the sky, comets and the colors of blood" (*algunas veces de tal manera, que cree lo que no hay, haciéndole parecer en el aire una espantable figura, fuegos encendidos, pelear hombres armados, tres soles, aperturas del cielo, cometas y colores de sangre*) (263).

This "terrifying figure," these "burning fires," and this "battle of armed men" already connect, in their echoes of *Quixote*, to a new symptom of the modern scientist: insanity. Daza warns against visionaries, against crackpots, against anyone who makes an immoderate use of technology. He then enters the realm of one of the most fruitful literary figures of these decades, that is, the virtuoso, a figure that was already familiar in Italy with savants like Leonardo Fioravanti and Giambattista della Porta. This figure would begin to take root in the Iberian Peninsula through inventors and collectors like Jerónimo de Ayanz and Juan de Espina, as we will soon see. The doctor's final assertions have enormous importance from an epistemological perspective, in that they connect the telescope with the Baroque concern with appearances and truth, and specifically with the controversial standing of judiciary astrology. If Baroque spyglasses show "distant things close-up," as the doctor concludes, they will also help define the perceived object under a new moral fabric: "the large, small; what is above, below, and vice versa." This inverted world, this carnival of things barely made out, will soon become the stuff of novels, as I argue later. However, I first would like to examine what the figure of the scientist in its many forms meant for writers of fiction, and what the standing of that figure did for the discussion of scientific exploration. The influence of certain Italian cases, as I show in the next chapter, is of vital importance in this dialogue between fictional forms and the so-called new science.

II

Galileo and his Spanish contemporaries

✷ 2 ✷

Foundations

Minima cura si maxima vis.
Take care of small things if you want the greatest results.

MOTTO OF THE ACADEMY OF LINCEI

¡Oh, maligno mirar!
Oh, maligned gaze!

BALTASAR GRACIÁN

Science (and) fiction: Elements for a new mechanics

The dissemination of scientific ideas and technical instruments in seventeenth-century Spain was determined in large measure by Spain's extremely complex relations with the Catholic Church. This complexity was not, however, a new phenomenon in the Iberian Peninsula: toward the end of the Middle Ages, judiciary astrology—which had been dealt with in texts such as *Libro cumplido de los juicios de las estrellas* (Complete book of the judgment of stars), translated by Alfonso X in 1254, began to find itself substituted by what we might refer to as a "literary" astrology in pieces by authors such as Francisco Imperial (ca. 1350–ca. 1409); Íñigo López de Mendoza, Marqués de Santillana (1398–1458); and Juan de Mena (1411–1456).[1] As we know from numerous extant sources, this form of astrology was characterized by its complex intersection with areas such as magic and the supernatural, both of which boast a tradition in early modern Spanish literature that is as rich in examples as it is complex in sources. Classic scenes such as the magic cave of Salamanca (*cueva de Salamanca*), and literary figures such as Merlin, Luzbel (i.e., Lucifer), and the "stone guest" (*convidado de piedra*) shape textual traditions that extend through the centuries, reemerging with success in all manner of genre and style. These variants alternate between the serious and the playful in their internal dialogue, at times following the taste or caprice of the author, at times the tendencies imposed by the aesthetic trends of the time. It is not necessary to move much beyond the masterworks of the period to see how these mechanisms operate in Spanish literature: in Lope de Vega's *El caballero de Olmedo* (The knight of Olmedo, ca. 1620), for example, one sees what the language of theater can offer when we pair it with Francisco Antonio de Monteser's 1651 burlesque version; and *Don Quixote*—with the character's oracular head or his bleary-eyed and gaunt embodiment of Belerma—is in itself a rich catalog of both respectful and irreverent attitudes with respect to a rich tradition with which it is in frequent dialogue.

In the most canonical works of the late medieval and early modern literary archive in Spain there emerged a broad esoteric current that was of great interest to the reading public. From the magnetism exercised by historically documented figures such as the necromancer Enrique de Villena or fictitious creations such as Celestina, to the fascination with the occult in the first plays devoted to magic that closed the seventeenth century (and went on to enjoy great success in the following century), Spanish literature acquired particularly suggestive tonalities thanks to works that connect, over the course of two centuries, with practices and rituals associated with witchcraft, visions, alchemy, demonology, and possession.[2] The problem of vision and the perception of that which is secret play a central role in these works, and these are frequently accompanied by complex optical mechanisms and games of perspective. One could speak, in modern terms, of the emergence of a kind of local "science fiction" through which meander all sorts of fantastic visions and in which, as part of these journeys through the unknown, celestial exploration likewise finds expression. We may take as a common example the figure of the astrologer, repeatedly represented in numerous plays of the period.[3] Alchemy also finds expression in many different genres, the alchemist being one of the urban archetypes most frequently ridiculed by Menippean satire. In 1630s Madrid, the alchemist Vincenzo Massimi conducted secret alchemical experiments in the residence of the Hermitage of St. John the Baptist that became something of a scandal; in the end, even the Count-Duke of Olivares became implicated, as he was likewise very interested in the lexicon of the occult. Necromancy would also find numerous detractors and denouncers, as evidenced by Martín de Castañega's *Tratado muy sotil y bien fundado de las supersticiones y hechizerias* (A very subtle and well-founded treatise on superstitions and sorcery, 1529), Pedro Ciruelo's *Tratado en el cual se reprueban todas las supersticiones y hechizerias* (Treatise in which all superstitions and sorcery are debunked, 1529), Benito Perera's *Adversus fallaces et supersticiosas artes, id est, de Magia* (Against the fallacious and superstitious arts, that is, on magic, 1603), and Martín Antonio del Río's *Disquisitionum magicarum libri sex* (Six books of magical investigations, 1624). These works represent only a small slice of a widespread concern with matters that exceed the parameters of my analysis but whose mere mention reveals to us both the porous boundaries between the licit and the illicit, and what seemingly disparate materials could come to serve as fictional material for the writers of the time (López Piñero, *Ciencia y técnica* 67–73). The great quantity of printed pages related to vision and the occult indicates that, despite the fact that these matters were repeatedly condemned or relegated to secret study, they were nonetheless highly seductive not only for readers but also for popular theater audiences and those who had begun to enjoy the technical advances associated with performance spaces such as the coliseum. The dramatic discourse of this period offers numerous cases in which the faculty of vision plays an absolutely decisive role, particularly in the context of scrutinizing the illicit or merely observing that which cannot or should not be seen. In this way, the oblique gaze and sleight of hand would have great success as visual mechanisms and, in the end, as epistemic values insofar as they reflect on the limits of sight and knowledge. There is an extensive bibliography associated with these processes, and I therefore move on.[4] Much can still be written, however, on this particular crossroads of moral, juridical, and aesthetic expression.

If the inquisitive eye moves us into new realities, language should be that which gives materiality to these visions. The link between the two must always be, however, a flexible and porous one, open to every means that might render it opaque, ambiguous, and provocative, expanding with it its range of signification, both eliminating and confusing, expressing half truths. In the field of inquiry that interests us, the play between *antojo* and *anteojo* (lens), or the many valences of the word *antojo*—not so much signifying "caprice," as we understand it today, but rather "madness" or "extravagance," as it was then used (or also *antojado*, an adjective that modifies a kind of moral myopia, when it does not simply mean "insane")—became so frequently employed that it essentially rose to the status of a commonplace. We also know that the term would be used in underworld slang as a synonym for shackles (*grilletes*), which helps reveal how tremendously popular this wordplay was as an instrument of quotidian use in the culture of early modern Spain.

Many literary texts also engage in the play between *anteojo* and *enojo* (meaning "nuisance," "irritation," or "pain"), and this is especially the case with poetry.[5] The first registered appearance of the word *antojos* linked to optics takes place in the *Poesías* (Poetry) of Alfonso Álvarez de Villasandino (1340–1424), which are contained in the *Cancionero de Baena* (Songbook of Baena, ca. 1426–1430) and were composed sometime between the final years of the fourteenth century and the first decade of the fifteenth. There is included within the *Poesías* a poem with the epigraph: "The aforementioned Alfonso Álvarez composed this poem for our Lord the King" (*Este decir fizo e ordenó el dicho Alfonso Álvarez para el Rey nuestro señor*), which begins: "I hear poorly and I don't see well. / Look, my Lord, what two nuisances! / What a sin! Without glasses / I can neither write nor read" (*Mal oyo e bien non veo. / ¡Ved, señor, qué dos enojos! / ¡Mal pecado! sin antojos / ya non escrivo nin leo*). And in the *Coplas de Vita Christi* (1467–1482), Fray Íñigo de Mendoza, says of the Virgin Mary:

You are a sacred woman
In whose womb the Trinity
Quiets its displeasure and
Further corrects the flaws
Of the most ancient host;
For you God loses the anger
That He has against us;
You are the rich lenses
Through which our eyes
Were able to behold God.

Tú eres sacra doncella
en cuyo vientre apacigua
la Trinidad su querella
y más repara la mella
de la hueste más antigua;
por ti pierde los enojos
que tiene Dios contra nós;
tú eres ricos antojos
por cuyo medio los ojos
pudieron mirar a Dios.

Here one finds a reference to *antojos* that is very close to its etymology, namely, a mediating object that facilitates a different vision; in the second case, in fact, it underscores the mediating character of the Virgin and the possibility of seeing God through her. This is a fascinating reference insofar as it uniquely binds the techno-scientific element to a search for religious experience: the Virgin is not only the poem's object of praise; she also serves as an intermediary to God through her condition as *antojo*.

The alteration that one finds in Villasandino would endure throughout the sixteenth century and reach great levels of popularity in the seventeenth. Lope de Vega, for example, characterizes Fabia in *El caballero de Olmedo* as a carrier of "rosary, staff, and eyeglasses" (*rosario, báculo y antojos*); and in Sonnet 22, "A dos niñas" (To two girls), included in his 1609 collection *Rimas humanas*, he speaks of the possibility of crying twice as much due to wearing glasses, playing at the same time with the disemia of *antojos* and with that of *niña*, "girl," and "pupil" (of the eye), which is further complicated by the play between *niño* (boy) and *Cupido* (Cupid), who is also referred to as *niño Amor*:

"To two girls/pupils"

To take vengeance on my disdain,
Love took away my girls (eyesight),
with which I commonly saw
all things in equal temperance.
I am at least familiar with change
in the restlessness of my vision;
from one day to the next I never stop;
that which time reaches flees from it.
The souls of my girls seem placed
in my eyes, and tender weeping bathes them,
Oh girls, Cupid, childish madness,
childish desire that costs me my life!
But should one with four *niñas*
in his eyes be forced to cry so much?

A dos niñas

Para tomar de mi desdén venganza,
quitóme Amor las niñas que tenía
con que miraba yo como solía
todas las cosas en igual templanza.

> *A lo menos conozco la mudanza*
> *en los antojos de la vista mía,*
> *de un día en otro no descanso un día,*
> *del tiempo huye la que el tiempo alcanza.*
> *Almas parecen de mis niñas puestas*
> *en mis ojos, que baña un tierno llanto.*
> *¡Oh, niñas, niño Amor, niños antojos,*
> *niño deseo que el vivir me cuesta!*
> *Mas, ¿qué mucho también que llore tanto*
> *quien tiene cuatro niñas en los ojos?*

If eyeglasses here offer an innocent game, the use of their long-distance sibling would situate early modern poets in much more dangerous, slippery territory. With the goal of dodging the censors' yoke, writers in seventeenth-century Spain found themselves forced to make use of a series of techniques that were often inherited from the literary tradition itself. For example, the voyage of the reader to the world beyond the grave as well as to the celestial infinite consistently employs a high degree of ambiguity, irony, and humor, to which is added (at times in the final lines of the narrative) a brief clarifying note to protect oneself from any possible blame. The writer adroitly manages in this way to put distance between himself and his characters, without abandoning any secretive winks to this or that innovative aspect of scientific activity that was then placing in check the established cosmology. This is, in the end, a fundamental technique of "saying without saying," proclaiming something without really committing oneself: the author drops in this or that piece of information, and the possibility of believing it resides within the reader, as does the possibility of taking up a certain measure of complicity with the material, of investigating a bit more deeply. In this sense, I am much more interested in the voyage than the destination and more intrigued by the means by which the path is facilitated than by the new space that is reached. This is a path richer in mechanisms than in landscapes, and what matters to me now are the trajectory and the means of travel rather than the destination per se. As I argue throughout this book, this work of reflection, at times taken up from silence and deliberate absence, is symptomatic of how and why scientific advances made their way into seventeenth-century Spanish fiction.

❋

The figure of Ptolemy served as a foundation for many centuries of history during which literary works also reflected and shaped anxieties with respect to the mysteries of the firmament. Poetry from the turn of the seventeenth century also offers examples of unquestionable merit and interest. An early text such as the *Tratado poético de la esfera* (Poetic treatise on the sphere), whose 258 verses Alonso Jerónimo de Salas Barbadillo inserted playfully into his long poem *Patrona de Madrid restituida* (The patroness of Madrid restored, 1609),[6] is of great importance given that—compared to the brief references seen in the theater—it is entirely dedicated to the constitution of the heavens. It contains twelve books in 733 octaves, in which he narrates the events that inspired the construction of the altar of Our Lady of Atocha; also included (and dispersed throughout the narrative) are the 258 verses that make up the poem. It was published one year before the telescopic observation of Galileo, and in it Salas Barbadillo still demonstrates a preference for the Ptolemaic system, situating the sun in the fourth sphere, in the middle of the eight "planets" (that is, the moon, Mercury, Venus, the sun, Mars, Jupiter, and Saturn). Above these rest the fixed stars, the primum mobile as the outermost sphere, and finally the empyrean. The entire theory of the poem has its source, as was the case with other cosmic journeys of the time, in the *Tractatus de sphaera* (Treatise on the sphere) by Johannes de Sacrobosco (ca. 1195–ca. 1256). The influence of Sacrobosco's text was still very much evident in the seventeenth century, as it was required reading not only at the University of Salamanca but also at other centers of higher learning.[7] From the medieval period, it had formed part of the content in classes that were taught in European universities under the rubric of "the sphere," which included the celestial as well as the terrestrial. For Sacrobosco, this sphere was nothing but the sum total of the created world—the first region, perfect and unalterable, was made up of nine heavens or spheres: seven that carried the planets; the eighth (or firmament), which contained the fixed stars and signs of the zodiac; and the ninth (or crystalline), the general source of movement. From the high spheres one descends to the elemental region, which contained the four elements placed in concentric spheres. The teaching of astronomy was complete with an anonymous text from the middle of the thirteenth century called *Theorica planetarum* (Theory of the planets), which offered a synopsis of Ptolemaic theories along with Alfonso X's *Tablas* (Tables, 1283), which were the most popular astronomical text in Spain until the publication in 1551 of Erasmus Reinhold's *Prutenic Tables.*

Salas Barbadillo, who had studied astronomy in his youth before devoting himself to literature, maintains in his poems a traditional vision of the universe, evidenced by verses of obvious meaning such as "at the center of the beautiful firmament / rests the earth like a point within" (*de centro sirve al firmamento hermoso / la tierra que parece un punto dentro*), at the same time that "the twelve beautiful signs that in one year / with lovely circuit the sun visits" (*los doce bellos signos que en un año / con hermoso rodeo el sol visita*). Within the poem, in fact, there are echoes of Ptolemy's *Almagest*, specifically his belief in the 1,022 stars, when he affirms: "Of the fixed stars, the smallest / covers more distance than the earth. / There are one thousand twenty-two, dazzling power; / the hand of God has created so much" (*De las estrellas fijas, más distancia / la menor comprende que la tierra. / Son mil y veintidós, poder extraño; / tanto la mano poderosa cría*). In this way, we see how deeply rooted traditional science was for some writers during the first years of the seventeenth century. Salas Barbadillo's composition is a product of its time, and it reveals a desire to connect his learning with the broader reading public. We know through the work of scholars such as Eugenio Garin (48–49) that during the first half of the seventeenth century, there were numerous relations between astrology and other forms of Renaissance knowledge, many of which even affected the daily behavior of people, such as medicine, science, philosophy, political theory, propaganda, and religion. With respect to these last three items, Luis Miguel Vicente García ("Lope" 219) reminds us:

> There was in principle no reason for Christianity to have an antagonistic relation with astrology; on the contrary, the latter allowed the former to become universal, which was the aspiration of cutting-edge humanism: incorporating all that science and reason were then saying, in the end, about the existence of the one true God. In all the orders of thought and art, astrology secured its roots. And even with respect to the thorny topic of prognostication, its prestige and utility, especially with respect to medicine, guaranteed its presence at court.

> *No había en principio razón para que el Cristianismo antagonizara con el saber astrológico, sino al contrario, permitía universalizarlo, como fue la aspiración del Humanismo más avanzado, incorporando todo lo que la ciencia y la razón venían a decir en último término sobre la existencia de un Dios único. En todos los órdenes del pensamiento y del arte la Astrología afianzaba sus raíces. Y aún en el espinoso tema del pronóstico su prestigio y utilidad, especialmente su vertiente médica, aún les garantiza la presencia en la corte.*

Vicente García continues, touching on early modern Spanish fiction, arguing that:

> It is necessary to consider that astrology is an organic mode of understanding the universe, interrelated in all of its parts like a living being (*anima mundi*). According to this, that which occurs in one place will be reflected in another and might assume, therefore, the hermetic principle of correspondence: "above as below," inherited from the oldest cosmologies of the classical and preclassical worlds, consecrated by the prestige of Aristotle, Plato, and Ptolemy. And all this so as to bear in mind how incorrect it is to pass judgment on the stance of Lope de Vega or Cervantes vis-à-vis astrology as mere superstition, fruit of the scientific backwardness of the period, of the weakness of old age or of bad feminine counsel, as has at times been suggested.
>
> *Es fundamental considerar que la Astrología es un modo organicista de entender el Universo, interrelacionado en todas sus partes como un ser vivo (*anima mundi*) que permite que lo que ocurre en una parte se refleje en otra, y que asume, por tanto, el principio hermético de Correspondencia: "arriba como abajo", heredado desde las más antiguas cosmologías del mundo clásico y preclásico, consagradas por el prestigio de Aristóteles, Platón y Ptolomeo. Y ello para tener en cuenta lo poco acertado de enjuiciar la postura de Lope o de Cervantes frente a la astrología como meras supersticiones, fruto del atraso científico de la época, de la debilidad de la vejez o de los malos consejos femeninos, como ha veces se ha dicho.* (219)

These two giants of Spanish literature offer fascinating insights for scholars interested in science and technology in seventeenth-century Spain, generally in the form of brief asides that many times resort to humor and ridicule. Although we know little about Cervantes's scientific education, his writing seems to suggest a certain sense of fatigue with respect to inherited teachings.[8] On this matter, Chad Gasta has written:

> There is no way of knowing for sure what Cervantes thought about the great scientific advances of his own day, but the many scientific references in his works suggest he was at least conversant about what was transpiring, and understood how such discoveries were shaping the manner in which people viewed their own world. . . . The novelist was aware of contemporaneous science even if he relayed that knowledge in an often funny or irreverent way, yet still firmly grounded in sound scientific principles. (79–80)

Mariano Esteban Piñeiro ("La ciencia" 34) has gone even farther in affirming that Cervantes possessed "ample astronomical knowledge, knowledge that went far beyond that held by the average person or, in fact, the majority of enlightened humanists" (*amplios conocimientos astronómicos, conocimientos que no se corresponden a los que disfrutaba sobre esa materia el hombre común de la época ni, tampoco, la mayoría de los ilustrados en humanidades*). In the exemplary novel *La gitanilla* (The little gypsy girl, 1613), for example, one finds a popular ballad (*romance*) to Queen Margaret of Austria, who died in childbirth in Valladolid in 1611, and whose childbirth mass was celebrated on May 31 of that year. The poem constructs a series of mythological identifications between the members of Philip III's cortege, sung as "sun of Austria" (*sol de Austria*); his eldest daughter, Anne of Austria, is "the tender Aurora" (*la tierna Aurora*), while "ancient Saturn" (*anciano Saturno*) is the cardinal bishop of Toledo, Don Bernardo de Sandoval y Rojas—his nephew, the Duke of Lerma, appears as Jupiter. The following allusions, to the moon and Venus, are relatively conventional, but it is telling that Cervantes goes on to write that: "Little Ganymedes / cross, pass, return and spin / fastened at the waist / to this miraculous sphere" (*Pequeñuelos Ganimedes / cruzan, van, vuelven y tornan / por el cinto tachonado / de esta esfera milagrosa*) (36). Let us remember that Ganymede was the name assigned by Simon Marius to one of the four moons of Jupiter discovered with the help of the telescope. The two final verses are written, therefore, in an astronomical rather than mythological key, as they allude to the ecliptic, the imaginary circle traced out by the planets. This leaves us to two different conclusions: that Cervantes knew *Sidereus nuncius*, and that his reference to Ganymede inspired Marius when he later named Jupiter's four moons.

It would be most prudent to affirm, however, that in Cervantes we most frequently find evidence of an already celebrated writer with an occasionally ambivalent attitude regarding which authoritative astronomical sources to follow. Such is the case, for example, with Sancho Panza's well-known humorous manipulation of Ptolemy's name in *Don Quixote*. The famous dialogue in part 2, chapter 29, between the knight and his squire suggests once again Cervantes's preference for ambiguity with respect to debates revolving around the new heliocentric system, debates of which he must have been aware given, for example, the large catalog of astronomical terms contained in his oeuvre. When Don Quixote and Sancho arrive at the river in the episode of the enchanted boat, the former expresses the distance covered in Ptolemaic terms, to which his squire, playing with the phonetic possibilities of the words *Ptolomeo* (*meo*, "piss") and *cómputo* (*puto*, "queer"), responds: "My God, . . . but you're asking me to take the word of a queer and a leper, and on top of that a piss-pants, or piss, or I don't know what" (*Por Dios* [. . .] *que vuesa merced me trae por testigo de lo que dice a una gentil persona, puto y gafo, con la añadidura de meón, o meo, o no sé cómo*). The narrator continues, indicating that "Don Quixote laughed at the interpretation that Sancho had given of the name, calculations, and theories of the cosmographer Ptolemy" (*ríóse don Quijote de la interpretación que Sancho había dado al nombre y al cómputo y cuenta del cosmógrafo Ptolomeo*) (870). How do we read this exchange if not as more evidence of Cervantine irony? The dialogue between Don Quixote and Sancho is constructed from the artificiality of the terms that both manipulate to their respective benefit until arriving at the boundary of the absurd. It combines different astronomical and geographic notions by which one perceives a very suggestive oscillation between the ancient and the modern that perhaps reflected the internal tensions of a man living in a moment of rapid change. Let us remember, for example, that the publication of *Sidereus nuncius* coincides with the zenith of Cervantes's muse, halfway between the first and second volume of *Quixote*. As the novel parodies through its very language certain examples of chivalric romance it also does with authoritative scientific texts. As Wendell P. Smith has recently argued in his analysis of the episode: "What Cervantes sought was the clash of ideas, not to crown the winner" (69). It is very significant, for example, that *Don Quixote*'s (second) narrator presents Ptolemy not as an astronomer but as a cosmographer, an occupation that in Renaissance Spain had a rather broad range, encompassing both the art of navigation and geography. Cervantes likewise speaks in this same dialogue of the "globe of water and of earth" (*globo del agua y de la tierra*), revealing a very modern notion of the terraqueous globe as a three-dimensional solid with a varied surface of water and earth. And it is not that Cervantes is reluctant to use a term such as *astrologer*: we find it, for example, in the conversation that Quixote has with Lorenzo, the son of Diego de Miranda, which reveals a particular appreciation of mathematics; an appreciation that, in contrast, was quite common in a society that depended on this discipline to carry out its overseas conquests.

Sancho also offers a very telling reading of Don Quixote's broadside. Terms such as *piss-pants, leper,* and *queer* in the mouth of a character in no way connected to the academic culture responsible for scientific innovation should not necessarily be read as personal attack on Ptolemy but rather as a symptom of the fact that his untouchable authority—perhaps that *calculation* that is mentioned—was already a thing of the past in a moment of flux.[9] The key to reading this episode depends not so much, in this case, on our understanding of Ptolemy at the beginning of the seventeenth century as on always keeping present the changes taking place within Sancho (even though he is simultaneously reactionary to all novelty). The joke undermines, I would argue, the blind obedience to an *auctoritas* who presided over all judgment of knowledge but is now questioned by a character whose oral culture equally determines his use of language. Sancho's words, which situate the celestial explorer at the very bottom of human service, are clearly critical, since in the end we find ourselves faced with a peasant who depends on his oral culture for humor and joking—and this joke, with its wordplay, could not be more oral in that sense. Nevertheless, as a person who goes right to the heart of the matter and does not accept the verbosity of his master, Sancho embodies a very healthy attitude—and even a very modern one—with respect to opinions lacking in evidence, that same evidence that seems absent in Don Quixote's verbal edifice when he speaks of the arduous path covered in his bookish journey not through a landscape, but through the *Tratado de la esfera*: colures, lines, parallels, zodiacs, poles, solstices, equinoxes, planets, signs, points, measures, and so on. These are words that doubtless sound to Sancho as empty as the names that his master had invoked in the famous battle of the herd of sheep (I.18): Alifanfarón, Pentapolín, Micocolembo. If, as *Quixote* seems to argue through the continuous act of citation, chivalric romances are already obsolete by 1600, then the same can be said of Ptolemy, that Ptolemy whom Sancho refers to as a queer and a leper, taking up a tradition that since the Middle Ages had associated these two groups. This exchange seems to be telling us that at the beginning of the seventeenth century people were ready for change.

If Cervantes's modern gaze reaches its highest point of virtuosity in *Quixote*, no less useful in a survey such as the present one is the well-known dialogue in *Persiles y Sigismunda* (Persiles and Sigismunda, 1617) between Mauricio and Soldino, two judiciary astrologers who discuss the validity of astrology and how difficult it is to predict correctly given that they are always forced to "always judge very carefully and with little confidence" (*juzgar siempre a tiento y con poca seguridad*) (116).[10] Cervantes suggests in this way that the much debated and threatened discipline of astrology is nothing but a form of knowledge born of experience and contrastable to it rather than a reductionist superstition. He, in fact, offers numerous interesting examples of optical phenomena. I will not enter here into the complex subject of the social construction of sight in pieces such as *El retablo de las maravillas* (The tableau of the marvels, 1615) or paradigms of seeing and not seeing (or at least wishing not to see) in the Baroque, as these have been studied elsewhere in detail. Neither will I focus on the fascinating use of glass as a reflection on knowledge in its rational and intuitive processes in texts such as *El coloquio de los perros* (The colloquy of the dogs, 1613), although in the latter example we do come across something that at least merits mention here: speaking about the famous problem of longitude that so interested Galileo, one of the dogs humorously indicates: "I've spent twenty-two years trying to find the fixed point; I cross it out here and situate it there, believing all the while that I've found it and that it can't possibly escape me, when suddenly I find that I'm so far from it that I'm left to wonder" (*veintidos años ha que ando tras hallar el punto fijo; y aquí lo dejo y allí lo tomo, y pareciéndome que ya lo he hallado y que no se me puede escapar en ninguna manera, cuando no me cato, me hallo tan lejos de él que me admiro*) (619).

What I do wish to discuss here is Cervantes's interest in eyeglasses. One finds various examples of this, for example, in *Don Quixote*: in chapter eight of the first part, Quixote speaks of the so-called *antojos de camino* in the episode of the Basque, with the two friars of St. Benedict wearing eyeglasses made of rock crystal—these were attached to a piece of taffeta that covered the face to protect it during travel and did not necessarily seem to have any sort of graduation. In the second part we might turn to chapter 64, in which one witnesses in the ducal palace a procession of six ladies, four of them with eyeglasses: Doña Rodríguez, a central character during Quixote and Sancho's eventful stay, is described as *antojuna*, or bespectacled; and Sancho comments, not without certain sarcasm, "Love, I've heard it said, sees through eyeglasses that make copper appear to be gold, poverty resemble wealth, and eye boogers look like pearls" (*el amor, según yo he oído decir, mira con unos antojos que hacen parecer oro al cobre, a la pobreza, riqueza, y a las lagañas, perlas*) (786). This supports the idea of crystal as a manipulator of the senses, an idea that authors such as Francisco de Quevedo and Diego de Saavedra Fajardo likewise developed a bit later.

If the first generation of early modern Spanish authors inherits a Ptolemaic reading of the universe to which they respond more or less critically in their writing, the use of the motif of crystal is extraordinarily fertile and varied. Luis de Góngora, who as a young man sang of the "clear honor of the watery element" (*claro honor del líquido elemento*) in an early sonnet (1582), played with the poetic motif of crystal in various compositions, demonstrating his familiarity with optical advances of the period. For example:

> Oh, clear honor of the watery element,
> sweet brook of flowing silver,
> whose water meanders through the grass

with a gifted song, with slow steps!
She for whom I feel I am freezing and burning
(while she looks upon you), Love paints
the snow and scarlet from her face
in your tranquil and gentle movement,
go slowly as you do; don't leave slack
the flowing reign to the crystalline brake
with which you govern your rapid current;
for it isn't right that Neptune should so
confusedly gather up so much
beauty in his deep breast.

¡Oh claro honor del líquido elemento,
dulce arroyuelo de corriente plata,
cuya agua entre la yerba se dilata
con regalado son, con paso lento!,
pues la por quien helar y arder me siento
(mientras en ti se mira), Amor retrata
de su rostro la nieve y la escarlata
en tu tranquilo y blando movimiento,
vete como te vas; no dejes floja
la undosa rienda al cristalino freno
con que gobiernas tu veloz corriente;
que no es bien que confusamente acoja
tanta belleza en su profundo seno
el gran Señor del húmido tridente. (114)

Kirsten Kramer has pointed out this facet of the early Góngora's "crystalline" vision:

> The description that the sonnet offers of the formation of the portrait stems from . . . the specific procedures of artificial optical magic associated with the Baroque that are derived from the implementation of the technical medium of the mirror and contribute to produce multiple optical figures that, due to their visual appearance as "a present absence" . . . acquire a profoundly ambiguous epistemic status. (60)

The mature Góngora offers his reader a panoptic vision of great significance in his *Polifemo* (Polyphemus, 1613), as well as that which some have seen as an elaboration of telescopic and microscopic perspective in some passages of the *Soledades* (Solitudes, 1613) (Del Río Parra 223).

A great deal more can be said about Lope de Vega, whose relation with astrology, astronomy, and cosmography has generated rivers of ink.[11] As with Cervantes and many other of his contemporaries, Lope is cautious of the correlations between planetary configuration and personal and collective destiny, but the need to adhere to church doctrine and his own situation as a public citizen make him consistently adopt a decidedly cautious attitude. Edwin Morby reminded us that during this period, the term *astrology* "embraced not only the present meaning of the term but astronomy as well; or that insofar as it purported to go beyond the field of astronomy to predict events to come, its position was highly ambiguous and its adepts exposed to the active displeasure of the Church" ("Two Notes" 110). If Don Quixote mentions that his niece had been born under the influence of Mars but subjected first and foremost to that which "my will desires" (*mi voluntad desea*) (I.6), and warns Sancho of the malevolent influence of the stars (I.52), Lope will argue something very similar in his 1632 masterpiece *La Dorotea* (Dorothea), constructing a dialogic framework by which he might discuss very seriously the validity of astrology. In scene 8 of act 5, for example, the astrologer César argues against subordinating free will to astrological interrogation:

> I assure you that I have always been displeased by and considered unreliable predictions regarding what inscrutable God has prescribed in his eternal mind. I studied this in my early years with the extremely learned Juan Bautista de Labaña from Portugal, and I only judge out of curiosity, and for no other reason, occasional births; but I don't respond to questions of any sort. Man was not made by the stars, nor can his free will be governed by them.
>
> *Os aseguro que siempre me desagradaron y parecieron temerarias las predicciones de lo que Dios inescrutable tiene prescrito en su mente eterna. Esto estudié en mi tierna edad del doctísimo portugués Juan Bautista de Labaña, y solo tal vez juzgo por curiosidad, y no de otra suerte, algún nacimiento; pero no respondo a las interrogaciones por ningún caso. El hombre no se hizo por las estrellas, ni el libre albedrío les puede estar sujeto.* (460)

Lope adopts a very conservative stance in these lines, hewing closer to Catholic doctrine, but he also relies on serious science through the figure of Labaña and on his training in astronomy and math. Equally interesting is what we know about his cosmographic vision. As Antonio Sánchez Jiménez reminds us, Lope was a student of Labaña's between 1584 and 1587; he studied the astrolabe and the sphere in the Academy of Mathematics, and he likely received training in astronomy during his time in Alcalá de Henares. The interest he has in these matters is reflected in the numerous allusions that are found scattered throughout his work across genres. In *La Arcadia* (Arcadia, 1598), for example, he states that "the poet must know not only the sciences, or at least certain principles of all of them, but he must also have extensive experience with the things that take place on the earth and the sea" (*no solo ha de saber el poeta todas las ciencias, o a lo menos principios de todas, pero ha de tener grandísima experiencia de las cosas que en tierra y mar suceden*) (440). This is precisely the advice that Don Quixote, as we have already seen, gives to the young Lorenzo de Miranda. There is also in Lope another mention of Labaña, in this case in Sonnet 115 of the *Rimas* (Rhymes, 1602), in which he explains to his master the state of his relationship with Elena Osorio, disguised as Filis in the poem, by means of terms and ideas learned in Labaña's course on astrology. Through the traditional link between *engaño-desengaño-daño* (roughly, "deceit-disillusion-suffering"), the enamored student comically admits to his teacher the error of framing love in mathematical terms given the mutability and unpredictable nature of the lover:

> My teacher, see whether the movement
> that Filis's sun makes in parallel
> with her intent, passing through the year,
> has been but a common amorous deceit.
> I measured her height in this disillusion,
> and in my suspicion, an accurate instrument,
> by coronas I took account of her thought,
> and I was thus given the balance of my suffering.
> Either these arcs are not well described,
> (I mean, these eyes) or this limbo is a sign
> that I will turn to that ancient darkness,
> or do I poorly understand your writing:
> if Filis makes solstice in Gemini, then
> my zenith of Capricorn is unavoidable.

> *Maestro mío, ved si ha sido engaño*
> *regular por amor el movimiento,*
> *que hace en paralelos de su intento*
> *el sol de Fili, discurriendo el año.*
> *Tomé su altura en este desengaño,*
> *y en mi sospecha, que es cierto instrumento,*
> *por coronas conté su pensamiento*

y señalome el índice mi daño.
O no son estos arcos bien descritos,
(digo estos ojos) o este limbo indicio,
que a aquella antigua oscuridad me torno,
o yo no observo bien vuestros escritos,
que si hace Fili en Géminis solsticio,
no escapa mi Cenit de Capricornio.

Within this construct, the only thing real, it seems, is the suspicion of the other: that "accurate instrument" (*cierto instrumento*) of the second quatrain.

At the same time, the figure of Ptolemy appears in the theater of these decades in numerous pieces. It is believed that his work came to Lope through Girolamo Cardano's *Commentariorum in Ptolemaeum de astrorum iudiciis libri IV* (Four books of commentaries on Ptolemy's "On the influence of the stars," 1554), as Lope cites it in pieces such as *La boda entre dos maridos* (The marriage between two husbands) or *Servir al señor discreto* (To serve a discreet lord). What is clear is that terms such as *epicycles, equinoxes, solstices,* or *colures,* which stem from the same Ptolemaic astronomy that one hears from the mouth of Don Quixote, are frequent in Lope's work, even if he also speaks of objects as modern as lunar craters, mountains, and seas in the popular play *Los Tellos de Meneses* (The Tellos de Meneses); in the eclogue *Amarilis,* he speaks of the "strange areas" (*partes raras*) and of "mountains" (*montes*) in pieces such as *Los ramilletes de Madrid* (The bouquets of Madrid). He in fact describes the moon as "húmeda y fría" (humid and cold) in *El premio del bien hablar* (The prize for speaking well), in *El vaquero de Moraña* (The cowherd from Moraña), and in *El hidalgo Bencerraje* (The noble Abencerrage). He defends a geocentric universe, for example, in *La doncella Teodor* (The young Teodora), through the words of a character named Fabio who indicates "If the elemental fire / were terminative for us, / and this star were visible, / the one that keeps us warm, / never would the earth see / the sun and the stars it sees" (*que si el fuego elementar / terminativo nos fuese / y como aquéste se viese / que nos suele calentar, / jamás la Tierra vería / el Sol y estrellas que ve*) (ll. 357–62). Fabio is making use of, as Julián González Barrera indicates in his 2008 edition of the work, a clear "medieval cosmology" (210). *El arenal de Sevilla* (The sands of Seville) presents us with a fascinating case: "'Who believes in judiciary astrology? Women" (*¿Quién cree en la astrología / judiciaria? La mujer*). Here one finds an echo of Sixtus V's papal bull, which had mercilessly attacked women ("wanton little women," *mujercillas,* it reads) and accuses them of being superstitious tricksters. Even the references of the older Lope in *El castigo sin venganza* (Punishment without revenge) (ll. 1468–69, 1494–95, 1735–40) confirm that not even at the end of his life, when Galileo's work was fully circulating throughout Europe, did he modify his presuppositions.[12] One finds an entire array of citations likewise shared by authors such as Cervantes, who in *Don Quixote* (II.25) denies the fortune-telling abilities of monkeys and later introduces the humorous story of the little dog, following the idea that fortune-tellers should be punished except in cases related to agriculture, navigation, or medicine. In fact, the gullibility of the masses is a nearly constant theme in early modern Spanish literature; we may take as examples the famous episode of the bull purveyor in *Lazarillo de Tormes* and the inattentive public that Lope attacks in works such as *El último godo* (The last Goth), *La prueba de los ingenios* (The test of wits), and *Sembrar en buena tierra* (To sow in good earth). This running joke at the expense of astrologers also extends to poetry, as occurs in Lope's famous sonnet in *Rimas humanas* that opens with the verse, "Let the heavens and the planets have their influence" (*influya el cielo, influyan los planetas*), which he also included in his famous play *El duque de Viseo* (The Duke of Viseu).

The catalog of references I have just presented is by no means comprehensive, nor does it fully represent Lope's scientific baggage, much of which he must have acquired, as was the case for many of his contemporaries, through the intersection of oral and erudite sources. To these we must also add his own life experience and that of his contacts and friends. The hypotheses that many decades ago were held regarding the scientific education of Lope's contemporary, Tirso de Molina, who owed his technical knowledge of astronomy to a variety of worldly experience and contacts, could well be applied to Lope. As Frank Halstead suggested many years ago:

> We may hypothesize that he gained certain information through formal schooling; through personal encounters with scientists and with students; through occasional reference to printed material and published texts, and, in short, through the media of those channels of social communication which from time immemorial have supplied and do yet supply the public with its vast, miscellaneous knowledge of the world and the dwellers thereon. ("Attitude of Tirso" 422)[13]

It bears mentioning that the dramatic works that I have just cited do not constitute in any way a study of the cosmogony of the period or of the astronomical theories that Lope acquired in his youth. Nor do they offer an analysis of the complicated astrological panorama of this period, whether we are speaking of judiciary astrology or not. They are, generally speaking, sporadic references, allusions to character types or common practices that set the stage for larger interventions, with the purpose of showing a certain type of erudition or simply to function as mechanisms of humor, of dramatic tension, or—as we have also seen—as misogynistic attacks.

I do not wish to close this cursory survey without making a very brief mention of the different uses of glass in early modern Spanish drama, a genre in which there would appear just decades later the most surprising effects of magic and illusion.[14] In one of the best studies on the symbolic potential of glass in Baroque theatre, Melveena McKendrick has pointed out that:

> Mirrors talk to us about power, authority, social values, patriarchal attitudes as well as about people and their desires. Although a presumption of replication is essential to the plot, . . . mirrors do not operate through the visual details of the images they represent. These are irrelevant to both play-text and play-performance—in no play that I can think of is it essential that the portrait-likeness or mirror-reflection be actually seen by the audience. They function as phenomena, and as symbols—as concrete elements in the unfolding of a narrative, as icons of possession and desire, as metaphors for what the plot is saying about the social and human relationships it explores. Stage mirrors are impartial revealers of truths. They are magic mirrors that discover much more than physical images; their reflections, whether sought or accidental, encapsulate entire crises, lives and destinies. (280)[15]

We find that within a period of a few decades, as McKendrick points out, "silvered glass mirrors had supplanted mirrors of polished metal only in the closing decades of the sixteenth century, and were highly prized" (268). This change equally affects corrective lenses: until the entrance of the telescope to the literary sphere, the term *antojo* was used in a relatively conventional way as a synonym for lenses, as well as in its most familiar polysemy: *antojo* as caprice, as a trinket, even as a weakness. This is what occurs with the already-mentioned duality of *antojo* and *anteojo*, exploited in numerous plays to comic end. Tirso de Molina, to give a well-known case, exploits the dual valence of *antojo* in a marvelous scene in *Los amantes de Teruel* (The lovers of Teruel), in which the old Rufino is tricked by his daughter Isabel, who hides from him a note to her lover justifying the poor vision of her father. Rufino sees an *antojo,* something that does not exist, to which he adds that he has no need for an *antojo*—that is, eyeglasses—as "I have not yet lost my sight" (*aún no me ha faltado el ver*). When the paper suddenly falls to the ground and the old man picks it up and reads it, he laments his lack of cunning, that "foolishness of my poor vision" (*antojo de mi poca vista*), in this way playing with the opposite of the spyglass or telescope (*anteojo de larga vista*) that the theatergoing public must have by then known well: the absence of shrewdness in an elderly man as opposed to the faculty of perceiving everything. Such dual valence permits Tirso, as a colophon, to reflect on the dissolute conduct of young people and the powerlessness of their parents:

> Does this stupidity correspond to my many years
> Or is it a foreshadowing of my death
> And the death of my honor? What is this, ungrateful girl?
> What liberty is this? What papers?
> While I most desire to please you
> And find you honor, nobility, and gold
> You dig mines that insult my nobility?
>
> *¿Corresponde a mis años este antojo*
> *o es sombra de la muerte de mis años*
> *y de mi honor también? ¿Qué es esto, ingrata?*
> *¿Qué libertad es esta, qué papeles?*
> *Cuando yo más deseo daros gusto*
> *y buscaros honor, nobleza y oro,*
> *¿hacéis minas de afrenta mi nobleza?*

Tirso's *antojo* functions as a marker not only of spatial distance but also of temporal space, insofar as it reveals something very common in selected urban plays of the period, that is, the incapacity of parents to adapt to new times, to new and "harmful" customs. The *antojos* that are now required are not merely physical but also cognitive, of a form of cognition constructed as well from that "sixth sense" that permits one to see what is hidden, the new ways of young people looking for autonomy and agency.

In any case, the presence of new elements derived from optics should not be interpreted as proof of this or that stance with respect to the then-recent achievements of Copernican science; rather, they are evidence that Lope and his contemporaries lived in a historical moment in which a certain expectation for the findings of the "new science" was felt with increasingly greater intensity, an expectation that, as we will see, became an active curiosity both for the king himself and for educated elites in centers such as the Academy of Mathematics. And yet, little has been written on the confluence of theater and science—especially in poets as versatile as Lope or Tirso—and the fertile combination of two apparently dissociated languages. In a genre as popular as theater, science and poetic verse harmoniously construct a symbiotic relation whereby that which is presented by one is popularized by the other, and that *other*, that ever-protean language of drama, enriches itself through the techniques of the new. Elements such as the mirror and the portrait facilitate, as we will see in the following chapters, a much-needed political reading of the present:

> This was a period in which not only the theatre but the nation itself was preoccupied with the shifting interplay of illusion and reality, image and self-image, self-expression and socio-sexual role-play, and every mirror-image, after all, reveals a refracted truth of sorts, every portrait image certainly every sixteenth and seventeenth-century portrait is in some way an ideological statement. (McKendrick 268)

Throughout the seventeenth century, therefore, one perceives greater complexity in the treatment of certain elements of the new science by Baroque authors, the result of a slow and cautious evolution that runs parallel to the impact of the Scientific Revolution in Spain. If actual practice is difficult to represent, what can one say about those who engage in the practice? The literary image of the scientist is seasoned with burlesque and/or ridiculous tones, when it is not openly critical. As a result, the scientist is converted into a kind of sewing kit in which there is space for nuance and difference in the creation of different typologies. The astrologer, for example, is "fake" (*fingido*) for Calderón, "belching" (*regoldano*) for Vélez de Guevara, and "false" (*falso*) for Lope in *La niña de plata* (The silver girl). The category of "scientists" thus does not appear in literary works in any stable way; rather, it fluctuates between the orthodox and the heretical, between the permitted and the prohibited, or between territories, always placed into evidence, between sanity and insanity. Also present is the difference between the material and the supernatural, giving place to a very cautious attitude when it is reelaborated as a fictional character. This is one of the reasons references to the historical figures of the time such as Galileo are practically inexistent in the literature of this period, and from this will emerge, as we will soon see, an entire genealogy of titles close to what we today understand as science fiction. There also exist concrete historical events that hold back the dissemination of the new, as in, for example, the famous prohibition against publishing novels in 1625 by the recently created Council of Reformation, against which canonical Castilian authors of the period, with greater or lesser resignation, had to struggle.[16] Given this, if elements or proposals of the new science were introduced, they were reworked in such a way that at times they seemed unrecognizable; as a result, modern readers should operate under the premise that, on many occasions, that same environment of change, not the changes per se, is what comes forth in the aesthetic expression of the time. The inclusion of compound terms such as *science* and *fiction* in the present chapter, as well as the word *mechanics*, should not be understood as an incursion into some reflection on the nomenclature of a genre; rather, they should be seen as the opposite: a deliberate semantic broadening that attains the greatest reach in its definition once it is translated to the sphere of aesthetic creation.

There is, however, something more at work here. The extensive genealogy that I have discussed in the present chapter demonstrates also the fragility of the barriers between the accepted and the prohibited, between the unstable categories of "literary" and "scientific," between the serious and the humorous, between the controversial and the didactic. If certain archetypes such as the doctor, the witch, the smoker, the mathematician, the schemer (*arbitrista*), or the astrologer proverbially strike the critical match of the most talented writers of the period, their role is not limited solely to the comic. Through them, authors of the period also sought to expand the boundaries of what was permitted. There is much humor and impudence in the confusion generated by outrageous discourse, but there is also a curiosity that pulses behind it—generally very cautious—regarding the possible scientific sediment of all optical effects. At the center of this curiosity are the eternal questions: And if one truly could? And if this were truly so? The following chapters present possible answers to these questions.

✷ 3 ✷
Assimilations

Italian influence and the culture of knowledge

In the preceding chapters, I have examined the contours of the various rivalries that revolved around the development of science and technical knowledge in Europe from roughly the end of the fifteenth century—rivalries that often pitted one nation against another. In the present chapter, I turn to the numerous areas of contact between Spain and Italy and how these facilitate the transmission of ideas and achievements that manifest themselves, in different ways, in the field of literary fiction as well in the construction of scientific endeavor, that is, science as a practice that generates an imaginary situated between the real and the fictitious. This of course does not mean that such transmission was necessarily harmonious, a point especially worth keeping in mind in the case of early seventeenth-century Castile—a "politically tempestuous context," in the words of Juan Gelabert, in which even inquisitorial authorities often failed to reach a consensus.[1]

As I have pointed out in the preceding chapter, the circulation of ideas between Italy and Spain had been extremely rich during the sixteenth century, with the latter frequently trying to catch up with the former. As Dietrich Briesemeister has argued: "The first alarming indices of dissatisfaction with the general state of culture, language, and scholarly inquiry in Spain appeared during the initial moments of contact and exchange with Italian humanists and the Renaissance of classical letters; in Spain, such contact and exchange provoked a feeling of intellectual inferiority and backwardness (*Los primeros indicios alarmantes de un malestar por la situación general de la cultura, de la lengua y de los estudios en España se manifiestan en los momentos iniciales del contacto e intercambio con la Italia humanista y el renacimiento contemporáneo de letras clásicas que causan una sensación de inferioridad y retraso intelectual*) (40).[2] It is worth asking, in the context of an exploration of the literary universe of early modern Spain, how it was that novelty found itself transmitted in early modern

Spanish fiction. Which processes of rivalry or which mechanisms of tolerance vis-à-vis the "other" manifested themselves between Spain and those Italian territories in which the "new science" was flourishing? What doors were then opening to the circulation of ideas?

In what follows, I focus on two modes of transmission that are fundamental to an understanding of how ideas and literary motifs were received during the first decades of the seventeenth century. On the one hand, there is the effect that texts such as Tomaso Garzoni's *La piazza universale di tutte le professioni del mondo* (The universal town square of all the professions in the world, 1585) and Trajano Boccalini's *Ragguagli di Parnasso* (Newsletter from Parnassus, 1612) had with respect to the dissemination of knowledge regarding glass and lenses, as well as the idea of the so-called keeper of secrets and of the virtuoso. On the other hand, one can notice the influence that political satire exercises in the projection of certain literary devices linked to the spread of telescopic lenses in Baroque culture. The texts of Garzoni and Boccalini—the former through a loose translation penned by Cristóbal Suárez de Figueroa (1615)—are not in any sense paradigms that completely exhaust the matter; however, they are representative of two fascinating and important modes of cultural translation. Beyond this, they allow us to compare some of their more suggestive presuppositions with the literary work of someone who, thanks to his active involvement in the local scene, became one of the most curious and admired figures of seventeenth-century Madrid: the musicologist and collector Juan de Espina, proud owner of a Galilean telescope that intrigued and delighted his contemporaries. In this chapter, I bring to light new forms of inquiry with respect to the construction of the virtuoso in the Iberian Peninsula, a figure whom Espina embodied like no other, a figure that is central to any adequately contextualized understanding of the reception and conception of the image of the scientist during these early decades of the seventeenth century and the dissemination of technical knowledge in Habsburg Spain.

Seventeenth-century Spanish satire, as we will see, offers contemporary scholars a wide-ranging catalog of examples for developing a detailed understanding of the points of intersection between scientific activity and the writing of fiction. Fernando Pérez de Sousa's partial translation of Boccalini's *Ragguagli*—with the title *Discursos políticos, y avisos de Parnaso* (Political discourses, and news from Parnassus, 1634)—is extremely useful and informative for an examination of the process by which the *an-*

teojos literarios, or "literary eyeglasses," became popularized in Spain. Also important in this respect is Suárez de Figueroa's translation of Garzoni's *Piazza universale*, a miscellany that provides a wealth of information for readers interested in the scientific landscape of the time, and in which one finds a reflection on the power of sight mediated by the telescopic lens. Both texts, equally complex and even kaleidoscopic, reveal ways in which motifs such as eyeglass and practices such as glasswork found their way into the collective imaginary of early modern Spain.

It is worth pointing out from the outset that Sousa's and Suárez de Figueroa's texts are in no way isolated cases in the sphere of early modern Spanish letters. It is also true that they are not indicative of a single, homogeneous Italo-Spanish paradigm of cultural relation: the intellectual ferment of Tuscany, for example, was very different from that which obtained in Naples, Milan, Rome, or Venice, as the cultural situation of each of these urban centers was quite particular, as was the Spanish presence in them. It is also important to make clear that the circulation of ideas does not take place only between the two peninsulas; rather, within each there were also fascinating cases of what we might term *internal circulation*. The diffusion of *Ragguagli*, for example, transcended national borders as it was translated into many languages and gave rise to new models of critique influenced to a greater or lesser degree by Italian.[3] In Portugal, one might turn to midcentury masterworks such as Francisco Manuel de Melo's *Hospital de letras* (Hospital of letters, written in 1650–1654 but first published in 1721), in which Parnassus is presented as a sanatorium where Quevedo, Boccalini, Justus Lipsius, and Melo himself offer a comprehensive evaluation of Spain and Portugal's literary past and present. Another excellent Lusophone example is Melchor de Fonseca e Almeida's *Sueño político* (Political dream, ca. 1650), which begins with the narrator reading Boccalini before falling asleep and having his eponymous dream. In Matias Pereira da Silva's *Fénix renascida* (Phoenix reborn, 1716–1728), a five-volume anthology of sixteenth-century Portuguese poetry of fundamental importance for the study of Renaissance Portuguese lyric, one also finds various poetic "reports" that follow the style of *Ragguagli*.

Even taking into account other Portuguese examples, whether written in Portuguese or Castilian, Sousa's *Discursos políticos* remains one of the most historically significant—and inadequately studied—texts of the seventeenth century. This is the case in large measure because he wrote it during the period of Spanish Habsburg rule over Portugal (1580–1640), and it perhaps logically manifests an almost constant tension between what its author wishes to say and what he can actually say. We may add to all this the fact that Sousa, like Suárez de Figueroa in the case of Garzoni's text, repeatedly allows his reader to sense in a very deliberate way the difficulties that come with translation as a tool of knowledge within a complex political landscape of friendships, offenses, rivalries, and suspicion.

Ragguagli offers its Spanish translator a series of fascinating challenges. From its very beginning, in the "Approval of the Augustinian Father Ignacio de Vitoria" (*Aprobación del padre maestro fray Ignacio de Vitoria* [. . .] *del(a) Orden de San Agustín*), the translator engages in the sort of linguistic and thematic juggling act that runs throughout the text. Already in the front matter we read: "The exquisite imaginativeness of [*Ragguagli*'s] topics and the skill and vividness of its concepts undoubtedly requires greater eloquence of the translator, so that these may be imbued with their proper political appeal while domesticated in accordance with the character of our language, than would be the case if one were to translate a narrative story or any other text" (*la exquisita imaginativa de los asuntos, y el primor y viveza de los conceptos, pide sin duda más felicidad en el traductor, para dar estas sus sales políticas, domesticadas al carácter de nuestro idioma, que si se tradujera una historia, u otra cualquier obra que sería de más fácil hechura*) (4r). António Resende, the work's second inquisitorial censor, in fact spoke of how Sousa had found himself compelled to mute the tone of some of Boccalini's cutting remarks found in the first Spanish edition of *Ragguagli*. Resende thus exercised his censorial role to the best of his abilities, "attenuating" (*modestando*) Sousa's text in an exercise of submission that, in a way, mirrored the existing political asymmetry between one nation—and, perhaps, one literary history—over another: "This book was first published with a few bitter, less modest elements, and it is now free of these acrid weeds, to the benefit of its readers, insofar as its liberties have been somewhat attenuated without sacrificing any of the work's piquancy; it has been no small feat to leave this text both clean and witty" (*salió este libro a la primera luz con algunos resabios menos modestos, y de esta mala hierba se ve hoy libre, por quien le traduce, modestándole las licencias, sin malograrle los picantes, y no es poca destreza dejándole limpio, que quede donairoso*) (5r). This work of "cleaning" reveals an open sense of geopolitical supremacy in which the offensive joke is eliminated while Castilian forms of wittiness and piquancy are foregrounded.

Even so, and despite Pérez de Sousa's concern as a Portuguese writer to exclude materials that might offend his Castilian readers, his translation made a notable impact on his contemporaries. Two editions came out in Madrid (1634, 1653) and Huesca (1640), followed by a few more in subsequent years. Although Sousa was merely attempting a rudimentary translation from Italian to Spanish, his *Discursos políticos* confirmed, nonetheless, a growing fascination in Spain with the style and ideas of

Italian writers, whose works had been circulating in their original language throughout the Iberian Peninsula for many decades. It is also worth mentioning here that, perhaps because of the limits of Sousa's Spanish edition, *Ragguagli* continued to be read in the original Italian by many of its most avid Spanish readers.

Born in Loreto in 1556, Boccalini was a tremendously popular author in Spain from at least the second decade of the seventeenth century, thanks to his *Pietra del paragone politico* (Political touchstone, 1614), a theoretical reflection that was very much read and enjoyed, when not openly imitated (Etreros 189). Boccalini's modus vivendi, however, had not always been writing, and it was only near the end of his life that he began the brilliant literary career that would close with his monumental, if scarcely known, posthumous work, *Comentarii sopra Cornelio Tacito* (Commentaries on Cornelius Tacitus, 1669). He composed his *Ragguagli* while living in Rome—"a storehouse for diversion and stimulation" according to Robert Haden Williams (9)—and later moved to Venice to avoid possible reprisals. Venice had long been considered a model of fair government, and it contrasted sharply with Habsburg Spain, a locus for Boccalini of favoritism and corruption (*Ragguagli*, I.5). Boccalini, according to Williams, "had long admired the independence of the Venetian republic, whose example he repeatedly invoked while attempting to incite other provinces to revolt against Spanish oppression" (4). He expressed an open admiration for Spain's enemies—such as France or, more specifically, King Henry IV—and celebrated the figure of the Duke of Savoy as an Italian warrior par excellence.[4] It is perhaps not surprising that his death in 1613, while officially attributed to illness, be wrapped in a fog of mystery, the hypothesis being that he was murdered by a group of Spanish thugs hired by the government—his sudden death served to enhance the legendary profile of his work outside of Italy. However, extant literary sources from the time are less than enthusiastic: among the most hostile of Boccalini's Spanish readers was none other than Lope de Vega, who refers to him in his famous play *El desdichado por la honra* (The proud unfortunate) as "more envious than eloquent and learned" (*más envidioso que elocuente y doctor*); and in his sonnet "A los Raguallos de Bocalini" (On Boccalini's Ragguagli), found in *Rimas humanas y divinas del licenciado Tomé de Burguillos* (Human and divine poems of the licentiate Tomé de Burguillos, 1634), Lope refers to Boccalini as the "mouth of hell" (*boca del infierno*).[5]

Despite such attacks, *Ragguagli* and *Pietra* paradoxically inspired a great number of imitations in Spain, insofar as they were received as fertile narrative models and not just chauvinistic provocations: Juan Cortés de Tolosa celebrated Boccalini's genius in *Discursos morales* (Moral discourses, 1617), possibly imitating him—the landscape of both texts is quite similar—in "Carta de una dama a Apolo dándole quejas del mal que pretende hacerla" (Letter from a woman to Apollo presenting him with complaints about the harm he intends to cause her), in which the fictitious author complains to the Greek god about the amount of bad poets who follow him everywhere. In the second part of *El caballero puntual* (The punctual nobleman, 1619), Alonso Jerónimo de Salas Barbadillo inserts a short piece, *El curioso* (The curious one), in which his protagonist, bringing letters of recommendation from Boccalini to Tacitus, shows up on Mount Parnassus; the same author, years later, includes numerous scenes in his *Don Diego de noche* (Don Diego at night, 1623)—a Senecan tribunal, a diatribe against cuckolds, a defense of comedies—that very closely follow the structure and language of Boccalini. And there is also the case of Matías de los Reyes, who in his *Curial del Parnaso* (Curia of Parnassus, 1624) directly plagiarizes various sections of *Ragguagli*, as both Carroll Johnson and Caroline Bourland demonstrated decades ago.

The second quarter of the seventeenth century was equally intense in terms of influences and imitations. In his posthumously published *Coronas del Parnaso y platos de las musas* (Crowns of Parnassus and plates of the muses, 1635), Salas Barbadillo returned, in Boccalinian fashion, to Mount Parnassus for a last evaluation of the present; and Quevedo—who had just arrived in Naples when *Ragguagli* was published—would later recognize the influence of the famous satirist in *Cartas del caballero de la Tenaza* (The letters of the knight of the pincers, 1627) and *La Hora de todos y la Fortuna con seso* (The Hour of all and Fortune with sense, 1650). Two decades later, Baltasar Gracián would pay Boccalini homage—"Boccalini's crises (*las crisis del Boquelino*)—in the prologue of his *El criticón* (The critic, 1651), as well as in *crisi* 7, where the protagonist visits a Parnassus store in which are sold glasses and gloves, and also through allusions such as "the window to the human heart" (*la ventanilla del pecho humano*) and "the entire world's fair" (*la feria de todo el mundo*) inserted into *crisi* 13, to which he would add, in *crisi* 6, the complaint that in his time slaves were bosses and the "blind lead" (*ciegos guían*).[6] Even the novel *La dama beata* (The blessed lady, 1655), by the Italo-Spanish novelist José Camerino, displays a markedly Boccalinian flavor.

In its condition as a "best seller" read and admired throughout Europe, *Ragguagli* provided its Spanish readers with new linguistic and thematic tools for analyzing the cultural scene. It also animated existing debates regarding political theory, the national literary field, and the perception of Spain in Europe, almost always stemming from a comparison between the decadent Habsburg monarchy and the admired and effervescent Venetian republic. This hostility soon triggered numerous metaphors in which the uses of glass were paramount, as in Luis Vélez de Guevara's insinuation in *El diablo Cojuelo* that Venice "moves with whatever wind blows it" (*se vuelve a cualquier viento que le sopla*) (55). For Spain's European neighbors, Boccalini's work was thus instrumental as much in the representation as in the projection of Spanish culture. In the tribunal of artists imagined by Boccalini, the citizens of Mount Parnassus do not constitute an exemplary society, although they do possess a certain *virtù* in the purest Machiavellian sense, close to what Gracián would term in subsequent works of fiction *héroe* (hero) and *discreto* (discreet or smart). In the text, Aristotle, Tasso, Tacitus, Lipsius, Dante, and Ronsard stroll around Mount Parnassus accompanied by allegories such as the muses, the monarchies, and the republics, identifying the problems with which a good government should deal. Through his first-person judgments, Boccalini effectively constructs a new type of satire that combines the parodies of the famous gazettes or newsletters with comic fiction in the form of allegory. He offers in this way a very personal version of what Mercedes Blanco has termed *intellectual journalism* through the articulation of a miscellany of diverse styles seasoned with new ingredients and received with enormous success by its readers. As Blanco ("Del infierno" 172) puts it: "The pleasure and admiration provoked by *Ragguagli* prove that Boccalini had arrived at an elegant solution for many of the structural and stylistic problems that were then of concern; he also found pleasing and effective formulas to say what he judged was important to say" (*El placer y la admiración que provocaron los* Ragguagli *prueban que Boccalini había dado con una solución elegante para muchos de los problemas de estructura y de estilo que preocupaban entonces, y había encontrado unas fórmulas gratas y eficaces para decir lo que se juzgaba importante decir*).

Boccalini's eloquence did not overshadow *Ragguagli*'s biting criticism, and the text's traces of Menippean satire were assiduously imitated over the following decades. Through the direct influence of figures such as Juan Luis Vives and Erasmus of Rotterdam, we know that an entire tradition of Spanish writers was familiar with classics such as works by Lucian, Tacitus, Juvenal, and Seneca.[7] Baltasar Álamos de Barrientos, for example, published his *Tácito español* (Spanish Tacitus, 1614), in which he commented on and updated the political theory of his Roman predecessor. What is truly innovative in Boccalini, however, is his use of narrative devices such as the *hospital de locos*, or insane asylum, which he had inherited from Garzoni's *L'hospidale de' pazzi incurabili* (The hospital of the incurably insane, 1586), and which also would inspire Salvador Jacinto

Polo de Medina's famous novel *Hospital de incurables y viaje deste mundo y el otro* (The hospital for the incurable and voyage through this world and the other, 1636), as well as certain passages of Gabriel del Corral's lesser-known piece *La Cintia de Aranjuez* (Cynthia of Aranjuez, 1629). There is also the innovative foundation of a literary academy, a concept that would become so important for figures such as Lope, Vélez de Guevara, and Salas Barbadillo.[8] In Salas Barbadillo, in fact, the idea of the distribution of what we would now call cultural capital in line with poetic merit was transformed into the very framework for his aforementioned *Coronas del Parnaso*, in which the object of his praise was none other than the Count-Duke of Olivares, praised both through cliché (such as *olivo*) and exaggeration. As with the pilgrimage fictions of contemporaries such as Lope and Cervantes, Mount Parnassus offered the possibility for one to take the measure of Madrid's contemporary cultural field through a playful rereading that was likewise cautious of the present moment. It was necessary to rename those aspects of reality that could not be reduced to allegory but that in their structure broke with inherited models: if Boccalini divided his book between *ragguagli* and *centuria*, the Spanish opted for other discursive divisions that were no less ingenious but were nonetheless typical of the genres of satire and the picaresque: *crisi* and *primores* (Gracián); *estafas* (Alonso de Castillo Solórzano); *trancos* (Luis Vélez de Guevara); *bulcos* and *transmigraciones* (Antonio Enríquez Gómez); *droga* (Marcos Fernández); *errores* (Juan de Zabaleta); and *hora*, *puntada*, and *esperezos* (Francisco Santos), which replaced the classic chapter (*capítulo*) and dismantled with it, through self-parody, any serious notion of how and why to write a novel.[9]

If there was a structural motif that managed to transcend boundaries in the new literary cartography, it was the *occhiali politici*, or "political eyeglasses." Canonized as a satirical weapon by Tacitus used to criticize the customs of his contemporaries, it came to be an invention that traveled widely, in different formats, throughout the centuries and literary traditions. The late Middle Ages, for example, had already taken hold of these "new lenses" through which one could draw so much, thanks in large measure to contributions by figures such as Nicholas of Cusa (1401–1464), who in his *De beryllo* (On beryl, 1458) proposed through Cardinal Krebs the use of "spiritual eyeglasses" to gain knowledge of the greatest to the smallest thing—*beryl* here was a synonym for *eyeglasses*, making reference to the crystals that had traditionally been used to make lenses. It was already assumed during this early period that lenses were a device that allowed one to see with clarity, to access realities otherwise hidden thanks to the use of convex lenses—such as those already being used to arrest the effects of presbyopia—and concave ones. We have already seen how in Spanish literature one finds sporadic mention of these famous accessories in *cancionero* poetry, playing, as we have seen, with the rhyme *enojos* and *antojos*.

In Boccalini, the "political eyeglasses" motif appears for the first time in *Ragguagli* when the author mentions the great variety of magical glasses sold in a store on Mount Parnassus.[10] The first part of the book, in fact, has as its title "The university of all politicians opens a store on Parnassus, in which are sold various sorts of merchandise, all very useful for the modest and virtuous life of all learned men and well-dressed people" (*La universidad de todos los políticos abre una tienda en Parnaso, en que se venden diversas mercaderías, muy provechosas a la modesta y virtuosa vida de todos los hombres doctos y personas de prendas*) (1v). This "university" store introduces a moralizing and lecturing tone through its extravagant offerings, typical of the Baroque *Wunderkammer*, but it also alerts us to all that can be learned from these small local or clandestine markets. And among the first objects mentioned—with erasers and pencils—are eyeglasses:

> There is also an infinite array of admirable eyeglasses, with excellent virtues, because some correct the vision of sensual men, who in the furor of their clumsy desires lose their vision in such a way that they cannot tell honor from harm, friends from enemies, strangers from family, nor any other thing that might merit some respect. And these eyeglasses are so widely used that the corrupt politicians themselves also wear them, as it is well known that there are few men endowed with good vision concerning sensual things.
>
> *Hay también número infinito de admirables antojos, de excelentes virtudes, porque unos sirven para la vista de algunos hombres sensuales, que en el furor de sus torpes disgustos se les acorta de tal suerte, que no diferencian la honra del vituperio, el amigo del enemigo, el extraño del pariente, ni otra cosa que merezca se le tenga respeto. Y es tan grande el empleo, que los políticos mercaderes hacen de semejante suerte de antojos, que se ha venido a conocer claramente, que son raros los hombres, que en las cosas sensuales tienen buena vista.* (1v–2r)

If this first statement is somewhat generic, Boccalini then launches into an antimonarchic argument by means of a speculation regarding the sense of vision. He recommends, in fact, a certain measure of blindness in order not to see what is occurring and not to fall into the same hypocrisy of the person whom he criticizes. In this sense, wisdom achieved through the examination of reality aided by eyeglasses is of little use because it leads ultimately to disenchantment. It is better at times, argues Boccalini, to turn a blind eye and feign that one sees nothing:

> Here there are also sold other eyeglasses, which help one not to see. And the politicians affirm that these are much more necessary to every man (and particularly to courtiers) than those made to help one see far away. This is certainly so, given that the many unpleasant and vile actions of some powerful princes are frequently offered to our sight, and if we turn our back to them, it seems as though we are disapproving, which will stir the anger of these lords. Witnessing these actions is a difficult trial, and putting on such eyeglasses serves to liberate one from having to see the corruption of such a depraved century—a time when the ignorant, persuaded by what is before them, applaud and observe with great attention.
>
> *Véndense aquí también otros antojos, que sirven a algunos para hacer que no vean. Y los mismos políticos afirman que son mucho más necesarios a todos los hombres (y particularmente a los cortesanos) que los de larga vista; por razón, que muchas veces se les ofrece a ella mil desagradables y viciosas acciones de algunos príncipes poderosos, a que, si uno vuelve las espaldas, parece que las reprueba, granjeando consecutivamente el enojo e ira de los tales señores. Siendo, pues, el mirarlas un penoso martirio el ponerse en semejantes ocasiones tan admirables antojos, servirá de librarse de la penalidad de ver la corrupción de siglo tan depravado, cuando ellos ignorantes están persuadidos a que los están asistiendo, aplaudiendo y con suma atención mirando.* (2r)

The antimonarchic tone of this text is forged through the criticism of "powerful princes" whose corruption constitutes "a painful suffering" to behold. The phrase "depraved century" (*siglo tan depravado*), which also frequently appears in Spanish satire, presides over the citation, in which Boccalini likewise attacks the lack of diligence shown by the governor or patron whom the courtier serves. Vision has no escape; there is no other option for the courtier but to look straight ahead and observe the misery before him, given that any expression of disapproval, as Boccalini observes, cannot be contemplated.

In like fashion, the ingratitude of those who do not remember their origins and forget the means by which they came to be crowned by Fortune merits the fabrication of a different set of eyeglasses, one that would be, perhaps, *of another type*, associated now with a passing of time and forgetting:

> Other eyeglasses help to preserve the vision of certain cold people who, the first day they are favored by fortune and rise to the summit of greatest dignity, are so swelled by luck that they become wholly ungrateful. The corrupt politicians say that these glasses are manufactured from the precious materials of the tenacious memory of the benefits received and of the reciprocal love of old friendship.
>
> *Otros antojos sirven para conservar la vista de algunas personas poco amorosas, que en el primer día, en que favorecidos de la fortuna subieron a la cumbre de superior dignidad, se les engruesa de suerte que llegan desconocidos a los términos de ingratos. Dicen los políticos de la tienda que son fabricados de la preciosa materia de la tenaz memoria de los beneficios recibidos y del amor recíproco de la antigua amistad.* (2r)

Boccalini also provides a sanction, very typical for the period, for the stupidity of the "unfortunate courtiers," who upon being falsely favored by their lords end up unable to distinguish generosity from hypocrisy. In this instance, eyeglasses serve Boccalini—and by extension, Sousa—as a metaphor for the illusory "augmentation" that these parasites of the Court experience upon being touched by the sight of their lords. This said, the enslavement to which some of them are submitted also bleeds, in an oblique way, into the broader problem of patronage in light of the unproductive nobility. In this context, learned people must suffer all type of poverty and humiliation to maintain themselves under the care of their protectors. From their perspective, everything seems to be magnification, distortion, and diminishment, and nothing is valued as it should be. And in this unjust universe the crystal lens holds a thousand marvels.

The problem of vision, embodied in the "observed being," is associated in this context with hypocrisy—"with a happy appearance, though artificial and forced" (*con alegre semblante, aun que artificioso y forzado*)—and so underscores the theatricality of a universe of appearances in which vision forms an essential part of a complex ritual of gestures tacking between the high and the low. The eye gives life insofar as being seen signifies being recognized, being legitimized within a closed group, to exist. But Boccalini thickens his satire by mixing scientific reality with its commercial possibilities, constructing a brief costumbrista sketch that derives, ultimately, from a moral reflection not lacking in social commentary. In this way, the new Flemish telescope—which as we have seen, competes between 1609 and 1610 with Galileo's own in terms of quality and popularity—is exported throughout Europe as a gift for the wealthy or the well connected, finding their way also to those who do not necessarily appreciate them, given that their unmerited use makes people believe that they can see that which is beyond their reach, that they can be that which they are not or ever will come to be. It should not come as a surprise that Boccalini uses the term *desvanecido* ("dissipated" or "vanished"), the polyvalence of which includes the sense "removing something from sight"—literally from the Latin *removere, disparere*—and in its metaphoric sense, to "give occasion for presumption and vanity."

The last of the eyeglasses that Boccalini discusses is, for him, the most beneficial. This is so not because of their ability to provide access to what is other but because of their power to mediate a reflection on that which surrounds us and, even more important, that which we are. They facilitate a genuine and necessary examination of oneself, a calm "internal gaze"—the idea being that without knowing oneself well, it is not possible to interpret reality correctly. It is here, as Blanco has argued, that Boccalini appeals to the mythical space of virtue: "*Parnassus* is the name of the space of liberty for a person of genius and learning, from which it is possible to watch and judge, as long as one remembers that the reasonable attitude consists in maintaining an ironic stance before the privilege of force, and in resigning oneself to the incurable illness of the age" (*El Parnaso es, pues, nombre de un espacio de libertad para los hombres de ingenio y doctrina, desde donde les es dado mirar y juzgar, con tal de que sepan que la actitud razonable consiste en inclinarse irónicamente ante el privilegio de la fuerza, y resignarse a la enfermedad incurable del siglo*) ("Del infierno" 174). Here Blanco correctly underscores the term *enfermedad del siglo*, or "illness of the age" ("Del infierno" 175). Wisdom, it follows, comes at a high price:

> Besides this, in the same store (but at a very high price) are sold human eyes of marvelous virtue, because it is incredible how much one can improve his own lot by observing it through other eyes; and even politicians agree that there is no other instrument by which one might achieve the happiness of reaching the excellent virtue of *nosce te ipsum* so desired and sought out by important men as with this one.
>
> *Demás de esto en la misma tienda (pero a muy caro precio) se venden ojos humanos de maravillosa virtud, porque no es posible creer, cuanto algunos mejoran las cosas propias, cuando las miran con ajenos ojos, y aun los mismos políticos afirman que con ningún otro instrumento se podrá llegar a la felicidad de alcanzar aquella excelente virtud tan deseada y procurada de los hombres grandes del* nosce te ipsum, *como con éste.* (2v)

As an instrument that designates at once the ideas of vision and eyesight, Boccalini's eyeglasses become a very effective narrative device that simultaneously pays tribute to that which was then an important local industry. As I have already mentioned, eyeglasses had been invented in Venice at the end of the thirteenth century, and it was not long after this that the city became a commercial center from which the first smooth lens, called *cristallo*, was exported.[11] Glass was, as it is even today, Venice's identifying brand, inspiring in the literature of the period audacious Baroque metaphors revolving around the fantasies of its forms, the sensation of movement that these inspired, and the material's capacity for manipulation—a transitory material and, in a certain way, unreliable. Such was the case, as we will see shortly, with Quevedo, who harshly criticized the Venetian republic in several of his prose works. And he was not alone. Many other writers of the period, although without Quevedo's overflowing genius, provide evidence of what Mercedes Etreros, among others, has interpreted as a reaction to ongoing crisis and changes in a sphere of fiction:

> Satire is a reflection of a dialectic tension that develops in the seventeenth century, a conflict produced by political and social change that is perceived, although without visible structural alterations, in specific changes at the cultural level. It is a reflection of symptoms of weakening in the modern state, of a splitting of its structures.
>
> *La sátira es reflejo de una tensión dialéctica que se da en el siglo XVII, tensión de conflictividad producida por un cambio político y social y que, aunque sin visibles alteraciones estructurales, se percibe en cambios particulares a nivel, sobre todo, cultural. Es reflejo de unos síntomas de debilitamiento del Estado Moderno, de resquebrajamiento de sus estructuras.* (153)

Boccalini, however, remained closer to the classical model of the myth. As Garzoni had done in speaking of ancient dioptrics in *La piazza universale*, Boccalini gives the illustrious profession of glassmaking a somewhat more noble treatment in converting the *occhiali politici* into a fascinating narrative element that recuperated, in a certain way, two scientific and philosophical traditions: that corresponding to figures such as Salvino D'Armate (1258–1312) and Alessandro Spina (d. 1313), who are believed to have invented eyeglasses around 1284, and that linked to Roger Bacon (1213–1292), who in 1268 had carried out the first scientific commentary on lenses as correctors of vision, using the term *reflection* to refer to the knowledge of oneself (it is also said that Bacon was the one who first proposed the idea of using lenses to correct eyesight, suggesting even the possibility of combining lenses to form a telescope).[12] With this, Boccalini effectively captured the interest generated by Galileo's recent discoveries. Galileo himself was in the process of perfecting during this period his famous telescope in collaboration with glassmakers from Murano. Through the corrective power of eyeglasses, Boccalini created a literary framework for the intersection of science and moral theory that Spanish authors would later lend a particular gravitas that rested halfway between the dignified and the humorous: this is what one sees, for example, in the case of Raphael's portrait of Pope Leo X (1517–1519), myopic like so many other members of the Medici family;[13] and very similar is the case, in the history of early modern Spanish portraiture, of various important paintings: El Greco's portrait of Cardinal Don Fernando Niño de Guevara (1600), or Bartolomé Murillo's *Patrician John Reveals His Dream to Pope Liberius* (1665) in which, from this "exaltation of the gaze," the seriousness of the portrait subject remains ultimately overshadowed by the strange effect of his eyeglasses (figs. 6 and 7).[14]

A combination of science and moral philosophy rests also at the base of Boccalini's poetics. In part 2, chapter 89 of *Ragguagli,* Apollo advises a vain writer to scrutinize reality with "political eyeglasses" (*occhiali politici*), and "with the eye of Lynceus himself" (*con l'occhio dello stesso Linceo*), probably also alluding to the image of the lynx, as Vélez de Guevara would do later in *El diablo Cojuelo* (The limping devil), of the famous Academy of Lincei. What Boccalini argues here is that the events of the day, as much in the cultural as in the political sphere, must be analyzed in a considered and extremely self-conscious way so that, in the purest Baroque spirit, one arrive at the true essence of things. This is, in fact, advice that would be very closely followed by Boccalini's Spanish contemporaries, who, upon adapting the *occhiali politici* motif to the *anteojos de larga vista,* created one of the most original satirical devices of the period. To take just some examples, in Juan Enríquez de Zúñiga's *El amor con vista* (Love with sight, 1625), Mercury permits Dionysus to see from the heavens what happens within people's homes; in *El hijo de Málaga, murmurador jurado* (The son of Málaga, sworn gossip, 1639), signed by Salvador Jacinto Polo de Medina under the pseudonym of Fabio Virgilio Cordato, the author feigns that a statue gives him the ability to detect the vices of those who visit the fish markets of the city, following closely a scene at the stores of Parnassus in chapters 9 and 10; in *La torre de Babilonia* (The Tower of Babel, 1649), Antonio Enríquez Gómez avails himself of a "tower of disillusion" and of eyeglasses to identify and ridicule the weaknesses of his contemporaries; in Diego de Saavedra Fajardo's *República literaria* (Literary republic, 1655) the protagonist visits a Parnassus

FIGURE 6. El Greco, *Cardenal Inquisidor Niño de Guevara* (1600)

city in which he meets a group of poets, among them a Tacitus wearing "telescopic eyeglasses" (*antojos de larga vista*); and, toward the end of the century, the war veteran Francisco Santos re-creates in *El sastre del Campillo* (The tailor of El Campillo, 1685) an optical shop (*tienda de anteojos*) in Madrid in *puntadas* 5, 10, and 11, playing with the lexical proximity of *anteojos* and *antojos.*

In what remains of this chapter, I focus on the implementation of the *occhiali politici* motif in early seventeenth-century Spanish satire, dealing also with how its migration from Italy to Spain allows modern readers to

FIGURE 7. Bartolomé Murillo, *El Patricio Juan y su mujer revelan su sueño al Papa Libero* (1665)

consider a series of significant parameters related to human curiosity, fear, and the recourse to humor during the period.[15] Like Boccalini, Spanish satirical writers did not only use the *occhiali* with a moral purpose; they also adapted them to the social, economic, and religious realities that surrounded them. Some of them even converted this literary device into the very center of their work, as is the case with three of the satires that form the center of this study: Rodrigo Fernández de Ribera's *Los anteojos de mejor vista* (Corrective eyeglasses, ca. 1625), Luis Vélez de Guevara's *El diablo Cojuelo*, and Andrés Dávila y Heredia's *Tienda de anteojos políticos* (Store of political eyeglasses, 1673). What is worth mentioning with respect to these specific (and very different) works, which exemplify distinct appropriations of what I present as *anteojos de larga* (or *mejor*) *vista*, is that each develops the motif from the context of a discrete historical moment within a so-called century of satire that witnesses both its zenith and its decline.[16] Beyond this, these works also represent three of the most important reflections on the sense of sight in seventeenth-century fiction and the connections between its use (and abuse) and the ordering of a new urban society. They constitute, in sum, a unique sequence in which the use of eyeglasses for moral ends reveals in the final instance the many concerns that scientific discovery created in the public and private sphere. I argue, therefore, that upon examining the *occhiali politici* in these texts in light of certain writings on optics, we can delineate new connections between scientific advances and courtly conduct in early modern Spain. To be sure, this narrative resource manifests a double attitude of fascination and mistrust toward innovations in optics, touching many times on magic and the occult, in what Pamela H. Smith has termed, with great incisiveness, "the lunatic fringes of the investigation of nature"—that is, alchemy, astrology, and magic (353). If the mere act of wearing eyeglasses endowed one with an aura of respect, seeing better and beyond carried with it the seed of the esoteric, of the magical, and even of the heretical, bound together as they were in an age of intense surveillance. The *anteojos de larga vista* wound up being, in this sense, an element that revealed a great deal regarding literary art as well as the role of science and the efficacy of the state in its control and dissemination.

A year before the publication of *Ragguagli*, the very interesting *Plaza universal de todas ciencias y artes* was published in Madrid. This is, as we already know, a highly idiosyncratic translation—insofar as it omits material from the original and adds new chapters—of Tommaso Garzoni's *La piazza universale di tutte le professioni del mondo*, an authentic European best seller during the seventeenth century, published first in Italian and then in Latin and German, and of which there appeared no fewer than one hundred thousand copies. This is, as Mauricio Jalón has pointed out, a "paraencyclopedic" book, dealing with a wide range of materials, sciences, and professions that had turned Garzoni into one of the most read authors of the sixteenth century, as well as one of the most translated outside of Italy. Between 1585 and 1665, for example, it is known that as many as thirty editions were printed, and that thirteen of these available when the first Spanish translation appeared in 1615. But Suárez de Figueroa not only produced a brilliant version of *La piazza universale*, using language at once technical and innovative, but also very adroitly modified the text for a literary and political context that was quite different from that which obtained in Venice. Seeking to be an "antidote for the venom of crass ignorance" (*antídoto contra el veneno de la crasa ignorancia*), Suárez adapted Garzoni's data to his time and place, pruning from the original in some places and expanding it in others.[17] This rich and densely populated *Plaza universal* was reedited various times during the seventeenth century, and during the following decades it appeared reordered and expanded with the title slightly changed; although it suffered from a temporary decline during the nineteenth century and well into the twentieth, today the text can be enjoyed as a very complete manual thanks to the attention that has been given to it not only in the field of literary history—which continues to perpetuate the image of Suárez de Figueroa as gossipy and vile, involved in numerous polemics—but also, and more properly, in the history of science and studies of the history of encyclopedic projects. From this perspective, and as an essential part of the present chapter, focused on scientific curiosity, I approach this masterwork.

Plaza universal is a very useful source of information on the state of various disciplines of knowledge in the early moments of what we have come to term the Scientific Revolution. Jalón has defined the period in which Suárez worked as "a rather confusing moment of anticipation . . . , a world increasingly tired of the sixteenth century (*un compás de espera bastante confuso* [. . .] *un mundo cada vez más envejecido de los centones quinientistas*) ("El orden" 82). With the text one finds close to 550 professions in 107 sections distributed over 368 folios that make up, it is worth remembering, only one-third of the original work. One of the work's greatest scholars, Ángeles Arce, maintains that Suárez took on the translation of his own will (97n13). Whatever his motivation, Suárez makes it clear from the beginning that the project has its roots in his own enthusiasm for Garzoni's text, notwithstanding certain problematic areas:

> Being myself more full of faults than others and with less understanding than the most stupid man, having read through the book in Italian by Tommaso Garzoni titled, *Universal Plaza of All Professions,* I came to like its variety, judging it worthy of communication, although it contained some expressions not current in our own language. I did not include these in the translation, and I added other things where it seemed appropriate. It is now published, cut, and with additions.
>
> *Habiendo yo más lleno de faltas que todos, y menos entendido que el más rudo, pasado los ojos por el Libro en Toscano de Tomás Garzón, título* Plaza universal de todas profesiones, *me aficioné a su variedad, juzgándole digno de comunicación, como careciese de algunas cosas, por ventura no bien corrientes en nuestro vulgar. Éstas no puse elegida la traducción, y añadí otras donde pareció convenía. Publícase pues ahora traducido, cercenado, y añadido.* (35)

This passage reveals very specific aims with respect to the translation, aims that Jalón divides into three different levels: the philological, the historical, and the mythological. Although it is true that Suárez alludes to the heliocentric ideas of Aristarchus of Samos, it is evident from the date of the original model that he belongs still to a generation of thinkers with a very traditional understanding of astronomical theory. In essence, he follows Garzoni, who had based his own background on the Venetian edition of Ptolemy annotated by Girolamo Ruscelli in 1561, which was very popular at the time. If Garzoni seems to praise Tycho Brahe on at least one occasion (among, it is worth pointing out, a long string of names), in general he adopts an obsolete vision of the cosmos.[18] The astronomers on whom he depends, such as Johannes de Sacrobosco, Georg von Peuerbach, Giovanni Bartolomeo Marliani, and Abraham Zacuto, all belong to a period already very much in the past. We may take as an example discourse 39, dedicated to the astronomers, in which Suárez links astronomy with astrology: "astronomy, and with it natural astrology, which are like sisters, united and intertwined" (*la Astronomía, y junto de la Astrología natural, que son como hermanas, unidas y abrazadas entre sí*) (374). This apparent resistance to innovation is countered in Suárez's translation by the argument—by the seventeenth century a kind of commonplace—that all science is good provided that it is put to good use, a sentiment echoed by Cervantes in the epigraph that opens the present book:

> Natural and true astrology is an extremely useful science and thoroughly necessary for human life; but this does not prevent the appearance of many errors in the works of those who write on it, not to mention the seemingly infinite problems that make it suspect and give it an almost doubtful value, as occurs in all the sciences.

> *la Astrología natural como verdadera, es ciencia utilísima, y necesaria sumamente para la vida humana; mas no por eso dejan de hallarse muchos errores en los autores de la misma, sin infinitas repugnancias que la hacen sospechosa, dándola casi una engañosa estimación, como sucede en todas las ciencias.* (379)

With the goal of distancing himself from all "doubtful value," "infinite problems," "suspicion," and "error" (*engañosa estimación de toda infinita repugnancia, de toda sospecha y todo error*) in his description of the movement of the planets and the sun, Suárez adheres to Ptolemy, whom he cites in various passages (45, 379–80). He shows himself, in fact, openly anti-Copernican (380) and in favor of "the truth of judiciary astrology" (*la verdad de la astrología judiciaria*) even if he criticizes, as Lope and Cervantes had already done, the servitude of those of his time to the stars as determinant of human destiny (392, 395).

Very different, however, is his approach to the discipline of optics, which he tackles through arguments that are much better constructed. It is worth remembering, however, that Suárez is not completely original. He takes directly from Rafaelle Mirami's *Scienza degli specchi* (The science of mirrors, 1582), as well as from one of the most common sources of the period, Girolamo Cardano—in this case, book 5 of *De subtilitate* (On subtlety, 1550) and book 10 of *De rerum varietate* (On the variety of things, 1558). In moving from theory to practice, one also finds in Suárez extremely interesting material (albeit somewhat topical) in his commentaries on the art of glassmaking. In the chapter dealing with glassmakers and their trade, we read the following:

> Today the glass shop of Murano, a pleasant place next to Venice, exceeds all others in terms of perfection and quality; this is so in part because of the quality of the water which is very important for works such as this, the lack of dust that might harm the work, the quality of the local wood, which makes the fire extremely clear, and because in other places they do not use salt from the stone called soda as is done in Murano, which makes it so that there is formed extremely beautiful glass. . . . The making of glass is so highly developed in Murano and Barcelona that there is nothing imaginable that cannot be made there from glass and crystal; there have been made desks and even castles with towers, bastions, artillery, and high walls.
>
> *Hoy el cristalino de Murán, lugar ameno junto a Venecia, excede a los más del mundo, en perfección y bondad; parte por lo salobre del agua muy proporcionada a los labores de este género; parte por carecer de polvo que pueda hacer daño a las obras; parte por la comodidad de la leña forastera que hace clarísima la llama, y porque no se usa en otros lugares hacerse el sal de la piedra llamada soda, como se hace en el mismo Murán; causa de formarse allí bellísimos cristales.* [. . .] *Está hoy en Murán y Barcelona tan en su punto este ejercicio, que no hay cosa imaginable, que no se obre con vidrio, y con cristal; habiéndose hecho hasta escritorios, y castillos, con torres, bastiones, artillería, y murallas.* (505–6)

We see in this citation that the Venice-Barcelona axis is extremely important for the glass industry of the Mediterranean; perhaps Suárez here has in mind, as was the case of many of his contemporaries, the legacy of the Roget family of Girona. The citation also reminds us that a great amount of the value assigned to glass was because of its use in the art of war and in different architectural structures that with it could be improved; this includes, as he tells us, towers and walls. The chapter is, in any case, merely descriptive; in it Suárez copies directly from Garzoni the passages that explain how crystal is made, the ingredients it requires, how eyeglasses are made, and the foundations of the art of glassmaking (to which he dedicates relatively little space and with which he closes the chapter). He also announces, however, that which Jalón has defined as "an expansion that will give way in just a few decades to technological thinking" (*una expansión que dará paso en unas pocas décadas al pensamiento tecnológico*) ("El orden" 77), and therefore the selection of this text—a kind of hinge between the old and the new—is of great relevance to the present study. It is an effective form of *translatio* that is eminently faithful to the didactic spirit of the book.

The disquisition on optics does not end here, however. A bit later, in discourse 102 ("De los espejos, y sus artificios," On mirrors and their artifices), Suárez links the science of mirrors to astrology and natural science:

> [Mirrors are] extremely useful to astrology, to resolve many questions on celestial matters, such as, for example the dark spots on the moon, eclipses, and the proportion of rays. It is also of great benefit in natural philosophy to reflect on many impressions that are formed in the region of air, like the iris, and the heat generated by solar rays, and many other effects that natural philosophy examines and discusses with great excellence. . . . Mirrors serve finally to bring light to dark places, to invert certain kinds of shadows from the place in which they are, to measure heights, depths, and distances, to put objects into perspective, and all things related to it.
>
> *Es utilísima a la Astrología, para resolver muchas cuestiones en las cosas celestes, como por ejemplo la de las manchas de la Luna, de los eclipses, y proporción de los rayos. Es también de gran provecho en la filosofía natural para discurrir acerca de muchas impresiones que se forman en la región del aire, como el Iris, y el calor engendrado de los rayos solares, y otros muchos efectos sobre que la misma juzga, y discurre con mucha excelencia* [. . .] *Sirven finalmente de alumbrar los lugares oscuros; de volver al revés algunas suertes de sombras de aquel sitio en que están; de medir con la vista alturas, profundidades y distancias, de poner en perspectiva, y de todas las cosas pertenecientes a ella.* (699–700)

We are here perhaps before a very veiled description of the power of the telescope—though not necessarily Galileo's—through the correct use of lenses that make it up and that give it the capacity to illuminate lunar dark spots and other obscure regions, or to appreciate the sun not only for its light but also for its heat. The citation also offers a metaphorical reading, insofar as the new science unveils previously unknown aspects, in its attempt to "invert certain kinds of shadows," evaluating in a new light "heights, depths, and distances" (700). This is a text in which curiosity blends scientific hypotheses with established dogma, and this intellectual desire overlaps with the teaching in the plaza that gives the book its title. The discipline of optics, the book suggests, is in a prime position to produce new advances thanks to the sophistication of modern lenses produced throughout Europe. In this sense, it is fascinating, to give just one case, to examine Suárez's brief note on refraction: "There is a great variety of mirrors, and it can thus be said that the profession of mirror-making has never been so advanced as it is today; this is so because the glass that is produced in Venice is essentially faultless, as is German steel, not to mention that produced in other places" (*se puede decir no haber estado esta profesión jamás tan en su punto como ahora; porque cuanto a los de cristal son perfectísimos los que se labran en Venecia, como lo son también los de acero de Alemania, sin otros de muchas otras partes*) (703). The refracted muse arrives at this plaza from its birthplace in Venice, but it is later transformed in Barcelona and made to take root in the "autochthonous" textual culture of the Iberian Peninsula.

However, glass in the form of a mirror, as we see in many other examples from the period, is a double-edged sword: its purely esthetic properties (*industria, ingenio*) are praised, as is its metaphoric potential (*artificio*) and its capacity for marvel; however, Suárez warns his readers that the artifices of glass should not be the object of blind praise, given that the mirror reflects only vanity, undermining the honor of the individual. Everything is, as the end of the citation playing with the dual valence of the term indicates, a question of perspective, returning us to the fundamental thesis that pulses through all of these pieces, that is, that the marvel of science depends on the use that one makes of it: "The use of mirrors is for human adornment, although its artifices deserve little praise, given that their works are as fragile as glass; and its honor and glory is all merely apparent and sophistic, as are all things related to perspective" (*hácense los espejos de muchas maneras, según que también son ellos diferentes, interviniendo en todos, industria, ingenio, y artificio. El uso de ellos es a propósito para el adorno humano, aunque sus artífices no tienen de qué loarse mucho, por ser sus obras tan frágiles como de vidrio; y su honor y gloria toda aparente y sofística, como son las cosas de perspectiva*) (703).

In the end, this universal plaza of sciences and arts offers an extraordinary array of opinions about what was then beginning to take shape in European courts, particularly in relation to new powers of seeing. Practice mixes harmoniously with theory in these discourses, which reveal to us a highly developed encyclopedic zeal on the part of its two authors, who were each looking for a kind of platform from which to educate their reading public. This said, Garzoni and Suárez do little to help us identify concrete archetypes, names that might allow us to gain a deeper sense of the technical-scientific Parnassus of the period. This is so because what they offer are essentially catalogs of names that are but accumulations in which scarcely anything is offered beyond one's occupation, erudite lists with no deeper reading to justify them.

There are, in any case, exceptions that I do not wish to leave out of the present examination. Such is the case, for example, with Discourse 20, which deals with secrets and turns out to be surprisingly specific and revealing with respect to the scientific practice of the period. Citing Cardano once again, Suárez reveals to us that a secret is "an obscure and veiled thing, the reason of which ceases to be known by all, keeping within it some seedbeds of invention by which speculators might find whatever they desire . . for ornamentation, for material gain, to show what one knows, or to trick; a thing abhorred by virtuous men" (*es una cosa escura y velada, cuya razón deja de ser notoria a todos, reteniendo en sí algunos seminarios de invención, con que facilitan los especulativos el camino de hallar cuanto desean* [. . .] *para ornamento, para ganancia, para mostrar que sabe, o para engañar; cosa aborrecida de virtuosos*) (198–99). A secret, in a masterful intuition, is thus nothing but a tool employed to obtain that which is sought after, a weapon of power that on occasion can mislead or manipulate one searching for precise information. If that which is hidden many times possesses an esoteric character, the keeper of secrets cannot emerge unscathed. Reflection on the virtuoso is centered, generally speaking, in attacking magic and in casting out the occult, and for this reason the portrait that Suárez paints is less than favorable; he reduces figures such as Pliny, Albertus Magnus, Roger Bacon, Cardano, and others to mere "fathers of fraud" (*padres de embelecos*). He goes on conclude that from secrets and their professors come "more smoke than substance" (*más humo que sustancia*) (200).

Más humo que substancia—this phrase is enormously seductive. We find it transformed into literary satire just a few years later, for example, in the beginning of *El diablo Cojuelo*, in that equally funny and merciless *cuadro* on the "belching" (*regoldano*) astrologer from whose artisanal laboratory the devil is freed. We see it also in portraits (not always positive) of some of the virtuosi of the Hispanic Baroque, the keepers of those secrets that pass through the filters—some more smoke, others more substance. The discourse brings together figures such as Leonardo Fioravanti and Giambattista della Porta, who up to that point had been referred to in passing but who are absolutely central to any study of the development of academies and the new experimentalists—a very grateful Galileo will become a part of this last group upon his acceptance, as the sixth member, into the Academy of Lincei. Within Spain, and in a way that is more symbolic than real, one can add a figure such as Juan de Espina, a Spanish virtuoso who brings together, only a few years after the publication of *Plaza universal*, some of the most interesting literary encomia of the time. From the plaza of the arts, we now move to the domestic sphere of science.

✷ 4 ✷
Inscriptions

Visible intermittence: The voyage of the secret and the creation of the virtuoso

For centuries, the popular imagination has reveled in presenting archetypal figures endowed with esoteric knowledge. In Spain, these figures displayed many of the same traits, such as eccentric behavior, permanent seclusion, and the ability to create life. From the medieval tradition of historical characters such as the necromancer Enrique de Villena and fictional creations such as Celestina, to the fascination with the occult in the first *comedias de magia* in the late seventeenth century, early modern Spanish literature captured a vast array of practices and rituals such as witchcraft, visions, alchemy, demonology, and possession (Lima 15–27). Given this tradition, it likely comes as no surprise that scientific and technical innovation could be received with a high degree of suspicion. Consequently, the scattered references to Galileo that we find in early modern Spanish literature are typically the result of a very cautious approach by seventeenth-century writers, who for the most part employed the word *galileo* to refer to Jesus of Galilee—as in the demonym *el galileo,* for example. This caution also mirrored the troublesome existence of a number of historical figures who, though extremely popular at the time, were not always well received by their neighbors. As a result, the literary portrayal of the scientist in Baroque Spain was often anything but benevolent, no doubt because of the influence of a number of local characters associated with practices such as alchemy and magic.[1] If there is anything we can learn from the realm of fiction in these important decades, it is that this literary exploration into new territories ultimately reveals very delicate and unstable boundaries between the acceptable and the prohibited, thus weakening the categories of what we understand as fictional writing vis-à-vis scientific rhetoric.

The figure of the scientist appears everywhere in early modern Spanish fiction, from narrative (picaresque novel, *vejamen,* costumbrista novels, allegorical dreams, dialogues, and colloquies) to poetry (*poesía de circun-*

stancias in sonnet form, burlesque epic, mock hagiography), and even to theater (interludes, urban comedies, dances, and *jácaras*). Represented primarily in the form of alchemists, mathematicians, and astrologers—and sometimes even as fake ones, as in Calderón's famous *El astrólogo fingido* (The fake astrologer, ca. 1625)—they are part of a stock of characters that appeared with extraordinary frequency in the satire of the time to the point of becoming structural elements rather than simple thematic features. They were not only the objects of critical scrutiny but also the eyes through which one could witness the dynamics of the early modern city in its learning centers and domestic retreats. Along with doctors, dentists, and cobblers, they were frequently portrayed as owners of a secret craft that only they knew how to nurture and perfect. It is a craft, I would argue, generally defined by an aura of secrecy that simultaneously placed them inside and outside of the social machine, both above it and below it, and also, most significant, ahead of it, in an epistemological future riddled with fear and uncertainty. This uncertainty was captured time and again by the writers of the time, for whom journeying into uncharted territory did not belong only to the institutional work of scholars and professors but also to that of the so-called virtuosi as well, independent—and sometimes isolated—figures who were feared and revered, admired and despised, but certainly never neglected.[2] Although I would not like to stray off course here, I do think, however, that it is important to briefly examine the adoption of the model of the so-called virtuoso in Spanish intellectual circles, given that his presence in urban centers helps us understand how certain ideas coming from abroad were perceived locally, ultimately affecting the reception and adoption of tools like the telescope. The model of virtuoso that interests me here is the one popularized in works like Alessio Piemontese's *Secreti* (1555) and Giambattista della Porta's *Magia naturalis* (1558), sometimes filtered through the work of Tomaso Garzoni, and in particular through *Piazza universale di tutte le professioni del mondo* (1585). These are paradigms that were influential in their Spanish counterparts, particularly in the adoption of a public image in which certain traits, like secrecy itself, were of paramount importance.

As a fictional character, the rise of the virtuoso is symptomatic of a very particular time in Spanish history, defined by the rapid growth of the commercial metropolis, the "colonization of the night" (Koslofsky 158) with the arrival of artificial light, and the formation of a highly sophisticated court society in which the mixture of the local and the foreign facilitated the development of new fields of inquiry and the rise of literary characters such as the experimentalist and the collector.[3] In his influential study on secrecy and natural magic, William Eamon described the virtuoso as a "cultural ideal . . . born out of the crisis of the aristocracy . . . drawn mainly from the ranks of unemployed gentlemen with too much time on their hands," whose curiosity "was particularly drawn to the rare, unusual, and 'extravagant' phenomena that might entertain and delight as well as instruct" (*Secrets* 302–3). He also adds:

> The "secrets" of nature—in the broad, seventeenth-century sense of that term—fascinated and delighted him to the point that the virtuoso became the butt of endless jokes about the dilettante who dabbles in trifles but understands nothing of himself. Trifles or not, the virtuoso's secrets were the stuff of early modern experimental science. (*Secrets* 301–2)

The recurrence of these types in the fiction of Habsburg Spain invites us to consider who—or what—can ultimately claim the alleged sanity these visionaries seem to lack. How, in other words, are these opposing traits, like experimentation and self-ignorance, construed and performed within the forces of tradition and innovation, brought about by the thrust of scientific inquiry in early modern Europe? Undoubtedly, this image of the virtuoso works advantageously when refashioned as a literary artifact: like the doctors, dentists, and shoemakers of early modern satire, virtuosi succeed in fictional forms by exploiting the hazards of improvisation when practicing their craft. Unlike these same doctors, dentists, and shoemakers, however, their path to knowledge is frequently paved with an aura of mystery that sets them apart, as every step forward raises new challenges to authority, be it legal or religious. In their appealing eccentricity, they manage to test the limits of fiction and the reach of censorship—or vice versa, the reach of fiction and the limits of censorship, ultimately challenging both. Most important, however, they succeed in exposing the challenges of a language that must face the new and the bizarre by testing and expanding its own vocabulary, when not importing a foreign one. Think, for example, of the Americanism *tabaco*, which soon became, in the pen of a wordsmith like Quevedo, a playful tool that equates smoke with heresy in the dialectic *tabacano* and *luterano*—that is, "smoker" and "Lutheran."[4] This is, I would argue, what some of these inventors do best for literary expression: they expose, transform, and re-create. With the arrival of objects and products from distant lands, some of these virtuosi become the ultimate masters of trinkets and substances that produce new trinkets and new substances, of matters that need to be named, inventoried, and, eventually, understood; matters, in short, that must be assimilated into the marketplace of wonder in intellectual and academic centers like Seville, Valencia, Barcelona, and Madrid. Scientific practice thus performs an additional act of inquiry, as it transports linguistic and thematic exploration into the realms of the unknown. Such is, for instance, the case of substances like the foam of boiling chocolate or the nasal drippings of snuff, which first became unworded images of Amerindian contamination brought to Spain by merchants and black slaves, and, soon after, contentious objects of scholarship in the work of doctors and university professors like Bartolomé Marradón (Seville) and Cristóbal Hayo (Salamanca).[5] Beyond academies and universities, these cities also witnessed the rise of independent research by savants who surrounded themselves with unique artifacts in spaces that were frequently closed to the public. The practices surrounding these artifacts, in many cases, rivaled the work of official learning centers: Stuart Clark reminds us, "Natural magic, in particular, conceived as the study of nature's most secret processes and powers, invariably occupied a place in the many surveys and systematizations that occupied early modern academics and structured both their courses and their textbooks—and not just in faculties of philosophy" (80). These practices helped in the development of the *magia artificialis*, built on the controlled experiment, on exact observation, and on the use of devices like the camera obscura, the mirror, and the lens that tricked the eye. The literature on the early modern virtuoso, I would argue, is a literature of things, of strange objects that become familiar and familiar objects that can become weird and out of hand, like the famous beakers that release the devil, prompting a fantastic literary voyage in the aforementioned satire *El diablo Cojuelo*, by Luis Vélez de Guevara. As a result, if the object defines its owner to the point of supplanting him, as we will soon see, the collection does so exponentially: the scientist who is immortalized in these fictional accounts is celebrated for what he has more than for what he knows, as talent is measured by the display and the concealment of knowledge rather than by its acquisition—such is, after all, as José Antonio Maravall persuasively argued in his influential *Culture of the Baroque*, one of the tenets of Baroque ideology. Consequently, these instruments of knowledge are usually portrayed in isolation, resting in all their iconic power in dusty rooms and hallways as if they were always idle, untouched, and made to be worshipped. Their size and shape, along with their attention to detail, were also staples of Baroque ingenuity, triggering surprise and admiration on the part of the viewer, as Stuart Clark reminds us: "Natural objects were often described as curious by virtue of their smallness, exquisiteness of workmanship being exhibited more strikingly in miniature" (315).[6]

This view of the unseen inspires a literature of the collector and the museum, of the shelf and the cabinet, of the storage and the display, of the original and the counterfeit. Such is the case, for example, of the eccentric Lázaro in the work *El cortesano descortés* (The rude courtesan, 1621) who is first mocked for his obsession with accumulating and displaying hats in his private chambers, and finally tricked by his neighbors with a counterfeit one. Similar cases include those of the morose hidalgo in *El malcontentadizo* (The malcontent), in the miscellany *Fiestas de la boda de la incasable mal casada* (Festivities of the wedding of the unmarriageable poorly married, 1622), and of the angry hermit Doctor Ceñudo in *El necio bien afortunado* (The fortunate fool, 1621), whose tenebrous house is also his own private shrine, sealed from public view. These pieces allow the novelist Alonso Jerónimo de Salas Barbadillo to delve into a number of understudied perspectives such as the performance of male domesticity, the coexistence of chance and sociability, the dynamics of collecting, and the showcasing of knowledge, generally portrayed from a comical perspective.[7] What happens, however, when fictional forms depict historical characters? And what happens when these forms face the limits of secrecy and the allure of revelation?

Very few figures in Spain embodied the complexities and nuances of these experimentalists as thoroughly as the musicologist and collector Juan de Espina. Despite the (mostly descriptive) attempts of his most illustrious biographers, Espina remains largely unknown to this day, as do the existing literary testimonies about him.[8] In approaching him and his work, this chapter takes a two-pronged approach: first, it argues that his figure offers a unique opportunity to explore important facets of the culture of collecting and—because of the representational nature of such practice—of the politics of vision in early modern Spanish literature. Second, it suggests that his literary persona provides, like no other historical figure of his time has done, new perspectives on the complex relationship between fictional forms and scientific practice in the turbulent decades of the 1620s and 1630s.

Most of what we know today about Juan de Espina comes from what his neighbors wrote about him. Born to a noble family around 1563, he died in Madrid in 1642. A contemporary of Luis de Góngora, Lope de Vega, Tirso de Molina, and Francisco de Quevedo, he was a famous collector and musicologist, as well as a virtuoso on the lyre and the sixteenth-century vihuela, a guitarlike instrument with six double strings. In his *Grandes anales de quince días* (Great annals of a fortnight, ca. 1623), Francisco de Quevedo, notorious for ridiculing the pastimes and aptitudes of others, offered one of the most detailed portraits in existence. He began by praising Espina's parents—his father, Diego, worked for Philip II as an accountant—and spoke of his "exquisite, and shy persona" (*condición recatada siempre al trato vulgar, pero no desapacible*) (219). Juan played music, reaching unrivaled heights (*tocando prodigios*) (219) and excelling at painting as well. In Espina's massive collection, Quevedo explains, he graciously followed a system of order and harmony (*introdujo por la mayor gala la órden y armonía*) (219). As Quevedo puts it, Espina was always kind enough to ask his guests what they did for a living and what they liked, and according to their tastes, he would show them one thing or another. "I never heard a complaint, impoliteness, or a vice," writes Quevedo, adding that his friend "avoided vanity and pomp, and tended to seek solitude" (*Yo no oí jamás de don Juan queja ni demanda, ni inadvertencia, ni descortesía, ni vicio; ni le he conocido enemigo.* [...] *Aborreció con singularidad y virtud robusta la pompa* [...] *anduvo solo entre la gente*) (220). Along with Jerónimo de Ayanz—a soldier, painter, and musician called "the Spanish Leonardo" for his invention of a steam-powered water pump for draining mines, which earned him a patent by the Spanish monarchy in 1606—and Juanelo Turriano, famous for his *artificio*, an engine that lifted water from the river Tagus to a height of more than three hundred feet to supply the city and its alcazar, Espina was able to captivate the popular imagination and penetrate the realm of the fictional.[9] Unlike Ayanz and Turriano, he managed to amass an enviable collection of artifacts that made his home in Madrid the site of literary praise by all those who longed to see what was behind its doors. In a culture built on visual display and ostentation, Espina's power stemmed from all that was hidden from public view. Deeply immersed in this "culture of virtuosity," he was admired, envied, and even idolized for his collection of works of art, musical instruments, rare books, and taxidermied animals of all kinds, which he purchased with the stipend of five thousand ducats he received as a clergyman, and as Philip IV's *sumiller de cortina* (vice-groom of the king's bedchamber) (Caro Baroja 433). He was famous for having two autographs by Leonardo da Vinci—the two stored today in Madrid's Biblioteca Nacional under the names *Codex Madrid I* and *II*—which he had acquired through the sculptor Pompeo Leoni.[10] As testament to this "Baroque obsession with crisis" (*obsesión barroca por la crisis*) (Bouza, "Coleccionistas" 246), he also owned the knives that had been used to execute prominent figures like the corrupt statesman Rodrigo Calderón, whose blindfold and locket found with him at the time of his execution ended up in Espina's hands too, until they were posthumously given to the Marquis of Villanueva del Río.[11] This was not, by any means, an unusual practice, if we consider other well-known cases:

the Count of Villahermosa's coin collection, the Duke of Villamediana's jewelry collection, the Marquis of Montealegre's book collection, and the Marquis of Leganés's weaponry collection, just to name a few.[12] Espina's paradigm, as we will see in Vicencio Carducho's testimony and in Espina's own writings, is significant for the variety of objects he possessed and for the intricate ways he was able to find a specific narrative for each of them, building a collection that was highly attentive to its different elements.[13] It is believed, for instance, that he was a master at public relations and that he knew how to exploit the allure of the rare and unique by personally issuing invitations to study in his home.[14] By the time Philip IV came to power (1621) Espina had already become too famous for his own good, as proved by testimonies from the 1620s that referred to his many activities, such as the anonymous ballad *Relación de la fiesta que hizo D. Juan de Espina, Domingo en la noche, último día de febrero. Año 1627* (Chronicle of the party that Don Juan de Espina threw in his house, on a Sunday night, the last day of February 1627), which chronicled the abundance of magic tricks performed in his house from seven at night until three in the morning: "What happened in just one night / Lauso, you want me to sing, / but it was so much, that I doubt / I may be able to capture it" (*lo que cupo en una noche, / quieres (Lauso) que refiera, / pero fue tanto, que dudo / que haya atención donde quepa*) (112r). The poem's dedicatee, Lauso—pen name of Luis Vélez de Guevara—captured the astonishment produced by a mock bullfight, a puppet show with giants, and a three-hundred-course banquet in which food levitated and left the house through the window: "fruit, china, pastries, ceramics / all flew through the window" (*frutas, vidrios, dulces, barros / volaron por las fenestras*) (114r).[15] It is believed, as Julio Caro Baroja has suggested (429), that Espina was later prosecuted by the Inquisition and banned from Madrid for allegedly practicing white magic (*magia blanca*). Such unfortunate events, however, did not quench his thirst for knowledge and accumulation, as indicated in his *Memorial que Don Juan de Espina envió a Felipe IV* (Memorial that Don Juan de Espina sent to Philip IV, 1632)[16] in which he presented himself as a virtuous citizen whose name had been tainted out of envy, and as a wealthy man who, in his own words, "wanted nothing for he had more than he needed" (*a tanto ha llegado mi riqueza que me sobra todo*) (198).[17] And this was the case, indeed: when he died on the night of December 30, 1642 in his house in Calle de San José—close to Atocha, and near where the great Lope de Vega lived—a document was found in his pocket that reiterated what he had spelled out in his will: after his death his belongings were to be donated to the king, and all the wooden automata that populated the hallways of his house were to be destroyed.[18] As this final act of self-promotion suggests, he was until the very end a man concerned with the performative and material aspects of knowledge, with hidden information that led to ultimate revelations, and with the irresistible power of the secret and the double. This is a tension that was captured, as seen in accounts like those of Alonso de Castillo Solórzano and Anastasio Pantaleón de Ribera, in a language that sometimes mirrored, in its opacity and hollowness, the very same objects it aimed to represent.

Espina's house, a villa (*casa de recreo*) named Angélica worth thirty thousand ducats at the time of his death, possessed a distinct theatrical quality, as it included fountains and *burlas acuáticas* (hydraulic machines that created music and storms), revolving walls, and wooden automata that were thought to double as his butlers.[19] In his *Memorial* written to the king, he stated:

> When it comes to rare, curious, and beautiful artwork made by the most famous masters from these and other kingdoms and nations, my house in this court can compete with all the extraordinary [objects] worldwide, and even leave them behind, as the experts of all major disciplines have already certified in writing.
>
> *En materia de las cosas insignes, curiosas y primorosas del arte hechas de los más afamados maestros que ha habido en todos los reinos y naciones, tengo mi casa en esa corte que puede competir con todo lo excelente del mundo y dejarlo atrás, como lo tengo firmado, de todos los maestros de mayor nombre en todas las artes.* (200)

Espina's aim, therefore, was to rival the exuberance of his surroundings by collecting and displaying objects of all kinds, with the rooms of his villa the guiding thread for exploration and amazement. He not only stored items in his house but also transformed his house to better fit the uses of the items, much like his contemporary Francis Bacon had done with his fictional Salomon's House, defined by William Eamon as a "utopian research institute, an artificial world fashioned and crafted in imitation of the natural world" (*Science* 315). Through these inventions, Espina successfully carved out the image of an eccentric neighbor, a misanthrope who allegedly received his basic staples like food and drink from the exterior through a spinning wheel (*torno*) to keep his privacy intact—Charles I of England, who coveted Espina's Leonardo codices, called him a "foolish gentleman" (Bouza, "Coleccionistas" 250).[20] He lived in the center of the city, yet remained sealed off from it, nurturing a love story with a villa whose female name evoked the epic love affairs of the Italian Renaissance. This feminization of the property was not uncommon in the Baroque: "As nature was feminine, natural philosophy was a 'Male Virtu' whose 'curious sight' followed nature 'into the privatest recess of her imperceptible Littleness'" (Eamon, *Science* 315). Espina thus shaped a vision of male domesticity that had already been cultivated by some of his contemporaries, such as the aforementioned Salas Barbadillo, who—perhaps inspired by him—depicted masculine interiors in great detail when creating small fraternities with the goal of entertaining other men, as was the case of his novel *Casa del placer honesto* (The house of honest pleasures, 1620), in which these *placeres honestos* (theater, music, and dance) already resembled those chronicled in Don Juan's 1627 lavish party. Whether conceived as an exclusive place or a welcoming one, Espina's was a house that elicited enormous curiosity, as proved by the unusually high number of writers who, despite their name and fame in the literary field, gave in to the humility of public praise.

The Juan de Espina who has survived in Spain's cultural memory is that of the necromancer, a negative portrait heavily mediated by the impact of the two eighteenth-century *comedias de magia* by José de Cañizares, *Don Juan de Espina en su patria* and *Don Juan de Espina en Milán* (Don Juan de Espina in his homeland, and Don Juan de Espina in Milan). I do not delve into these two plays here, for they have already been thoroughly analyzed by Susan Paun de García in an excellent critical edition of Cañizares's works. I focus, instead, on a number of pivotal testimonies from the years 1625 to 1641 that trace what I consider the literary making of Juan de Espina during his lifetime, and I offer critical information on the issues I have outlined here. This textual genealogy is shaped by a fascinating shift from collector to necromancer. His *Memorial* to the king thus plays a pivotal role for the modern reader, as it carefully crafts the image of a studious musicologist and mathematician widely respected by his peers, a paragon of decency who is "executioner" (*verdugo*) and "the finger that points to all" (*dedo malo de todos*), and of a humble vassal and exemplary neighbor who has been the victim of "the war of gossip" (*la guerra de la murmuración*) (199–200): "the utility of sciences can't even be shared with those who practice them if they are not rigorous enough" (*lo útil de las ciencias no se puede hablar aún con los que tratan de ellas si no son muy científicos*), Espina wrote to Philip IV (193). In the end, Espina's *Memorial* reveals a monarch that was not only fluent in the nuances and technicalities of new scientific trends but also influential in the development and acquisition of private collections.[21]

❋

One of the most important testimonies in existence is that of Alonso de Castillo Solórzano, who wrote the poem "A Don Juan de Espina" (ca. 1623), later included in his collection *Donaires del Parnaso* (Graces of Parnassus, 1625).[22] As Fernando Rodríguez Mansilla has argued, Castillo Solórzano's piece is an exercise in self-fashioning in which he presents himself as a well-connected poet of academies who merits a visit to his house. There is no shortage of praise for himself or for Espina, whom he calls a "Spanish Phoenix" (*Fénix español*) in the opening lines of the composition. Espina is a "subject who represents / the most praiseworthy and unique [traits] / of skill and grace" (*sujeto que comprehende / lo más célebre y más primo / de ciencia y agilidad*). The uniqueness of being *primo*, "without equal," is highlighted two lines later with the word *peregrino*. Espina is also lauded for his "eminent and erudite wit" (*ingenio / eminente y erudito*) (15–16), and for his rarities, or *caprichos* (18). Castillo Solórzano then moves on to his *mansión* (21), where everything can be found:

> Because neither sciences,
> nor the most mundane professions
> have manuals and instruments
> that can't be found in your house;
> for your generosity
> has allowed it [your house] to be
> a source of all knowledge,
> and the place for all to practice.
>
> *Porque no tienen las ciencias*
> *ni aun los comunes oficios*
> *de su práctica instrumentos,*
> *de su teoría libros,*
> *que en tu estancia no se ostenten,*
> *que tu providencia ha sido*
> *maná de todos ingenios,*
> *taller de todo ejercicio.* (23–30)

The terms *manna* (*maná*) and *office* or *workshop* (*taller*) not only stress the abundant resources of Espina's house but also highlight its many practical purposes, with all the disciplines that can be learned. Castillo Solórzano's poem, for instance, makes reference to his collection of glassworks: "Is there any extraordinary invention in glass forged by the Venetians that cannot be found [in your collection]?" (*¿Qué invención extraordinaria / forjó el veneciano rico / uniendo los elementos / que no la tengas en vidrios?*) (45–48). It also mentions Espina's knives—like the aforementioned ones used to execute Rodrigo Calderón—when he speaks of "sharp blades" (*agudos filos*) and "steel" (*acero*). Espina's musical talent is praised through his instruments (53); his collection of local and imported animals is sung of soon after when speaking of "feather, hair, or scale / in cave, water, or nest" (*pluma, pelo o escama, / ya en cueva, en agua o en nido*) (65–66), as well as his abilities as an art collector: "you display original works of art" (*muestras el pincel valiente / no imitado, sino vivo*) (67–68). Castillo Solórzano also confirms the belief that Espina allowed entry only to those who were educated enough (*sin habilidad / a nadie se abren tus quicios, / que es de ellos tu rectitud / querubín del paraíso*) (77–80). He then moves to present himself as a famous poet (81–162), before closing the composition with a request, that he opens him the doors of his house (*os pido me concedáis / lo que tanto he pretendido. / Valga aquesta habilidad, / valgan los deseos míos / para que de vuestra casa / queráis abrirme un postigo*) (163–68). Although it certainly is one of the most precise accounts of Espina's tastes and interests in terms of collecting unique objects, the poem still relies heavily on the clichés that were part of his public image in the early 1620s. By placing the narrative voice outside the house, this piece ultimately reveals more about Castillo's curiosity and sense of humor than about the specificities of the collection itself.

A few years later, Anastasio Pantaleón de Ribera celebrated the wonders of Espina's house of tricks in his poem "A la curiosa y celebrada casa de don Juan de Espina" (To the unique and celebrated house of Don Juan de Espina) (ca. 1627), in which he shared some of the images and rhetoric previously employed by Castillo Solórzano.[23] Pantaleón de Ribera's piece is a celebration of ingenuity, a "well-wrought urn" that marvels at other fellow urns because of their uniqueness: espanto, éxtasis, peregrino, culto, rico, curioso, argumentoso . . . The poet saturates his composition with variations of the same idea that are captured in the architecture of the mind, the body, and the home. The term *culto*, in particular, is significant in that it points to both rare and learned, as it describes a house that will someday elevate his stature from man to icon (*fama después venerará futura*). The sonnet, probably built on hearsay, ends with a petition by its author to have access to the wonders of the prodigious, to enjoy, in the words of Fernando Bouza, the wonderment of things (*la experiencia de las cosas*) in this universal encyclopedia (*enciclopedia universal*) ("Coleccionistas" 249) of which nothing else is revealed.[24] Profoundly Baroque in form and content, it certainly resembles Castillo Solórzano's view from the distance, but ultimately tells us more about the ends than about the means of how this wonder is achieved. Ribera's poem is, as a result, as bombastic as it is vague.

This representational—and physical—distance would be overcome in two pivotal accounts. The first one is by the art theorist Vicente (Vicencio) Carducho. Impressed by Espina's art collection after an eight-hour visit on April 10, 1628, Carducho wrote about feeling overwhelmed by the display of rarities in front of him and about the need to spend more time to enjoy such collection fully (38–39). His chronicle is truly unique for two reasons. First, because it was included by Espina himself in his *Memorial* as a way to legitimize—and put a price tag on—his possessions to Philip IV, in a critical move in which he revealed himself not only as a collector of artifacts but also as a gatherer of testimonies, which helped him build a specific image of his house and his persona. But it is truly exceptional because his guest Carducho gives us a sense of the quantity and, most important, of the quality of Espina's collection: it is expensive (*costosísimo*), unique (*singular*), and all authentic (*modelos originales*). It is, above all, a feast for the eyes that leaves the viewer speechless, as he is himself unable to find the proper words to depict such wonder: he not only feels that his report "didn't do justice" (*he andado corto*) to what he saw but also regrets that he was left at a loss for words (*no sé con qué palabras encarecer ni ponderar*) when writing it. Espina's gesture is, once again, typically Baroque, as he hides behind someone else's voice when crafting his self-portrait. In fact, what Carducho actually wrote in his masterpiece *Diálogos de la pintura* (Dialogues on painting, 1633) was slightly different, focusing on the existence of the famous manuscripts by Leonardo da Vinci:

> He has the rarest objects, worthy of being examined by a learned and curious visitor (in addition to all the artwork), because he always considered himself exceptional and unique, which drove him to find and acquire, no matter the price tag, the most beautiful and strange things. That's where I saw two sketches, handwritten by the great Leonardo da Vinci, of unusual beauty and contents.
>
> *Tiene cosas singularísimas, y dignas de ser vistas de cualquiera persona docta, y curiosa (demás de las pinturas), porque siempre se preció de lo más excelente y singular, que ha podido hallar, sin reparar en la costa que se le podía seguir, preciándose de coger lo muy acendrado, y extraordinario. Allí vi dos*

libros dibujados, y manuscritos de mano del gran Leonardo da Vinci, de particular curiosidad y doctrina. (438)

The second testimony is by Juan de Piña, a close friend of Espina's, who in his novel *Casos prodigiosos y cueva encantada* (Prodigious cases and the enchanted cave, 1628) included a detailed assessment of his house. The narration begins with Piña expressing his gratitude for this unexpected honor, which is bestowed upon him at night to further enhance the theatrical nature of what is to come (*hacía estos favores de noche, y dejando a los que como yo lo entraban a ver y admirar, que así, de más de haberlo visto, lo entendí del padre de mi fortuna*) (258). Once inside, this "prodigious case" unfolds with relentless precision: a room devoted to glassworks from Venice and ceramics from China of unparalleled quality: "a hall with such beautiful china and ceramics, that [it looked like] he had plundered Venice from the most admirable and precious [items] of its workshops, and China from its tableware and other wonders, and Spain from its most renowned goods" (*una sala de vidrios y barros con tal compostura, adorno y riqueza, que había saqueado a Venecia de lo más admirable y dorado de sus fábricas, y a la China de sus vajillas y maravillas y lo que en España tiene mayor nombre*) (259); numerous mechanical clocks and distorting mirrors; a scale—invented by Jerónimo de Ayanz[25]—so precise it could weigh the leg of a fly; two-hundred candles in silver chandeliers that illuminated its rooms; paintings and tapestries by world-class artists like Rubens; and a ship with shooting cannon that sailed in a sea of mercury, achieving the effects of tides and waves in what was technically known in theater as *burletes* and *rivetas* by the choreographers of the time. Don Juan de Espina, we are told from the start, is a wealthy aristocrat who decided to invest his fortune in his cabinet of wonders (*que siendo mucha la renta del caballero, toda la gastó, no al desperdicio, sino en cosas de curiosidad*). His was a cabinet of wonders that, through the use of images like the "plunder" from the Venetians and the preserved birds from the Indies, reflected a very personal quest not only to engage in dialogue with the past but also to build an idea of national patrimony that fit very well with the practices of the seventeenth-century virtuosi. In their analysis of Piña's ties to Espina, Both Cotarelo y Mori (35–37) and Caro Baroja (181–82) convincingly argued that this account was the result of a true event, and certainly the details of its depiction shouldn't lead us to think otherwise. It is, in sum, a long and detailed chronicle in which Piña builds an image of Espina's house that will be adopted by future writers as a sort of master narrative, a cornucopian text from which to borrow freely.

These two accounts are of enormous importance in the creation of Espina's public standing. They unveil his prized collection to contemporary readers, thus offering a wide array of objects and practices from which to build a specific image of the virtuoso. All the aesthetic traits that have come to define the period of the Baroque, of which scholars like William Egginton (*Theater* 4) and Gregg Lambert have reminded us recently, are here: anamorphosis, *mise-en-abîme,* and trompe l'oeil, the juxtaposition of disparate terms (*coincidentia oppositorum*), the proliferation of décor, and a conscious embrace of artifice. In addition, Espina's aesthetics is that of the still life, the *naturaleza muerta,* in which species of all kinds were displayed in a deliberate narrative. Perhaps most important, these characteristics also facilitate the transition from public citizen to literary device, as proved by a number of mentions to his persona in the 1630s and 1640s. Still during Espina's life, a handful of new texts came to portray him archetypally, thus showing that by this period he was such a well-known figure that he hardly needed any elaboration as a fictional character.

This shift in characterization appears in the work of four well-known *ingenios* of the time: Tirso de Molina may have borrowed the idea of the spinning wheel, or *torno,* in his famous cloak-and-dagger play *Por el sótano y el torno* (Through the basement and the turnstile, ca. 1630) from the one Espina had at home to have his food delivered, but what we know for certain is that he admired certain qualities of his house, as he wrote in his piece *En Madrid y en una casa* (In Madrid and in a house, 1635):

MAJUELO: This house is haunted!
ORTIZ: It is not possible, for Don Juan de Espina has not been its guest.
MAJUELO: That is absolutely true.

MAJUELO: *¡Válgate al diablo la casa!*
ORTIZ: *No es posible, que no ha sido Don Juan de Espina su huésped.*
MAJUELO: *Verdad dueñísima has dicho.* (549)

Another of Espina's contemporaries, Antonio Coello, wrote *Vejamen que se dio en el certamen del Buen Retiro* (Pun written for a contest held in Buen Retiro, 1638), in which he dreamed he visited the house of the "illustrious and never sufficiently praised Don Juan de Espina" (*insigne y nunca bastantemente alabado D. Juan Espina*) (315–21). He used Espina as interlocutor in a dialogue in which he mocked some of the most notable aristocrats of the time, celebrating his friend's command of the lyre, a command so impressive that, in the words of Espina himself, "with it in my hand I make miracles, such as grow wheat in fifteen minutes" (*con ella en la mano hago yo Milagros y me atrevo a hacer crecer el trigo en un cuarto de hora*) (317). It was, no doubt, a satirical portrait, but one that also hinted, as Tirso had also done, at the supernatural, God-like powers of this maligned figure. And a similar brushstroke, one could argue, was given by the court dramatist Diego Hurtado de Mendoza when he attacked his contemporary Jerónimo de Villaizán by comparing him to Espina and his knowledge of necromancy:

> Who could have such thing
> that, even if Don Juan de Espina
> had it in his office,
> neither could it smell Celestina,
> nor see Tamorlan?—Villaizán.
>
> *¿Quién la cosa peregrina*
> *que, a tenerla en su oficina*
> *el señor don Juan Espina,*
> *ni la oliera Celestina*
> *ni la viera Tamorlán?—Villaizán.* (83)

However, the most daring testimony of the time was undoubtedly penned by Luis Vélez de Guevara, who praised his famous astronomical observing chair in the 1641 Menippean satire *El diablo Cojuelo* (The limping devil) as a response to Espina's ballad. Vélez de Guevara's journey is a rich exploration that looks at Madrid's streets and buildings through the gaze of Cojuelo and his student companion Don Cleofás Leandro Pérez Zambullo, pushing the boundaries of Spanish satire by including opinions on contemporary astronomy from the point of view of the aficionado (García Santo-Tomás, "Fortunes"; Zuese). He reflects on a number of current affairs regarding some of Galileo's most significant discoveries: his work on the telescope in 1609–1610, his argument against the incorruptibility of the sun, and the publication of his groundbreaking *Sidereus nuncius* in Venice a few years before, where he had divulged his findings regarding matters of interest to Vélez de Guevara, like the surface of the moon. It is then that the limping devil praises Espina but considers the current fashion of stargazing to be worthless:

> Don Cleofás, our fall [with Lucifer from heaven] was so swift that we had no chance to see anything; and I swear that if Lucifer hadn't brought with him a good third of the stars, as we frequently hear in the allegorical plays at Corpus Christi, astrology would have even greater means than it now

has to play tricks on you. I say this with all due respect for Galileo's telescope and the one used by Don Juan de Espina, whose famous house and extraordinary chair are the results of his own imagination; for I speak of this with great caution.

Don Cleofás, nuestra caída fue tan apriesa que no nos dejó reparar en nada; y a fe que si Lucifer no se hubiera traído tras de sí la tercera parte de las estrellas, como repiten tantas veces en los autos del Corpus, aún hubiera más en que hacernos más garatusas la Astrología. Esto, todo sea con perdón del antojo del Galileo y el del gran don Juan de Espina, cuya célebre casa y peregrina silla son ideas de su raro ingenio; que yo hablo de antojos abajo, como de tejas. (75–76)

This passage is remarkable not only because of its boldness in including the name of Galileo—at a time when he was being prosecuted for his refusal to give in to the pressures of the Vatican[26]—but also because Espina's name is also included in this sequence with the adjective *gran*, as well as through the mention of an alleged telescope and a precious chair. Vélez is here referring to a gyrating chair that allowed Espina to stargaze in comfort from multiple angles, a chair that was equipped, according to the Spanish historian Nicolás García Tapia, with all the instruments needed for it (*todos los instrumentos necesarios para ello*) ("Los códices de Leonardo" 387). As one of the most cherished items of his collection, and one that triggered great interest on the part of his guests, Espina's famous chair was donated to the king after his death—and right after Vélez's novel was published. These last references, as brief as they are, complete the portrait of a collector who managed to become a quasi-mythical figure in his lifetime. They are the last touches of a slowly evolving canvas that took shape over fifteen years, offering a view that became clearer and clearer until, one could argue, it had nowhere else to go. If the making of Espina benefitted greatly from fictional portraits, it also suffered its inevitable undoing when falling into the trappings of overkill: as the testimonies accumulated, Espina's persona became a simplified caricature of his past self. The texts studied here frequently blur a distinction that, as some critics have recently observed,[27] is more imagined than real. This is, for example, the case of Piña's detailed account of Espina's collection: although it is framed within a larger fictional narration, it also displays certain features—like a careful eye for order and taxonomies—that we see in some forms of scientific rhetoric. With so much clandestine experimentation and so much prosecution, the writer of the time is left to wonder: What if? What if all this were possible?

In his monumental study *The Emergence of a Scientific Culture: Science and the Shaping of Modernity, 1250–1685*, Stephen Gaukroger has argued:

> The "Scientific Revolution" of the early-modern West breaks with the boom / bust pattern of all other scientific cultures, and what emerges in the uninterrupted and cumulative growth that constitutes the general rule for scientific development in the West since that time. The traditional balance of interests is replaced by a dominance of scientific concerns, while science itself experiences a rate of growth that is pathological by the standards of earlier cultures but is ultimately legitimized by the cognitive standing that it takes on. (18)

It is precisely the "dominance of scientific concerns," more than the presence—or the awareness—of a revolution per se that I have addressed here through one of Spain's most fruitful case studies. If world-class universities like those of Valencia and Seville, and learning centers like Madrid's Academy of Mathematics, demonstrated that there was a voice for serious scientific endeavor, the existence of these private collectors with ties to the king himself, in the form of the *donativo* (gift), also unveiled very significant forms of intellectual exploration. In this age of curiosity and curiosities, Juan de Espina embodied the experimentalist, the heterodox (*heterodoxo*) (in the words of Menéndez y Pelayo), yet his house was located at the very center of the city, near the core of scientific activity. He was, by all means, a well-connected, central figure, situated at the crossroads of literature, art, music, kinematics, and architecture. As the "invisible neighbor," he was equipped with so much information regarding his many talents, his many possessions, and his many eccentricities that in the end, all this abundance paradoxically turned into a simplistic image of his persona and his surroundings. His objects rested away from the public eye, yet they were known beyond the physical and chronological boundaries of his time, as proved by the aforementioned plays by José de Cañizares. The fact that Cañizares devoted two of his most important *comedias* to Espina should not come as a surprise: by outfitting his own house as a place in which anything was possible, Espina successfully built his own stage, equipped with all the technical innovations brought to Philip IV's Madrid by famous engineers and designers like the Roman Giulio Cesare Fontana (1622) and the Florentines Cosimo Lotti, Pietro Gandolfi (1626), and Baccio del Bianco (1651). He purposely became a man of theater and magic, a multi-instrumentalist whose interests—much like those of other European figures like Galileo—put scientific endeavor in dialogue with the performing arts.[28] He formed part of a tradition of collectors that culminated in the seventeenth century with figures like the Aragonese Vincencio Juan de Lastanosa (1607–1682), famous for his cabinet of curiosities, his chemical laboratory, the garden of his palace, and the books, manuscripts, and maps of his library.[29] After all, the virtuosi believed, as Eamon reminds us, that openness to novelty and rarity "was the starting point of scientific inquiry" (*Science* 316).

In the end, the "making" of Don Juan de Espina, which runs hand in hand with the creation of the virtuoso in Spain, was a very nuanced process, going back a few decades in time through layers of fictional experimentation and conceptual wrestling. An analysis of his reputation in Madrid's court society can also open the door to important insights into the processes of reception, accumulation, and dissemination of those scientific instruments and ideas that were reshaping early modern Europe, at a time in which Spain was caught between strong currents of tradition and innovation.[30] This tension is proved by the extreme care with which some of the writers of the time wrote on Espina's achievements as much as the careless scorn his detractors displayed when he was seen as a necromancer. Espina's slow downfall helps us see the Baroque *Wunderkammer* in its dialogue with the literary Parnassus and the political arena; and if all his ventriloquizing objects and architectural wonders do tell us of a sustained personal endeavor, they also succeed in illuminating new details on the complex development of scientific inquiry in early modern Spain.

III

The science of satire

✷ 5 ✷

Situations

The refracted space of the city

The previous chapters have shown how the shadow of Galileo repeatedly projects itself onto Spanish costumbrista literature and theater without being alluded to explicitly. This projection occurs at times through the mediation of concrete objects such as the telescope, and at others it manifests through a kind of mystical discourse that was, by then, already revolving around his persona. The work of the reader in the seventeenth century, as well as today, largely consists of knowing how to read between the lines so as to discern Galileo's (mediated) presence in texts of a wide range of genres and subgenres. Throughout the previous chapters, we have also seen that Galileo's shadow tends to be projected onto an urban setting, and that the scenes in which these projections take place, especially in a genre such as satire, are relatively similar. The urban space, as I have pointed out in previous studies (*Espacio urbano* 131–85; *Modernidad* 57–90), exercises the double function of serving as a literary theme and a catalyst for fiction-as-critique; that is, it stands as a scenario in continuous transformation that is at once a reproducible motif and an incitement or source of inspiration for further discursive innovation. To remain faithful to the dominant imaginary of the present book, one might argue that the urban space functions as a kind of prism that transforms reality into fiction through its situated refraction of the everyday. Put another way, one might say that the city effectively transforms perceived experience into new experience. Importantly, this operation—a fully processual cartography—does not cancel out the ongoing working-through of Baroque experience in early modern Spanish fiction; rather, it refracts it into something new, with a different point of departure that generates a new angle and a development that is distinct without being completely separate.

The new "refracted writing" to which I have just referred proposes a reality composed of recognizable materials that are simultaneously made strange through a series of inherited discursive tools. The delimitation of allegorical or symbolic spaces within the city and the simultaneous redefinition of those spaces through the use of hyperbole, animalization, and the carnivalization of everyday reality redefine the urban context; this in turn generates a compact but also kaleidoscopic repertoire of situations. The development of genres and subgenres such as the picaresque, the novella, the earliest scientific and technical manuals, political schemes (*arbitrios*), the mirrors of princes (*espejos de príncipes*) that had so much success in the sixteenth century, miscellanies such as the aforementioned *Plaza universal* by Suárez de Figueroa, and even certain forms of comic theater, as well as the influence of classical authors recuperated in these pieces, provide the necessary materials—language, situations, characters, tone, and the like—for different types of satire that would become predominant during the middle of the seventeenth century. One of the main common threads that run through these satirical works is the persistent act of situating characters within new angles of observation, or what the film critic Christian Metz referred to many years ago as "scopic regimes." This placement of characters into a position of perspectival (visual) privilege had the effect of expanding, as the century wore on and the imperial decline was made more evident, the array of allegorical denouncements of a reality that was progressively more sordid and disenchanting. The use of different forms of glass serves to gain new perspectives, and the privileged vantage point—in all its diverse variance—constitutes in the texts studied here the optimal platform for engaging in pointed social commentary even while it permits, from its physical and moral height, the use of the long-distance lens to observe horizons that were previously unreachable.

If the situation in the Iberian Peninsula was not completely one of closure and condemnation, Madrid had made a significant contribution to the development of scientific knowledge thanks to the scholarly and pedagogical plans of its prestigious Academy of Mathematics and the Imperial College. Seville, for its part, had done much the same through its university chairs and mercantile institutions, as well as through its active peerage culture of celebration and patronage. These are two cities, as we have already seen, that provide a kind of chaotic allure: the misadventures, mishaps, and calamities suffered in them at the turn of the century seem almost to have been even more numerous than their inhabitants; this abundance is somewhat misleading, however, as one finds ample evidence of agency as well as oppression—that is, the environment is frequently made to turn on itself. In part 2, chapter 58, of *Don Quixote* for example, there is the brief, tormented narrative of Doña Rodríguez, who speaks of a Madrid that is at once unpredictable, violent, and almost comical; through the tragic family episodes that condemn her to solitude and abandonment, alone in the dukes' castle and separated from the world by her own eyeglasses, she is nonetheless able to provide Cervantes's readers with a correct and mordant diagnosis of the duchess within the perverse space of the palace. Cervantes, who had already lingered over a ruthless Seville through the marginalized characters of Rinconete and Cortadillo, Escalanta, and Gananciosa, nonetheless chose to rescue through the old *dueña* the narrow *calle* Santiago and the "useless people" (*gente baldía*) of the Puerta de Guadalajara from the leisurely gangs that would later be cultivated with cruel detail in the costumbrista and theatrical fiction of the seventeenth century. In the end, it is not surprising that the urban environment should be experienced in this literature through a voracious form of eyesight endowed with a seemingly unquenchable appetite for continuous pleasure, stupor, and surprise. For the foreigner, noise is seen as much as smoke; and then there is the texture of fabrics, often enough presented as a visual image, or the power of exotic and unknown dishes, which also seem to enter through the eyes. The city becomes an imposing theater and facade, a visual saturation that also cultivates an inner world of secrets, precarious answers, and an entire visual catalog of clandestine uses.[1] For Fernando R. de la Flor:

> The Baroque gaze, through which one echoes or deforms the reality of the world, represents the capacity to alter, diffract, or erase the natural; in this way it places a veil over the world, an enigmatic and artificial note placed on top of the earlier creative pretensions for a pure mimesis and a direct and sympathetic correspondence between the eye and its object.
>
> *La mirada barroca, sobre la que o bien rebota, o bien se deforma la realidad del mundo, lo que representa es la capacidad de alterar, difractar, borrar lo natural, poniendo así un velo, una nota enigmática y artificiosa en las anteriores pretensiones ingenuistas de una pura mimesis y una correspondencia directa y simpatética entre el ojo y su objeto.* ("El *Quijote* espectral," n.p.)

The "visual" literature of the Baroque city is, from this "enigmatic and artificial note," a phenomenon apart in which the visible is manifested in its refashioned edifices, raised in all their majesty, danger, beauty, oppression, and brilliance—from the darkness of its haunts and dives to the verticality of its towers with their filigrees. What emerges is a new urban cartography that allows one to see much and nothing at all. Vision also implies the syncretism of the natural and the cultural—of the person and the means—in two important derivations: first, the movement of people from outside to within and vice versa (or perhaps already always irremediably *within* the city), and second, perspective (whether this means to see without being seen, to see nothing while one is being observed), distortion, or even blindness. The human body moves here within an economy of hidden and covered things, creating its own poetics and industry, its own incipient voyeurism. Such visual saturation leads also to a now-common urban phenomenon that is absolutely new in the seventeenth century: anonymity.

With respect to the visual, the fiction of these decades has a double function insofar as it documents this scopic quality of the urban while at the same time illuminating it in surprising ways. To illuminate also implies a hiding away, because one's focus in writing is not unlimited. The city dweller is hidden within the furtive and the prohibited, or in that which is new and not yet codified; leisure, a central matter in these urban-themed pieces, articulates itself as an original rather than normative activity, and as often as not, as a detour into crime, snares, or previously unassimilated personal uses of space. A painting such as Bartolomé Esteban Murillo's *Four Figures on a Step* (1655–1660) captures an interesting Sevillian scene in which a family seated in the hallway of their house observes something that has just occurred outside the home; it calls attention not only to the boy lying down with part of his body exposed (in a way that strongly suggests Quevedo's picaresque creations) through ripped pants but also to the mother, who is delousing him with the help of eyeglasses, helping us to see the extent to which these were woven into the daily life of the lower social strata (fig. 8).

The use, abuse, and misuse of objects, fundamental for such conduct, operates as a symbol and a sign. In theater, materiality itself is saturated with functions, often converted into a kind of exemplar: boards, powders, and disguises are put on before the eyes of the public to make it see new and more profitable dramatic possibilities. To what extent does language in theater illuminate the hidden only to once again obscure it? How is text interposed within the urban landscape? How does it act as a filter? How are edges perceived, the silhouette of the hidden? Baroque theater transforms reality into fantasies that permit the spectator to escape from the realist spaces of the picaresque or from the beautifully stylized spaced of the courtly novel to enter into the masque itself along with the larger allegorical and oneiric game.

For the moment, I wish to distance myself from Baroque theater to attend to more popular and bawdy (when not marginal) forms of literary expression within metropolitan centers in continual transformation. In

FIGURE 8. Bartolomé Esteban Murillo, *Four Figures on a Step* (ca. 1660)

the pieces I examine in this chapter, there begins to insinuate that which Fernando R. de la Flor has called

> the construction of a true history of the "productive eye," a subtle instrument that destabilizes reality, introducing alarming doubt, into its relation to the exterior (primitively considered to be "direct"); perhaps also it introduces concern or, in any case, ontological instability and establishes, to the contrary, the spectral order of beings.
>
> *la construcción de una verdadera historia del "ojo productor", instrumento lábil que desestabiliza lo real, introduciendo en su relación con el exterior (considerada primitivamente como "directa") la temible duda; tal vez el desconcierto, también, en todo caso, la inseguridad ontológica, estableciendo, al contrario, el estatuto fantasmal de los entes.* ("El *Quijote* espectral," n.p.)

Taking this phantasmagoria as a point of departure, I begin my examination in Seville. My principal focus is Rodrigo Fernández de Ribera (1579–1631), a writer active during Galileo's lifetime who has remained largely unknown to the present day. Of particular interest is his prose work *Anteojos de mejor vista* (Corrective eyeglasses, 1623), which provides a useful entry into one of the most attractive thematic currents of Baroque satire.

Watchtowers, visions, horizons

Fernández de Ribera was a writer of many registers. He was very much aware of trends in the world of letters, and he was also well known and respected by more than a few of his fellows. Lope de Vega, for example, praised his poetry in the second *silva* of his panegyric *Laurel de Apolo* (Laurel of Apollo, 1630). However, such textual praise does not come close to reflecting the interest that readers had in Fernández de Ribera's satirical work. Admittedly, the stature of this oeuvre has never risen to the level of satires written by other authors such as Quevedo and Vélez de Guevara, even though these possess a tone and hold objectives very similar to those found in this novelist. Little was known of him, in fact, until the work of Joaquín Hazañas de la Rúa more than a century ago, and little has been done since.[2] A lover of religious poetry and cultured panegyric, Fernández de Ribera's satirical texts include *Epitalamio a las bodas de una viejísima viuda dotada de cien escudos y un beodo soldadísimo de Flandes, calvo de nacimiento* (Epithalamion for the wedding of an extremely old widow with a one-hundred-escudo dowry and a drunk soldier from Flanders, bald from birth, 1625) as well as his more popular *El mesón del mundo* (The inn of the world, 1632), which includes representations of a large number of social types from the period; the type of inn, as Edward Nagy reminded us in his edition of the text, that would be transformed into one of the favorite scenes of satire from the seventeenth century up to the satirical-costumbrista narratives of Francisco Santos.

It is believed, as Richard F. Glenn has pointed out (123–33), that Fernández de Ribera never would have produced his *Anteojos de mejor vista* had his friends not encouraged him to publish it. Beyond the author's intentions and those of his readers, it is clear that *Anteojos de mejor vista* is an enormously interesting text, as much for the themes that it explores as for the vivacity and spark of the language with which this exploration takes place. The piece begins, like many satires of the time, with a voyage that is ultimately of central importance not just from a structural perspective but also from a somatic one. Under the pseudonym Miser Pierres, the author narrates the return to his much-desired Seville by way of an eventful journey to Majorca.[3] The initial scenes, peppered with scatological elements linked to examples of comical cruelty, immediately suggest certain passages of *El buscón* (The swindler, 1626) and other coetaneous picaresque novels such as *La hija de Celestina* (Celestina's daughter, 1612) by Salas Barbadillo or Cervantes's *Rinconete y Cortadillo* (Rinconete and Cortadillo, 1613), which perhaps served as models for Fernández de Ribera: falls, filth, the presence of hanged men at the entrances to the city, and other elements of the genre project the darkest, most brutal, and primitive side of what was then the most fascinating city in the Iberian Peninsula while also establishing the tone of what would come in subsequent pages. As a cure or remedy for so much misery, Miser Pierres decides to stop at the cathedral, which serves as a center for news as well as a spiritual refuge. The first function, however, very quickly displaces the second; he strikes up a conversation with a cultured boor, a compound (and openly paradoxical) figure whose physical aspects remind one of Quevedo's masterful creation in *El buscón*, the infamous Dómine Cabra:

> His entire face was reduced to the point of his nose, sticking out of what must have been a face; the latter was situated between two curtains of dark brown hair that served, in effect, as a pen for sows, a limp collar that has long been out of style, and a horse-sized pair of eyeglasses that covered his entire forehead. Above this was a felt hat that captured his very essence, and even then there would be much hat left over. I later came to learn from all these parts that the man was a cross between cultured and rough, and by the latter I mean not unrestrained and vulgar but rather circumspect and respectable.
>
> *Reducíase toda su cara a un pico de nariz, asomado por dos cortinas de cabello castaño oscuro a uno que debía ser rostro, abrigado en un pabellón de cerdas, entre una valona opilada—que ya no hay celos, ni se usan aun en esto—y un anteojo de caballo en que traía encajada la testa, o un morteruelo de fieltro que le recogía el meollo, y aún debía sobrarle mucho sombrero* [. . .] *Lo que después vine a entender de este compuesto fue que el tal señor era un mixto de culto y bravo, no de lo desgarrado y vulgar, sino de lo circunspecto y respetable.* (35–36)

Fernández de Ribera describes Miser Pierres's interlocutor as respectable; however, the reader quickly perceives that the man in question is in fact an old-fashioned pedant, halfway between the *philosophus gloriosus* and the *miles gloriosus* who filled the steps and plazas of Spanish cities with their war stories while searching for favors or simple mercy. The opening of the conversation, in fact, approximates the style of local gossip, of the "news" (*noticias*) and "dangers" (*peligros*) of the day. To the lightness and even absurdity of the content of *Anteojos de mejor vista*—which Fernández de Ribera adorns with numerous linguistic artifices very typical of the genre—he also adds ambitious and grandiloquent notes corresponding better to the *arbitrista*: "He distanced himself from the gazettes, but not the sins of the statesman, which, having been purged of their choler, were still as disparate as they had been with it" (*Deslióse de gacetas, no sin sus pecados de estadista, que, a ir purgados de cólera, fueran tan dispares como con ella*) (37). As it turns out, this learned man is extraordinarily rigid in his ideas, incapable of opening himself up to outside judgment, and the new friendship quickly finds itself in need of an escape narrated in a tone not lacking in sexual ambiguity: "I saw that he was as closed in his opinions as a silkworm, even to the death; and I did not contradict him so that he would not close himself up or close up on me" (*advertí que se cerraba en su parecer como gusano de seda, para morir en él, y no le contradije para que no se encerrara o cerrara conmigo*) (37).[4] What most bothers Miser Pierres about his interlocutor is the latter's poorly understood erudition, accumulated as it is within an infertile vessel and so lacking in any kind of intellectual dividend. Through this false erudition, Fernández de Ribera parodies and criticizes a literary type that had by that time become a caricature of itself: the bad Latin, gratuitous citation, and an insufferable petulance complemented by a lamentable narrowness of vision. Unable to accept the more playful side of the conversation and of language in general, the learned pedant becomes the very object of the game, quickly becoming relegated to the margins of the work's true narrative engine.

The most important part of *Anteojos de mejor vista* begins when Miser Pierres and his companion climb to the highest point of Seville's Giralda tower. There they encounter the third and final character of the satire, Master Disillusionment (Maestro Desengaño), who is just as strange as the pedant. Once again, an emphasis on the material, and especially on the quality of clothing, points to the need to rescue only the voice and not the body, the word and not the substance that pronounces it. Everything thus descends into a great Baroque artifice that serves to place the veracity of what is recounted in doubt, thus undermining the narrator's authority. Both the cultured lout and Master Disillusionment—and, to a certain extent, Pierres himself (about whom we know scarcely anything)—are essentially oracles without bodies, voices without context; Master Disillusionment is defined as a machine whose echoes reverberate from within the boundaries of his clothing and dirty, tangled hair.[5] The first part of the citation is in fact a clear recentering of Quevedo's famous famished "blowpipe priest" (*clérigo cerbatana*) with his "eyeglasses riding bareback on a hook of a nose, which was poorly nailed to something like a face" (*anteojos que traía a la jineta sobre una alcayata de nariz, que tenía clavada en uno como rostro*) (46).

Master Disillusionment is found up in the tower scanning the horizon with his moral eyeglasses, taking in a landscape that is completely different from the one he had seen without them. He decries the blindness of his two new colleagues, and these dutifully begin to check their eyesight. Fernández de Ribera's satire is in essence a direct attack against certain social types that were commonly perceived to be vain and hypocritical, and it is thus not surprising to see Master Disillusionment sharing his magical eyeglasses with his new companions, telling them, "You will see things just as they are, without illusion clouding the light of what is most important to see" (*verá las cosas en el mismo ser que son, sin que el engaño común le turbe* la luz *de la vista más importante*) (47, emphasis mine). The narration then transports the reader through several areas of Seville and, in the process, mercilessly dissects an entire series of social types: where before there were doctors, one now sees executioners; where once there were scribes, one now finds vultures; where once there were ministers of justice, one now finds predators. Hens stand in for the valorous, horned oxen for husbands, horses for horsemen, carnival heads for potentates. The eyeglasses allow one to make out an entire inverted Noah's Ark floating through the narrow streets of Seville: swine, bugs, crows, eagles, pigeons, hedgehogs. With their help, Miser Pierres is able to see a completely different city: all that once merited praise has now been transformed into a grotesque zoological imaginary, into a saturated canvas that will also make its way into the future satires of writers such as Antonio Enríquez Gómez and Francisco Santos. In image after image, Fernández de Ribera surprises his reader with monstrous figures that remind one of Hieronymus Bosch's *Garden of Earthly Delights*, a work of art that had deeply influenced Spanish writers for decades (Salas; Levisi, "Hieronymus Bosch"; Iffland, *Quevedo* 1:128–30, 2:43–49).

The authentic Seville, which we now see through illuminating eyeglasses, is nothing but a tableau of misery visible only from an anamorphic perspective, from a fresh, new angle.[6] Through the recourse of allegory, Fernández de Ribera's lens works just like Galileo's telescope insofar as it not only brings distant objects close but also allows one to determine their true nature, offering a new and bare reality that tears down all mistaken interpretation. This new perspective, gained from a high vantage point and mediated by eyeglasses, enables the street and the plaza to raise new suspicions in the reader before offering their teachings: where there had once been notaries, procurers, businessmen, and ministers of justice, Miser Pierres now sees vultures, crows, kites, eagles, and pigeons, "all mixed together" (*todo barajado*) (48) in an indistinguishable mass. Constables are now referred to as "false devils" (49); what was once a doctor on a mule is now an executioner entering the city "on some executed man's ox" (*en un jumento de algún ajusticiado*) (50); bulls are now cuckolded husbands; criers are now hedgehogs wandering aimlessly on the steps of a church, their heads indistinguishable from their tails; coaches do not transport people but rather garbage; pedantic poets, or *cultos*, have been converted into cacophonous hogs or dogs. Here all sensuality is corrupt and dirty, much as it is in other satirical tableaux from the century. Following artists such as Holbein and Bosch, Fernández de Ribera likewise presents us with an abject mix of the animal and the human when he speaks of centaurs surrounded by hounds who wander through the city looking for prey; and like Holbein and Bosch, he delights in the fantastic, describing bugs that emerge from their mud-covered eggs fully grown.

The sensation of movement in Seville's human stew invades everything, making the work of distinction and classification difficult. Fernández de Ribera goes so far as to speak of worms that emerge from within (whatever is meant by *within*), thus offering a strong image of moral putrefaction that spreads throughout the city. This space is, in the end, both infernal and apocalyptic: "mountains of fire" (*montañas de fuego*) appear for backbiters, and even the pedant, when Miser Pierres observes him through eyeglasses, ends up not "a man with an apparatus of words" (*un hombre de con tanto aparato de palabras*) (54), but rather exactly the opposite: "a man-sized hen, such that all that remained of what he had once been was a sword, a dagger, and a guitar-bridge mustache; and he even had another two chicken wings on his feet" (*una gallina de su tamaño, sin que le quedase, de lo que antes era, más que una espada, una daga y unos bigotes de puente de vigüela, y tenía más de gallina otras dos alas en los pies*) (54). And if lenses employed from a high vantage point serve to strip urban social types of their patina of false dignity, we also find a persistent critique of their customs. Such is the passage in which the narrator depicts an impossibly large Paschal candle with a wick so long that it took an entire ship's worth of Indian cotton to make it. The broader politico-economic situation is likewise addressed, such as when the narrator laments what he sees at the port of Seville; especially troubling are the foreign ships: "these people seem to do nothing but pull fish from this city and all the others in Spain" (*que no hace esta gente, sino pescar de esta ciudad y de las demás de España*) (62). In similar fashion, Fernández de Ribera criticizes the pernicious effects of the gaming house as a waste of money and a source of corruption. He also describes with great mastery the players as bees in their hives, watched over by constables transformed into bears that steal the honey.

The surreal urban landscape offered up by *Anteojos de mejor vista*—a paradigm of ignorance, conformity, and brutality—merits a final reflection. In the first place, if no one is truly safe from the book's cruel attack, one may well wonder why there is no mention of women or students, two very common targets in the satire of the time. This is so because, as the author indicates at the beginning of the text, they happen to be a lost cause and unworthy of any further mention whatsoever. And beyond this blanket condemnation of women and students, Fernández de Ribera likewise constructs his entire text as an a priori denunciation. For him, all humans live in complete blindness, and given that the entire world is blind and it is therefore the blind who guide the rest, vision has no purpose: "The entire sense of sight, and with it the others, is reduced to touch: today we judge blindly, we cure blindly (although few are actually cured), and we live blindly" (*Todo este sentido, y aún los demás, están reducidos al tacto: hoy se juzga a ciegas, se cura a ciegas (aunque sanan pocas) y se vive a ciegas*) (64). As a result, as Glenn reminds us: "In Fernández's exposé of society, voluntary moral blindness and the hypocrisy it engenders is the root of all evil. Greed, adulation, fear, desire, jealousy, pride, and lust are all the motives for self-blinding, and each must be reckoned with as man—purposely ignoring the falseness of his reality—pursues his livelihood" (130). *Anteojos de mejor vista* advances, in fact, a gradual erasure of the senses, out of which the only one that survives is touch, which together with smell was considered the basest of the five.[7] The faculty of vision is placed in doubt, the sense of smell serves only for the use of tobacco (which Fernández de Ribera condemns on at least two occasions), and taste is irredeemably compromised by the devastating power of the crier's "volcano of the tongue" (*volcán de la lengua*) or by the "vegetable sellers" (*verduleras*) who "not only make one blind, but also mute" (*no sólo hacen ciegos, pero mudos*) (54). The brushing together of bodies, whether sensual or violent, is that which in the ultimate instance defines the whole. It is the identifying mark of this macabre urban vision.

Similar human degradation is achieved through what Martin Jay has identified as an "overloading [of] the signs in a painting, producing a bewildering excess of apparent referential or symbolic meaning" (51). Within this framework, everything becomes saturated, full of sense(s), and excessive; however, thanks to the mediation of corrective lenses, it can all be dissected in great and reliable detail. When Miser Pierres, aided by his eyeglasses, wonders aloud where all that he sees is found, Master Disillusionment responds: "In your rational mind" (*en su juicio*) (47). The irony lies, evidently, in that this new Seville is not an imaginary scene, but rather the real, authentic city, revealed thanks to the mediation of the *occhiali politici*. With them on, vision is, as we have already seen, at once moral and political. And while this Babylon might logically merit being the main protagonist of the piece, Fernández de Ribera teaches us that the observed object is not as important as the means by which it is observed.

The reality of Seville could be that of any other city, given that what is sought after is the unveiling of the

> theater in which is represented the comedy or farce of human life; and the dressing-room is the earth, from which we emerge dressed as human beings: this is the king, that is the shepherd, the other is a merchant and, in this way, each one has a role; and while we all look at one another, we are blind and so see nothing nor know what we are until we once again return to our earthy dressing-room and to the nothing that preceded us, the common resting place of these clothes until the day on which our actions are finally taken into account.
>
> *teatro en que se representa la comedia o farsa de la vida humana, y el vestuario la tierra, de donde salimos a representar vestidos de hombres: este es el rey, aquel el pastor, el otro el mercader y, así, cada uno su figura; siendo los que miramos unos a otros, y todos ciegos, pues no vemos ni conocemos lo que somos hasta que nos volvemos a desnudar al vestuario de la tierra y al nada que antes, depósito común de estos trajes hasta el día en que se dé cuenta de la acción que a cada uno se entregó.* (61)

From this it follows that the true protagonist of this bewildering and thought-provoking piece is not the city but rather the eye that scrutinizes it and the means by which this examination is carried out, or rather the device that permits one to escape blindness and discern objects from a distance. As a result, *Anteojos de mejor vista* is nothing but the history of a revelation facilitated by the powers of glass not turned toward the firmament but rather toward the *ground* and the many social, moral, and political challenges faced by the early modern city dweller. In this, it seems to suggest that the edifice of progress should begin at its foundation, since the recourse to glass is central for any adequate understanding of the didactic project that informs the novel. As Fernández de Ribera puts it: "Well do I know that if one were to look into this mirror, it would be impossible to cover up, given that the mirror reflects everything completely, inside and out; mirrors were made, in effect, to see and compose oneself, and eyeglasses were made to see and come to know others" (*bien sé que, si se viera a un espejo que yo tengo, no se había de poder encubrir de sí, por ser capaz de verse un hombre todo, dentro y fuera; que, en efecto, los espejos se hicieron para verse y componerse a sí, y los anteojos para ver y conocer a otros*) (60). This idea is symptomatic of its historical moment, and it synthesizes an entire tradition according to which glass had alternately served as a vehicle of knowledge and of introspection. What is truly original here is that Fernández de Ribera combines these two types of optical phenomena in one sentence. Writing thus becomes a kind a fellow traveler with the material of glass, and we could also say that it effectively generates the act of writing. As a result, Fernández de Ribera's eye is the "eye of the mind," the eye that penetrates into true knowledge, that strips down one's neighbor to reveal his or her most human essence. Master Disillusionment's eyeglasses offer us a radical novelty that Fernández de Ribera exploits to the maximum: "These eyeglasses were fashioned by experience, the glasses made of truth; because, although Venetian glass is very clear, it is overly subtle, and as all eyeglasses are made of ambition, they ultimately cloud the vision to a great extent. Nobody wears these, because everyone is guided by their own" (*Estos antojos los labró la experiencia, el vidrio es de la misma verdad; porque, aunque el de Venecia es muy claro, es demasiado de sutil, y allí, como todos los anteojos son de ambición, turban la vista mucho. Nadie usa de estos, porque todos se guían por los suyos*) (60). The purpose of *Anteojos de mejor vista* is not, therefore, to offer an interested portrait as Venetian writers might do, nor is it merely a political text or an ambitious search—as the citation makes clear—of the universe; rather, it strives to serve the common good, a goal that was much more pressing for Fernández de Ribera and his readers. It is not surprising, in fact, that in this meditation on the power of sight, we should find a brutal critique of astrologers:

> They are more blind than anyone, because they not only see what they don't see, but they also say that they see what hasn't yet been seen (nor, in the majority of cases, will be seen), trying to convince others that they are able to stop the sun in order to measure its circumference. It should not surprise us that those who speak with such certainty should be so blind and wrapped in shadows.
>
> *Son más ciegos que todos, porque no sólo ven lo que no ven, pero dicen que ven lo que aún no se ha visto, ni se ha de ver las más veces, dando a entender que hacen parar al sol para tomarle la medida. Y no es mucho que estén tan ciegos los que, aun en las tinieblas, comunican siempre con tanta luz.* (67)

Fernández de Ribera is ultimately a cautious narrator writing at the beginning of a new century and observing with great prudence the breakthroughs then taking place in Europe. Rejecting those who "see what hasn't yet been seen" and assuming the idea of a sun in movement, Fernández de Ribera focuses his interest on improving what is before him and most urgent, and it follows from this that he considers such debates to be much more important than those then taking place on the firmament. *Anteojos de mejor vista* is, in the end, a reflection on not seeing and a meditation on blindness. This is so because within it human perception is whittled down to and wholly dependent on touch, which would seem to produce beings that could not grow outward but would rather circle back to the womb. For Fernández de Ribera, the *occhiali politici* are destined for much more modest tasks, for undertakings supported by greater evidence. And this "political eye," a Spanish version of its Venetian predecessor fashioned to save humans from their blindness, is what gives sense to the entire text.

Other lesser-known texts offer a different moral vantage point. In *El hijo de Málaga, murmurador jurado* (The son of Málaga, sworn gossip, 1639), signed by Salvador Jacinto Polo de Medina under the pseudonym "Fabio Virgilio Cordato," the author imagines that a statue bestows him with the power to detect the vices of those who visit the city's pier, following very closely the thematic-contextual motif of Parnassus in chapters 9 and 10. This is a novel, however, that doesn't differ greatly from *Anteojos de mejor vista*, and therefore I won't delve into it. In what remains of this chapter, I wish to contrast Fernández de Ribera's text with a brief account of a piece that I find impossible to omit from my survey, insofar as it offers a different vantage point, more symbolic than material—a vantage point with a twist, we might say—but no less fascinating for the content that it presents. Charged with a strong feeling of resentment provoked by exile and prosecution, Antonio Enríquez Gómez's *La torre de Babilonia* (The Tower of Babel, 1649) offers an innovative development of the themes upon which I have been focusing, given that the objective of his critique is a direct reflection of a singular phenomenon in early modern Spanish letters. Modern scholarship devoted to Enríquez Gómez has consistently insisted on the Jewish sources of his work, insofar as he situates his diatribes against the Inquisition and his social and literary critiques within allegorical dreams, imagined academies, or the fictional journey.[8] Dreams and picaresque narration constitute, as Northrop Frye pointed out decades ago, two fundamental elements of Menippean satire, although to this scholars such as Matthew D. Warshawsky have also linked them—in Enríquez Gómez—to "new Christian protest literature" (*Longing* 45) suggesting that:

> it is clear that *El siglo pitagórico y vida de don Gregorio Guadaña, La torre de Babilonia,* and *La Inquisición de Lucifer* reflect the adverse conditions in which they were written. A descendant of crypto-Jews persecuted by the Inquisition, Enríquez Gómez saw in letters a means to express the injustice of his and his family's suffering, and resist inequities in general [. . .] the author's own experience of financial woes at the hands of creditors and irresponsible borrowers may motivate the depiction of money as a caricature of religion. (*Longing* 25, 203)

With his personal mix of the marvelous and the realistic, the fantasy of Enríquez Gómez finds expression in important texts of the middle of the seventeenth century, such as *Inquisición de Lucifer y visita de todos los diablos* (Lucifer's inquisition and the visit of all the demons, 1642), *El siglo pitagórico* (The Pythagorean century, 1644), and the aforementioned *La torre de Babilonia* five years later.

In *transmigración,* or chapter, 11 of the *El siglo pitagórico,* Enríquez Gómez speaks of an *arbitrista* in relation to contemporary problems of astronomy: "the first judgment that he rendered was to stop the sun; after this, he placed a new law over the moon" (*el primer arbitrio que le dio fue estancar el sol; asegundó con otro y puso un nuevo derecho sobre la Luna*) (18). With respect to dreams, Enríquez Gómez comments in *Inquisición de Lucifer,* for example, that "there is no inquisitor like one's dreams, as these make our tongues confess the heart's heresies without resorting to torture" (*no hay inquisidor como el sueño, pues hace confesar sin tormento a la lengua las herejías del corazón*) (18). Dreams are instruments, therefore, that permit Enríquez Gómez to reveal to his reader—and to a certain degree to himself, as they almost certainly serve as a personal catharsis in the face of a merciless reality—the state of anguish and resentment in which he lived, always on guard and always informed of the ever-changing religious policies that ultimately determined his family's possibilities. The dream motif, according to Warshawsky, permits Enríquez Gómez to achieve this desired objective insofar as

> the Baroque dream also permits distance between the author and the subversive, degraded world of his creation. . . . The narrators they create feign impartiality, when in fact they are the means of criticism and parody. As intermediaries between the authors and their degraded characters, these narrators also assure that a distance remains between the two. Another example of distance is a geographical one: the dream works create settings whose absurdities and stock characters caricature elements of actual societies. (*Longing* 68)

This is a writer who presents us with an extraordinarily interesting paradox within the broader scope of early modern Castilian literature. He is at once a novelist fully situated in his daily experience—communicated through a very personal voice—and one that exists in a kind of limbo or "living death" given all the veils that he constructed to hide his existence. Enríquez Gómez left Spain in 1636, and a great percentage of his literary production took place during his exile in France, first in Bordeaux and later in the bustling city of Rouen. Critics have likewise puzzled over the fact that some of his works—and others that were not his—were signed "Antonio de Zárate" and highlighted the famous and tragic episode that took place upon his return to Seville: while living under a false identity, he witnessed the burning of his own effigy, his own auto-da-fé carried out—officially, at least—in absentia. A deep concern with life and death and the development of a form of writing that looks to overcome this dialectic in the face of a situation that obliges him to erase and reinvent himself continuously are constant features of his writing. In the second *sueño* of *Inquisición de Lucifer*, he in fact makes this point explicitly: "Dreams, perhaps, are living works, while those [works] of the day are dead works" (*los sueños, tal vez, son obras vivas, y las del día obras muertas*) (48).[9] And if the actual daily *works* of Enríquez Gómez have come to us only in diffuse form, his written work is a major and urgent contribution, as alive today as it was in its moment of production, overflowing still with relevance and timeliness.

Enríquez Gómez's narrative landscapes are almost always marvelous, deliberately confused and ambiguous, and his written voice consistently benefits from the resulting interstice within which he situates his readers: he is at once innocent and guilty, magnanimous and merciless, content and resentful. Victim and executioner, he employs writing as a weapon both for assault and for self-defense; his attitude before the Inquisition is one of attack and denunciation, of sophisticated social critique and sincere human testimony. The inquisitorial tribunal is nearly ubiquitous within Enríquez Gómez's oeuvre, and he often contrasts Spain with Zion in his personal dialectic between Christianity and Judaism. To give but one example, the "Nimrod" who repeatedly appears in *La torre de Babilonia* as the builder of the famous tower, the tyrant who robs and abuses, is the spirit and the body of the entire repressive apparatus that his writing denounces. With masterful sarcasm, he goes on to say: "The Inquisition is a theater of the divine; it begins as a comedy and ends in tragedy" (*la Inquisición es una farándula a lo divino; empieza en burlas y acaba en tragedias*) (79). For Enríquez Gómez, the Baroque sense of spectacle and showmanship combined with an institutionally sanctioned hunger for the visual rendering of a body submitted to the law and tortured regardless of age or sex—in his narrative there appear beautiful young women who are submitted to inquisitional torture—is a deplorable form of theatricality that is also necessary; it is, one could argue, a weapon that has its effect as a legal tool but also as a narrative resource to be exploited fully. It comes as no surprise that at one point he makes the claim that the Inquisition consists of "inquisitors made eagles, prosecutors made sparrow hawks, commissaries made crows, lay officials made gyrfalcons, witnesses made wolves, notaries made kites, informants made storks, receptors made ostriches, sheriffs made mastiffs, qualifiers made basilisks, secretaries made owls, and the executioners made vipers" (*las águilas de los inquisidores, los gavilanes de los fiscales, los cuervos de los comisarios, los gerifaltes de los familiares, los lobos de los testigos, los milanos de los notarios, las cigüeñas de los malsines, los avestruces de los receptores, los alanos de los alcaides, los basiliscos de los calificadores, las lechuzas de los secretarios y las víboras de los verdugos*) (85). After alluding to this string of animalized characters already seen in Fernández de Ribera, the narrator wonders, "And if this peacock with his rich plumage were to attack so many birds of prey, how would he retain even a single feather?" (*si a este pavón lleno de pluma acomete tanta ave de rapiña, ¿cómo le ha de quedar pluma a esta ave?*) (85). In the face of this question, which expertly underscores the bivalence of the term *pluma* ("feather" and "pen"), it is worth wondering whether it is not Enríquez Gómez himself who is speaking here in a spark of sincerity, effectively revealing himself in the face of censorship, running the risk of becoming, quite literally, *desplumado* ("unpenned" or "unfeathered"). To read Enríquez Gómez is to read a small inquisitional archive, in large measure because in his prose the Spanish Inquisition functions both as a heuristic fiction and as a historical source.

La torre de Babilonia is unquestionably one of the most fertile testimonies of the literary device of "corrective lenses" in early modern Spanish letters. And if the book's invented landscape portrays much of the personal and the unique, it is also true that it contains a very generous dose of erudition. Taken from the eighth through the eleventh chapters of Genesis—which, as we know, describe the end of the flood up to the tower of Babylon episode—it also serves as inspiration for Calderón de la Barca's famous *auto sacramental* (i.e., a religious short piece) of the same name. The tower of Babylon likewise finds its way into other works penned by Enríquez Gómez: in the unpaginated prologue of his epic *Sansón nazareno* (Samson the Nazarene, 1656), for example, he confesses, "If I enter into the tower of Babylon, it is to remove documents that cause confusion" (*si entro en la torre de Babilonia es para sacar documentos de confu-*

sión) (9). In the second section of his *Academias morales de las musas* (Moral academies of the muses), he includes a "song for the ruin of an empire" (*canción a la ruina de un Imperio*) in which the tower serves as an image that represents the ruins of a better time.

La torre de Babilonia presents its reader with a kind of allegorical dream divided into *vulcos* (turns), an explicit allusion to the turns (*vuelcos*) one makes in bed while having a nightmare and that suppose a change in the thread of that which is dreamed. What is certain is that the work depicts an authentic nightmare vision—as is the case with Fernández de Ribera's *Anteojos de mejor vista*—that begins when the narrator finds himself the owner of a luxurious palace and married to a beautiful woman. Glenn F. Dille describes the text as a "*converso*'s dream of utopia" (90), a vision of a peaceful past, of an uninterrupted flow of time that is abruptly cut by the separation of his family—part of it remained in Spain after he emigrated—and the need to go into exile. It is important to note also that Enríquez Gómez's Babylon points not only to an existential disorder, to a cacophony of the new, but also to a continuous linguistic challenge, as Dille pointed out: "the story of the Tower of Babel must have seemed quite immediate to Enríquez Gómez as, with a sensation of confusion and alienation, he entered various foreign cities in his exile years. The incomprehensible chatter of the people involved in the streets would have been disheartening to an author so involved in the artistic nuances of his native language." (94) The morally praiseworthy use of sight governs some of the considerations with which the text opens. In the prologue, for example, we are warned: "Read with care and pay attention, as those who read with good sense have eyes, and those who do so without it have only eyeglasses" (*lean con cuidado y miren en él, que los que leen con seso tienen ojos, y los que miran sin juicio, antojos*) (337). The text's critique of authority represented by *antojo* as a mark of distinction is thus guaranteed, in this insistence on seeing things as they truly are, to see people not according to their social position. We know that the book's protagonist is a "young pilgrim" (337) whose trajectory has no defined structure but rather a reiterative character; in fact, the term *pilgrim* (*peregrino*) here refers to someone who wanders rather than a person who moves deliberately toward a specific destination. In this way, Enríquez Gómez points to his own unpredictable movement from one place of exile to another, embodying the figure of the wandering Jew. *La torre de Babilonia* begins with a novelized re-creation of the story of Genesis, with the expulsion from the Garden of Eden serving as the platform for a pessimistic vision of the present that is not unlike that offered by Francisco de Quevedo in many of his creations. In the second *vulco*, when the pilgrim arrives at the tower's entrance, a nymph emerges from a trapdoor such as those used in the theater of the time and sings to him a long ballad that synthesizes the misery that the narrator will face, gradually rolled out over the course of the rest of the text, in one vignette after another. For Enriquez Gómez, individuals are governed by almost absolute arbitrariness, a vision that is very far from what one finds in Gracián's *El criticón* (The critic) or Juan Martínez de Cuellar's *Desengaño del hombre en el Tribunal de la Fortuna y Casa de descontentos* (Disillusionment of man in the Tribunal of Fortune and House of the Discontent, 1792); in fact, if a specific source is to be discerned, it is likely Suárez de Figueroa's *Plaza universal*. In this way, after the tenth *vulco*, we are taken through the marketplace that contains all of the stores that the text will visit: there is, for example, the store of Necessity; then in the thirteenth *vulco*, the protagonist and his companion visit the store of Common Sense; and then because they are in Babylon, they visit some science classes—accompanied by the Marquis of Villena, with whom they had spent time in the previous section after the former had left his flask—attending classes taught by alchemists, theologians, philosophers, politicians, astrologers, doctors, anatomists, botanists, surgeons, and lawyers.

In her edition of the text, Santos Borreguero writes that "the vision is satirical, and the Marquis is the one in charge of, at the end of each of the classes, presenting the truth, criticizing exaggerations, and pondering the positive aspects of each of the sciences" (*la vision es satírica, y el Marqués es el encargado de, al final de cada una de las clases, sentar las verdades, criticar las exageraciones y ponderar lo positivo de cada una de las ciencias*) (98). Without being dealt with explicitly, the use of the *occhiali politici* in fact largely defines the narrative structure of *La torre de Babilonia*. Enríquez-Gómez, as Kramer-Hellinx reminds us, was well aware of the success of the literary gazette in Spain, and he had definitely read Boccalini's invectives, which he combined with a personal narrative "universalized through the use of biblical myth (the creation, Noah's Ark, Babel, and Babylon) in order to relate the work to the entire history of man" (114).

It is in the ninth *vulco*, however, that Enríquez-Gómez presents the passage that defines the book and the philosophy that underlies it, insofar as it is here that he most powerfully interweaves sarcasm with the most disarming sincerity in an explicit assessment of the reality that surrounds him:

> I never knew that Babylon was the honorable place that it is. May God bless you for the excellent republic that you are! Is it possible that honor has come to such a miserable state that the liar supports it with his lie,

the trader with his betrayal, the bad judge with the bribes he accepts, the woman with her body, and the man with his sin? No, my friend; these honors are more public dishonors than private honors. The true honor is virtue, jewel of the spirit and sun of human action. The good man is he who looks to true North to reach the port of salvation. Honor consists in the purity of life, and the good life consists of the justification of one's works.

Nunca entendí que fuera Babilonia tan honrada casa como es. ¡Válgate Dios por república y cuál estás! ¿Es posible que la honra haya venido a tan miserable estado que el mentiroso la sustente con su mentira, el malsín con su traición, el mal juez con el cohecho, la mujer con su cuerpo y el hombre con su pecado? No, amigo; estas honras más son deshonras públicas que honores propios. La verdadera honra es la virtud, adorno del espíritu y sol de las acciones humanas. El buen hombre es aquel que mirando el norte de la verdad alcanza el puerto de la salvación. El honor consiste en la pureza de la vida, y la buena vida en la justificación de las obras. (530–31)

It is toward this port of salvation that Enríquez-Gómez's pilgrim is headed, oriented as he is to that True North that cannot be justified except through good works. The vision that Babylon and its tower lookout affords us is worthy of mention in the genealogy of Menippean satire in which journey, dream, and explicit didacticism form a constellation toward which past and future texts gravitate. *La torre de Babilonia* serves also to connect my arguments up to this point to a series of coetaneous testimonies that will form the center of my analysis in the following chapter, testimonies that have much in common with some of the passages that I have commented on in the present chapter—for example, the fabulous voyage of the main characters in *El diablo Cojuelo* (The limping devil). The Marquis of Villena offers a kind of homage to this journey in his own short autobiography, which serves to lift the roof off the houses of Madrid in order to show the misery of their inhabitants. Equally curious is the childbirth imagery in the thirteenth *vulco*, where the marquis initiates his story. He begins by confessing his personal debt:

I, my dear Sirs, after the miraculous genius of don Francisco de Quevedo left me in his flask as minced meat, decided to leave it on Tuesday, July 21, 1647. Thank God I liberated myself from my mother's belly! Blessed be he that freed me of that dowser of guts and explorer of stews! Many will be envious of my birth, as if there were much difference between woman and glass. . . . I was born made of glass from a Venetian belly.

> *Yo, señores míos, después que el milagroso ingenio de don Francisco de Quevedo me dejó en su redoma hecho gigote, salí de ella un martes veinte y uno de Julio, año de mil y seiscientos y cuarenta y siete. ¡Gracias a Dios que me libró de barriga de mujer! ¡Sea bendito el que me libró de zahorí de tripas y explorador de cuajares! Envidiarán muchos mi nacimiento, como si hubiera mucha diferencia de mujer a vidrio* [. . .] *Soy* parto cristalino de una panza veneciana. (586, emphasis mine)

It is impossible, once again, to escape the influence of Venice and its glass, as both yield so much in terms of metaphor within the satires of this period. Enríquez-Gómez's debt to Quevedo is more or less common knowledge, as is the case with Luis Vélez de Guevara: two authors, at the end of the day, who masterfully integrate glass, mirrors, and eyeglasses into their work. I turn to Quevedo and Vélez de Guevara in the next chapter, but I wish to close the present one by underscoring once again how eyeglasses open and close Enríquez-Gómez's exploration, a set of political eyeglasses that distance themselves from the disciplinarian uses of the Inquisition that he critiques in order to offer a vision that is radically novel, shaped to a great degree by his experience as a prosecuted writer. It is here that he lays out the reasons we should identify true virtue beyond the uses that condemn difference. *La torre de Babilonia*'s lens is, therefore, one that is simultaneously similar and different, specific and universal, and through it we can approach the refracted space that is at once a particular city and all cities, that is a particular Spain and a universal Spain outside of time.

6
Explorations

The social critique in the universe of glass

The challenges that new scientific advances posed to orthodoxy, to dominant truths personified in the early seventeenth century by Aristotle and Ptolemy, were tremendously attractive for writers looking to expand the thematic, contextual, and linguistic horizons of Baroque fiction. There were a number of changes (albeit in small doses) within Spanish satire produced after the advent of the telescope. The goal of these satires is not to denounce madness, eccentricity, or the strange; rather, they present marginality in a positive and not exclusive sense, making it almost normative. And insofar as these texts problematize what their readers assume madness to be and what is accepted as sanity, their misanthropic protagonists are more often considered visionary than insane.

The shadows of the domestic space, the *studiolum* littered with apparently useless objects that nonetheless hold life within them (we will see an example of this in *El diablo Cojuelo*, where a simple flask opens up an entire novel), finds itself converted in seventeenth-century Spanish satire into an ideal frame for the examination of extreme and eccentric forms of conduct. These forms of conduct and their satirical examination are of course not without certain idiosyncrasies; the feminine, for example, seems to have no place. Put another way, the early seventeenth-century urban misanthrope does not participate in the masculine codes of his social milieu; rather, he creates his own, and these tend to rest outside of the canons of gender. For these satirical figures, honor scarcely matters and decorum is all but unobserved; their ethical codes emerge from the dictates of conduct, routine, and personal manias that—just as with the genre of satire itself—participate tangentially but significantly in court society. The invisibility of the home is also that of the individual separated from society—both are hidden and closed off.

The individual in satire is also, significantly, the carrier of nonnormative forms of knowledge and is occasionally ahead of his time. And if light manifests itself in the cavernous spaces of his lived interior, we are to imagine it always entering diagonally—as is also the case with the telescope and the angle of vision that this physiognomy imposes—tearing away the veil of the clandestine while leaving its bottom half in shadows. This has its ekphrastic foundation: shadows, as historians of Baroque art have taught us, are as important as the illuminated portions of the canvas, in that they invite us to investigate, to find true knowledge. And it is precisely the act of hiding, the interruption of sight, as Roland Barthes reminded us in *A Lover's Discourse,* that serves to increase desire.

The desire to see the intermittent, more than the invisible, is the central concern of the present chapter. In focusing on this desire, we see how the privileging of the eye to see what is prohibited both on earth and in the heavens provides a kind of outline for the critical fortunes of the *occhiali politici* when it overlaps with science. To this end, I also explore some of the "maladies" of the Baroque subject, who was consistently so in need of escapist pastimes such as judiciary astrology; my goal here is to examine the extent to which new technical inventions supported the complete jouissance of sight for this voracious Baroque eye, an eye, in the words of Fernando R. de la Flor (*Pasiones* 47), *sin párpado* (literally, "with no eyelid").

The remarkably fertile eclipse of vision in seventeenth-century letters initiates a specific trajectory. Given that this book focuses on, as its very title indicates, the Spanish contemporaries of Galileo, I do not include in my survey texts belonging to the sixteenth century, during which were written Menippean satires of great importance: the *Somnium* (Dream, 1520) of Juan Luis Vives;[1] the *Diálogo entre Mercurio y Carón* (Dialogue between Mercury and Charon, 1529) by Alfonso de Valdés; and Cristóbal de Villalón's *Crotalón* (Crotalon, 1553) being perhaps the best-known and most studied examples.[2] Together with the spatial motif of the high vantage point, which served as the focus of the previous chapter, we know that in seventeenth-century satire the aerial journey is likewise a common presence, manifesting itself through dreams and reveries.[3] The characters in these texts consistently initiate their voyages in a liminal state that can make it difficult to place them within a specific situation until well into the novel; this tendency can generate for the reader a generous dose of confusion and fantasy that, at the end of the day, is very much a hallmark of the tastes of the time. The observed landscape is thus wrapped in an air of marvel that importantly does not prevent the reader from perceiving specific, concrete references in the underlying critique—these appear as characters or topical situations that many times allude also to a reality mediated by flesh-and-blood people who pass through the novel in their more uninhibited form and their eminently "human" comportment. The city is the quintessential field of observation, a *champs* in which the master lines are traced out that reconstruct a disenchanted reality. And the use of glass, whether as eyeglasses or telescopes, or as mirrors that reflect and refract, returns us once again to the conceptual and magic richness of this crystal universe—a universe that without glass cannot be presented in all of its fullness, ambiguity, and play.

The complex characters of seventeenth-century Spanish satire, together with the mysterious and disorienting territories presented in their lunar fantasies and dreams, demonstrate that when Galileo's work was beginning to be widely known, there was already in Spain an ideal foundation into which his celestial observations could be integrated. I will say very little in this chapter about the genesis of political satire in the seventeenth century, as this topic already boasts an ample bibliography;[4] rather, I enter directly into those examples that are strictly contemporary with or come just after the publication of Galileo's *Sidereus nuncius*. Ramón Valdés ("Rasgos") has argued that Menippean satire in early modern Spain "takes place primordially through contact with the other world, or through fantasy, allegory, or recourse to magic," and that "it is not limited to repeat passively the model of Seneca's or Lucian's satires; rather, it picks them up with changes shaped by humanism and actively introduces further changes" (*se realiza sobre todo mediante el contacto con el otro mundo, o a través de la fantasía, la alegoría, de recursos mágicos*; and *no se limita a repetir pasivamente el modelo de las sátiras de Séneca, o de Luciano, sino que las recoge con los cambios operados por el Humanismo e introduciendo activamente nuevas mutaciones*) (202, 198). It is these changes to which Valdés refers that form the focus of the present chapter. In this way, I open up a path of critical inquiry that examines the genesis of a kind of "science fiction" in the literature of early modern Spain, a genre nurtured through the interweaving of autochthonous traditions with echoes and presences in Spain of the Scientific Revolution then taking place in neighboring countries to the north.

Among the writers who elevate satire to new levels of originality at the start of the seventeenth century is the Granadan playwright and satirist Luis Vélez de Guevara. His classic prose work *El diablo Cojuelo* has been amply studied over the past three decades with respect to its genre, its alleged theatricality, its vision of the urban; however, it has been only more recently that modern scholars have begun teasing out the work's dialogue with the scientific landscape of the period.[5] The allusions scattered throughout the work to "the Baroque gaze" and advances in cosmography point to the idea that *El diablo Cojuelo* offers its reader an image of Spain itself at that moment, a moment in which the eye is, once again, the essential protagonist.

Vélez de Guevara's narrative journey is a rich exploration that touches numerous themes and covers too much geography to be dealt with in any detailed form here, in part because the eye of the observer is an itinerant and double one—a loquacious and dialogic eye, we might say—that changes according to whether it is the student or the devil Cojuelo that projects it. In Velez de Guevara's proposal, the reader is invited to observe from a privileged perspective the streets and buildings of Madrid through the vision of a (limping) devil and his friend, the student Don Cleofás Leandro Pérez Zambullo. Through a mediator, then, two eccentric characters are brought together as they run from their respective enemies on an aerial journey that lays out the miseries of their time, an uncovering that in some cases is literal and serves to lay bare what lies within the scene under analysis. In his denouncement of the madness of others, Vélez de Guevara authorizes his characters the power to open up the suffocating space of Madrid in summertime. Following the *Icaromenippus* of Lucian of Samosata, and passing through the *Anteojos de mejor vista* of Fernández de Ribera, Vélez de Guevara stages his narrative perspective from above, portraying his creatures in constant movement as if they were in an anthill. Worth noting is that the use of eyeglasses (*anteojos*) is here replaced by the keen visual powers of the devil, whom Vélez de Guevara converts into a kind of "literary lens." Through this lens, the reader is able to take in a clandestine urban landscape, a landscape closer to the by then almost extinct picaresque setting than that presented by the so-called courtly novel so in fashion during these years. *El diablo Cojuelo* is, in this sense, a very particular case for its time.

The aerial journey is perhaps the piece's most suggestive device, since it is through the journey of Cleofás and the devil that the reader manages to share the perspective—double in this sense, as it is literary and also physical—employed by Vélez de Guevara. We cross over Madrid through the sky and land before we find ourselves situated in events that take place in Andalusia, only later to return to the capital in the final sections of the text. But the journey contains its own dynamic, its own vision, and this is achieved fundamentally from on high, as if Vélez de Guevara were looking for a globalizing perspective. From the perspectival privilege given to Cleofás by demonic powers—wielded by a mischievous devil fleeing from his own persecutors—the structural figure par excellence is the escape, and it is only in Andalusia that the protagonists touch the ground, abandoning their condition of privileged spectators.

In their dialogic journey, Cojuelo and Cleofás manage to see things that nobody else can: a vain courtier in the privacy of his room removing his many accessories—a wig, a false mustache, a wooden arm; a woman giving birth with her cuckolded husband by her side while the child's father sleeps peacefully on the other side of the city; a witch preparing a potion to restore young girls' virginity; a pair of pretentious *madrileños* living in their coach as if they were "turtles under their shell" (*tortugas bajo su caparazón*) (27), and so on. A great part of the novel's originality, in fact, rests in this active uncovering: in this way, the first chapter of the book closes when "raising the roofs of the buildings through demonic art uncovered the many thin layers of Madrid's meat pie just as they were; and with the great summer heat many windows were left wide open, and there was such a greater variety of rational beasties within this ark of the world that, compared with it, the biblical ark seemed of little distinction" (*levantando a los techos de los edificios, por arte diabólica, lo hojaldrado, se descubrió la carne del pastelón de Madrid como entonces estaba, patentemente, que por el mucho calor estivo estaba con menos celosías y tanta variedad de sabandijas racionales en esta arca del mundo, que la del diluvio, comparada con ella, fue de capas y gorras*) (20).

Vélez de Guevara transfers all of the symbolic potential of the meat pie to a new aesthetic dimension that lacks the moralizing charge of his contemporaries Francisco Santos and Juan de Zabaleta. For readers at the beginning of the seventeenth century, such grotesque and even ridiculous scenes are no novelty, and the meat pie in particular was commonly associated with the picaresque genre. We see this, for example, in Mateo Alemán's *Guzmán de Alfarache*, when the young Guzmán abandons his house with no explicit destination; we likewise perceive it in all its repugnance in the basest episodes of Quevedo's aforementioned *El buscón*. The strange treatment that Vélez de Guevara gives an element so frequent in this and other urban settings serves to construct a space of play where the outer shell hides the inner fruit, an amorphous mass of varied ingredients plural in their movement, a microcosm delimited by an external border to which we have access only through the aerial vision made possible through lenses. This resource facilitates the writing process; it makes it possible, and in the end, it legitimizes it as a portrait, as a tableau of customs that permits the entrance, from this grotesque and hyperrealist aesthetic, of new elements such as humor or caricature, exaggeration or animalization, and above all, the play of combined and interwoven senses: total and blinding vision, the smell of sulfur, the feel of the air at high altitude, the sounds of the coaches, of commerce, and so on. And, like one who uncovers a beehive in the full light of day, this urban re-creation reactivates the narrative tension in the opening of the following chapter, with poor Cleofás "absorbed by that human hodgepodge of so much variety of hands, feet, and heads" (*absorto en aquella pepitoria humana de tanta diversidad de manos, pies y cabezas*) (21). This "stew" establishes an imaginary parallel with the "hodgepodge" of human flesh that is packed into Vélez de Guevara's portrait of Madrid. For example, at the beginning of chapter three, the city awakens suffocated by the summer heat:

> Now men and women were beginning to boil up within the human stew of the court, some bubbling up, others down, and others across. They crossed each other to the tune of their own confusion, and Madrid's ocean seemed filled with four-wheeled whales, which are also known as *coaches*, engaging in their daily battle, each one with a different design and matter to attend to. They tried to fool one another, raising a dust cloud of cons and lies that prevented a single eye from uncovering the least bit of truth.
>
> *Ya comenzaban en el puchero humano de la corte a hervir hombres y mujeres, unos hacia arriba y otros hacia abajo y otros de través, haciendo un cruzado al son de su misma confusión, y el piélago racional de Madrid a sembrarse de ballenas con ruedas, que por otro nombre llaman coches, trabándose la batalla del día, cada uno con disinio y negocio diferente, y pretendiéndose engañar los unos a los otros, levantándose una polvareda de embustes y mentiras que no se descubría una tizna de verdad por un ojo de la cara.* (33)

Through this evidently critical and bitter tone, Madrid's "ocean" also announces a saturated urban landscape in which the configuration of new and strange elements (as in the case of the coaches) constructs an image that is nevertheless tremendously familiar: "Each day . . . there are new things in the court" (*cada día* [. . .] *hay nuevas cosas en la corte*) (34) claims the surprised Cleofás. The sense of novelty, as I have already indicated, permeates this image of the urban; however, different from other, more benevolent visions of Madrid, Vélez de Guevara endows this dynamic of additions and supplements with a transcendent sense of excess that results in a very pessimistic vision of the human condition. Disorder and the new capacities and infrastructures required by courtly life construct a landscape at the breaking point, a stew pot whose lid scarcely contains the bubbling of disparate and conflict ingredients. The long string of carriages that inundate the Calle Mayor in the text's seventh chapter are depicted "as many figures as different roles represented in the Theater of the World" (*tantas figuras como en aquel Teatro del Mundo iban representando papeles diferentes*) (94), and sometimes comically portrayed as "four-wheeled whales" (*ballenas con ruedas*) (27) that scatter dust throughout the city due to the maneuvers of their unscrupulous drivers:

> That married couple, so in love with their coach that all that they should have spent on clothes, shoes, and setting up their house had been used on it, even though they now had no horses to draw it; and they eat and sup and sleep inside of it without ever having left their reclusion—not even for their corporal necessities—in the four years since they bought it; they are encroached and sandwiched, and it has been so their habit not to leave it that the coach serves them as the shell serves turtles and giant tortoises, who when caught with their head stretched outside the shell place it back inside as one who sees exposure as something unnatural, and they catch cold even placing their foot, leg, or hand outside this very strict religion.

> *Aquel marido y mujer, tan amigos de coche, que todo lo que habían de gastar en vestir, calzar y componer su casa lo han empleado en aquel que está sin caballos ahora, y comen y cenan y duermen dentro dél, sin que hayan salido de su reclusión—ni para las necesidades corporales—en cuatro años que ha que le compraron; que están encochados, como emparedados, y ha sido tanta la costumbre de no salir dél, que les sirve el coche de conchas como a la tortuga y al galápago, que en tarascando cualquiera dellos la cabeza fuera dél, la vuelven a meter luego como quien la tiene fuera de su natural, y se resfrían y acatarran en sacando pie, pierna o mano desta estrecha religion.* (27)

The expression "boiling pot of people" manifests itself through a description of the suffocating heat of summer: "near the end of July in Madrid, when the clock strikes eleven at night, an unfortunate time for the streets because, since there is no moon, they become the jurisdiction and terminus for all manner of stupid flirtation and deadly triviality" (*daban en Madrid, por los fines de julio, las once de la noche en punto, hora menguada para las calles y, por faltar la luna, juridición y término redondo de todo requiebro lechuzo y patarata de la muerte*) (11). Playing with the bivalence of the word *río* ("river" and "I laugh"), Vélez de Guevara ridicules the Manzanares River and its bathers following an aesthetic employed by his contemporaries Quevedo and Luis de Góngora: "it's called a river because we rib those who bathe in it, since there is no water, and they are only bathed by sand; and on dark summer nights it passes like a counterfeit coin, being among the most visited rivers in the world" (*se llama río porque se ríe de los que van a bañarse en él, no teniendo agua, que solamente tiene regada la arena, y pasa el verano de noche, como río navarrisco, siendo el más merendado y cenado de cuantos ríos hay en el mundo*) (104). To the dirty waters of this poor river Vélez adds the humorous mention of the waters that are tossed into the street at night, dirtying the streets of the city with feces and waste. What emerges is an image of strange and grotesque domesticity that well could be the only one in existence, given that all the rest is pretension and lies. And behind all this humor, we find some unmistakably familiar targets: the specious nature of appearances, the perversity of ambition, and the prevalence of untrammeled hypocrisy. Anyone familiar with the carnivalesque in Renaissance Europe will know that the boiling pot of people to which Vélez de Guevara refers is in reality an invitation to read Madrid in summer as a kind of insane asylum, a locus for an eminently theatrical form of madness expressed through jingles, guitars, bagpipes, and other instruments mentioned in the novel's third chapter. It also finds expression through the *arbitristas*, who are "the most dangerous madmen for the Republic" (*que son los locos más perjudiciales para la república*) (38) in this "tableau of grief" (*retablo de duelos*) (39) that so surprises Cleofás. The activity of the human gossip mill on the steps of the Augustine Convent of San Felipe, the fruit market near the Convent of La Victoria, and the fountain of the Buen Suceso, praised for the beauty of its lapis lazuli and alabaster, all manage to reinforce the sense of spectacle on which Vélez de Guevara focuses the Baroque lens of his flying tandem.

However, the act of uncovering carried out by the diabolic lens eventually ceases to have a significant role; it is taken over by the optical play that is produced by the manipulation of (and by) crystal. Of great interest, for example, is the survey that the protagonists make on the Calle de los Espejos, or "street of mirrors," where the handsome and pretentious gesticulators submit their ridiculousness to a truly disorienting multiplication of images. The vision of the devil is, it follows, an expressive and structural means characterized by pessimism and bitterness, perhaps expressing the feelings of disillusionment felt by Vélez de Guevara with respect to the court in Madrid. It is also worth noting that the perspective of the devil is nocturnal in that it precedes and rivals that of the sun, a fact that permits him to move beyond mere appearances and enter into the obscurity of human nature through skylights and chimneys.

The novel also introduces a witch who goes up to the roof patio of one of the buildings on the Calle Mayor. Once there, Cleofás points out to her the parade of vanities occurring below; this vantage point is of course anything but gratuitous, since learned readers of the period commonly understood patios to be the spaces in which old women accused of witchcraft would gather. And from the tower of the church of San Salvador—

"Madrid's greatest tower" (*mayor atalaya de Madrid*) (20)—situated near the Calle Mayor and Plaza Mayor, Cleofás will see what takes place in the city because Cojuelo, through his dark arts, will raise the roofs of the houses of what Enríquez Gómez had called a "Spanish Babylon, that in confusion was equal to the other city with this name" (*Babilonia española, que en la confusión fue esotra con ella segunda deste nombre*) (20). There are, in fact, similarities of a personal sort that go beyond the literary, as Blanca Periñán has reminded us: "*Cojuelo*, closer to the scheme of the guided journey, presents the most detestable figures that operate in the court, and subliminally speaks, in an affable manner, from a diverse perspective that corresponds to the actual experience of the converso Vélez de Guevara" (*El Cojuelo, más cercano al esquema del viaje con guía, presenta las más detestables figuras que pululan en la corte, y subliminalmente habla, de manera afable, desde una diversidad que es la de las vivencias del Vélez converso*) (xvi). If there indeed exists a converso quality in social critique, this seems to be shared with other authors, such as Antonio Enríquez Gómez, examined in the present book.

El diablo Cojuelo presents us with a ferocious satire of the empty and pretentious urban character. The demonization of humanity to which the devil alludes when he tells Cleofás that "only you, because you're up so high, are safe from this demon that in a certain way is more of a devil than I am" (*que solamente tú, por estar tan alto, estás seguro deste demonio, que en algún modo lo es más que yo*) (32) crowns a gallery of types and customs that are examined with all of their miseries. In this way, in the final passages of the second chapter, a whole series of nocturnal characters passes before the eyes of the devil, who dissects them mercilessly in accordance with the canons of the genre, with their influences and subtexts, and with the folkloric tradition assumed by Vélez de Guevara. The principle *de turpido et deformitas* for laughter, the register *de ridiculis* shared with many other writers of the period, is fundamental for this "invention" of Madrid. As Periñán puts it, "the dynamic of its interior orientations maintains it within the structures of comicality to offer a new reflection on the world," through a "maieutics in reverse according to which a devil instructs a young man who does not see the deeper truths of the world, revealing to him the principle of reality" (*la dinámica de sus orientaciones interiores lo mantiene en las estructuras de la comicidad para llevar adelante su nueva reflexión sobre el mundo* [...] *a través de una "mayéutica al revés en la que un diablo irá enseñando a mirar a un joven que no ve las verdades profundas del mundo, revelándole el principio de realidad*) (13). The image of the deaccessorizing nobleman is telling in this respect, as he dismantles himself behind the scenes in a space to which only the devil and Cleofás have access.

✻

I wish to end this section with a brief but necessary mention of how early seventeenth-century astronomy finds its way into *El diablo Cojuelo*, completing in this way what I have already suggested regarding the possible relation between Vélez de Guevara and Espina. *El diablo Cojuelo* carries even further the expressive liberty of seventeenth-century satire by including certain opinions on astronomy from the point of view, we might say, of the aficionado, somebody still skeptical about what is occurring around him. We find ourselves faced with "a revisitation of the quest, an exploration of the folds of the world in a voyage of initiation" (*una revisitación de la* quête, *una exploración de las dobleces del mundo en un viaje iniciático*) (xiii), as Periñán has suggested. The sense of sight is once again the basis for all knowledge: the text is an exercise in dioptrics—the branch of geometry placed in charge of the formation of new images through lenses—since it permits the reader to analyze certain courtly social practices through a symbolic prism that conjures up an image only slightly different from that of earlier texts. Mirrors, lenses, telescopes, and even the retina of the human eye are converted into an essential part of the narrative when the two protagonists reflect on the unreliable nature of society in Madrid from the intersection of its two most festering sores: the appetite for capital or social distinction and the pseudoscientific practice of the *arbitristas*. The first phenomenon manifests itself in the aforementioned Calle de los Gestos, a narrow avenue in chapter 3 that is adorned with mirrors on both sides, and in which passersby see their own distorted figures, "becoming boogeymen to themselves" (*haciéndose cocos a ellos mismos*). When the student asks the devil what it is they are doing, the latter responds:

> This is the Street of Gestures, and only people from the court know of it; they come here to act as though they must go out that very day, and they walk around with a kind of palsy of beauty—some with swollen lips, and others with their eyes all sleepy, snoring beauty—and every one of them with their index fingers and pinkies raised along with other affectations of *Gloria Patri*.
>
> *Esta se llama la calle de los Gestos, que solamente salen a ella estas figuras de la baraja de la corte, que vienen aquí a tomar el gesto con que han de andar aquel día y salen con perlesía de lindeza, unos con la boquita de riñón, otros con los ojitos dormidos, roncando hermosura, y todos con los dos dedos de las manos índice y meñique levantados, y esotros de* Gloria Patri. (33–34)

The second case arrives halfway through the novel, when the two flying travelers, tired of "making like birds" (*pajarear*) fall into a meadow just outside of Écija (near Seville) to contemplate the starry nighttime sky. The conversation that they strike up is littered with technical terms that were known components of a certain scientific lexicon, already exploited by Cervantes, that readers in 1641 would inevitably considered affected. Cojuelo, who had been shut up in a flask in the home of a "belching astrologer and grafted necromancer" (*astrólogo regoldano y nigromante enjerto*) (43) until he was rescued by his friend, responds that the necromancers "sell imagination as truth" (*venden imaginaciones por verdades*) (75) and that astrology "wheedles" (*hace garatusas*) (76). We already know that there were seven heavens according to Muslim and Jewish natural philosophy, eight according to Aristotle, nine in Ptolemy, and up to eleven in Scholastic theology. The Scholastic theological framework, which was by far the most orthodox one during the early modern period, situated the firmament of the stars in the eleventh heaven. The imagined colures about which Vélez de Guevara speaks are the two imaginary circumferences that vertically divided the celestial sphere into four segments (*gajos*) that passed over the poles and equinoxes (equinoctial colure), and the poles and solstices (solsticial colure). According to the geocentric theory, as we also already know, the planets describe a double orbit: one around the earth, and the other—epicyclic—around the determined point of the first orbit. In this way, early astronomers explained the observed movements that did not fit with a simple orbit around the earth and which stemmed from, in reality, a heliocentric rather than geocentric design. Only certain heavens moved in the Christian-Ptolemaic system: the crystalline heaven (the ninth one) moved in a way that was almost imperceptible, trembling (trepidatious) and served to move the inferior spheres; the tenth heaven, the primum mobile, which was as light as human thought, completed its circuit in twenty-four hours. The eleventh heaven, the empyrean, remained immobile.

Vélez de Guevara displays a precise knowledge of the geocentric cosmogony that he inherited, but the doubt that he manifests in light of the change in paradigm indicates that he is not completely closed to the new; put another way, the "truth" to which he alludes when he criticizes astrology is as much the ancient one as that which had just been discovered. This is just one of many dialogues in *El diablo Cojuelo* that clearly manifest Vélez de Guevara's interest in astronomy, which he converts into narrative material without taking a position in favor of or against it—after all, he presents only the opinions of the devil. The game is carried out along three axes: the student observes, Cojuelo interprets, and the implicit author summarizes his cosmology; this places the reader in the role of judge who must determine, thanks to this generous maieutics, which side he or she is to take. From this narrative distance Vélez de Guevara's achievement is not only thematic but also formal: on the one hand, he manages simultaneously to ridicule and to elevate practices and beliefs related to identical aspects of what is still not completely understood, and which he takes on loan for his narrative; on the other hand, he successfully manages to construct a new satirical lexicon in which fantasy and hyperrealism harmoniously coexist. The final product is doubly meritorious: an idiosyncratic investigation of supralunar reality without leaving aside a critique of the earthbound landscape that surrounds him. As Periñán has indicated, his investigation effectively mixes a series of inherited elements that make it extraordinarily sophisticated:

> Vélez de Guevara reelaborates the cardinal principle according to which laughter is based in *turpitudo* and *deformitas*; he also observes that the humorous gains power when things surprisingly end up the opposite of what was expected, through the confounding of expectations. This illusion produced by recognition is in comedy the *pendent* of tragic vicissitude, requiring mediocre characters and style. Since ugliness and deformity can reside as much in the body is in the soul, and in that which is extrinsic to both, to this latter category is directed the attention of the comic writer, above all when reaching the triumphant moment of admiration. It is explicitly theorized that if to the *turpis* is added singularity and astonishment, one obtains a greater comic effect.

> *Se reelabora el principio cardinal según el cual la risa se basa en* turpitudo *y* deformitas, *observando además que lo risible resulta potenciado cuando las cosas salen, sorprendentemente, al revés de lo esperado, por medio de la desilusión de las expectativas; ese engaño producido por el reconocimiento, es en la comedia el* pendent *de la peripecia trágica, requiriendo personajes y estilo mediocres. Pudiendo residir la fealdad y deformidad tanto en el cuerpo como en el alma, y en lo extrínseco, a esta última categoría va la atención de la escritura cómica, sobre todo al sumársele la triunfante instancia de la* admiratio. *De manera explícita se teoriza que si al turpis se le agrega singularidad y estupor, se obtiene mayor efecto en la comicidad.* (x)

This is the epistemological perspective that this type of narrative assumes, which responds to a reality in which the sense of sight finds itself, we might say, threatened. The availability of eyeglasses and telescopes is a double-edged sword for these writers who, following the teachings of scholars such as Daza de Valdés, expound on the utility of glasses but also on the dangers that come with them. In Andrés de Dávila y Heredia's *Tienda de anteojos políticos,* for example, we will see how another devil complains about the pomp of the court. Equally vain are our two protagonists in *El diablo Cojuelo,* at least on two occasions: in chapter 9, Cojuelo gives his companion a pair of stolen eyeglasses "with guitar strings for the ears" (*con sus cuerdas de guitarra para las orejas*) (106) so that the student might seem more learned to the people they meet; in chapter 10 they participate in a literary academy and bring the fake eyeglasses to a reading in which Cleofás impresses those in attendance with his poetry. It is a caricature of gestures that also takes up the tradition according to which mirrors "of liquid mercury" (*de alinde*) attracted demons, thus giving sense to Cojuelo's particular construction.

The artful conversation demonstrated by Cojuelo, which Vélez de Guevara practices with such mastery, is purely traditional. It recuperates the famous cynic Menippus of Gadara, who was a principal character in many dialogues written by Lucian of Samosata, including the *Icaromenippus.* In this text, the reader observes the earth from the moon and encounters a biting satire of vain and ridiculous human interests. Cleofás's curiosity calls to mind this tradition of Menippean satire, in which a character tells another the truths of the universe, contrasting these with the foolish conclusions of scientists. What ends up being in a certain way novel is Vélez de Guevara's fascination with post-Copernican science: in speaking about what he has seen from above, the devil plays with the words *anteojo* and *antojo.* When the student hears Cojuelo for the first time, he is "leafing through the writings of Euclid and deceits of Copernicus" (*papeleando los memoriales de Euclides y embelecos de Copérnico*) (43). We have already seen, when thinking about Juan de Espina, how Vélez de Guevara recounted with extreme humor some of the recent discoveries, praising both Galileo and Espina in a very memorable phrase: "Galileo's telescope, as well as that of the great Juan de Espina, whose famous house and rotating chair are ideas stemming from his rare genius" (*el antojo del Galileo y del gran don Juan de Espina, cuya célebre casa y peregrina silla son ideas de su raro ingenio*) (75–76). Nevertheless, the second part of Cojuelo's judgment is a bit more ambiguous. Vélez de Guevara, master of the conceit, writes about "the optics of those capricious, telescope-assisted gentlemen, who have discovered a mole on the left side of the sun and have made out mountains and valleys on the moon, and they have also seen horns on Venus" (*la óbtica de esos señores antojadizos, que han descubierto al Sol un lunar en el lado izquierdo, y en la Luna han linceado montes y valles, y han visto a Venus cornuta*) (76). This is a clear allusion to the corruptibility of the sun, with solar stains presented as moles, to the existence of mountain ranges on the moon and the extensions of the Venusian cusps, which was exactly what Galileo maintained in his *Sidereus nuncius*. The term *lincear* ("to see far away") refers also to the Academy of Lincei, which had by then become a very controversial institution. The virtuosity of Vélez de Guevara permits him to express opposing ideas in a single phrase: on the one hand, he feels admiration for Galileo's telescope (*antojo*), expressed through a clear homage to the polemical *Sidereus nuncius*'s exposition on Venus and the sun; on the other hand, it argues against judiciary astrology, which is reduced here to a mere infantile caprice (*antojo*) in which planets such as Venus are caricatured and converted into faces (*rostros*) about which certain heretical astrologers could write their fantasies. A man of his time, Vélez de Guevara is simultaneously curious like the young student Cleofás and prudent like the old Cojuelo. But the Spain in which he operates is one of fear and suspicion, and through a Menippean veil, there moves through the narrative a particularly Baroque form of disillusionment: given that the word *antojo* is applied here at once to "eyeglasses" and "telescope" (Spanish, as I have already said, will not incorporate this word *telescopio* for another century), the ability to see the beyond acquires extremely disastrous features. *El diablo Cojuelo*, which questions the existence of newly discovered constellations, defends the idea that astronomy is an invention of the insane and that the telescope is as useless as the vain courtier's eyeglasses: both are for him "capricious" (*antojadizos*), for him the most censurable conduct. The political lens, Boccalini's *occhio linceo*, has essentially been turned against itself as a scientific instrument on the tight rope that Vélez de Guevara must walk in order not to offend his censors. His universe of glass is one open to many interpretations, as many as the effects produced by manipulated crystal itself. It is worth remembering, as a way of closing the present section, that the philosopher Francisco Sánchez, who was opposed to Aristotelianism and a supporter of the direct examination of reality by means of data culled from experience, exclaimed sixty years later in his *De multum nobili et prima universali scientia* (On the very noble and first universal science, 1581): "What can mirrors not invent!" (24). In the case of *El diablo Cojuelo*, mirrors invent an entire novel.

Dream/vigil: Moons, moles, and lunatics in the poetry of the Baroque

I have already pointed out that dreams were one of the most exploited literary motifs in early modern Spanish literature. In Baroque fiction, dreams are extraordinarily common, whether as a narrative frame or as a central aspect of a particular story. The new realities that lay beyond wakefulness delighted seventeenth-century readers, given their implementation of the free and the marvelous, the ridiculous and the tragic, the perverse and extraordinarily precise. This is demonstrated, to provide just two well-known cases, in the *Sueños* (Dreams, 1627) of Francisco de Quevedo and Don Quixote's dream in the Cave of Montesinos.[6] This highly animated and entertaining dream, so confusing and yet clear, was in fact a literary device that circulated at the same time as that of the magical or political eyeglasses, insofar as it served as a filter and point of access to truth, to the true reality and not simply to a scene or landscape. Dreams and glass came to be mechanisms for fiction that were in continuous dialogue with other literary devices developed during the sixteenth and seventeenth centuries, such as the famous university lampoons and parodies, either in zones of contact or embedded within broader narrative frames. Teresa Gómez Trueba has depicted these narrative dreams as a category that "is defined principally by certain semantic elements that do not necessarily exclude the presence of other elements considered particular to other genres, such as imaginary journeys, utopias, Lucianesque dialogues, Menippean satires, discourses, sermons, etc." (*que se define principalmente por ciertos elementos semánticos, que no excluyen necesariamente la presencia de otros elementos que se consideran propios de otros géneros, como los viajes imaginarios, las utopías, los diálogos lucianescos o sátiras menipeas, los discursos, los sermones, etc.*) (16).[7] This oneiric morphology incorporates certain features related with the use of glass, not only in narrative but also in poetry, and to a lesser extent in theater. And with these dreams, extraterrestrial space takes on a seminal role, especially other planets or satellites such as Mars and the earth's moon. From the Middle Ages onward, literary dreams had made ample use of the voyage to the moon, perhaps emulating the *True Fictions* of Lucian that, like his *Icaromenippus,* had so pleased its readers with fascinating lunar landings.[8] The "lunatic" was forged as a literary personage in these inhabited surfaces, the delirious space imbued with parallel realities, which on many occasions was likened to the bad poet: the tumult of the lunar landscape, its lack of evenness (*llaneza*), mirrored the *culteranista* (highly learned and challenging) verse criticized in these pieces. One could articulate a kind of inverted Parnassus in which the supposed ruggedness of the moon was converted into a symptom of the lack of common sense of its inhabitants. As an example, we might take the utopian dream of Juan de Maldonado (1532), which provides an interesting description of the lunar society that allows readers to reflect upon life on earth.[9] This cosmic fantasy was linked to a tradition no less seductive that had taken off many decades earlier, that is, that of the *lunarios,* who during the sixteenth century had produced up to thirty-four titles in Spanish, which reached a total of roughly 180 editions. These texts contained all manner of prediction; especially important were the works of Bernat de Granollach (1551), Joan Alemany (ca. 1565), Rodrigo Zamorano (1585), and Jerónimo Cortés (1594), which tended to mix meteorological questions with medical ones.

This "Galilean moment," understood as a historical time of controversies and condemnations that defines the first third of the seventeenth century, produced in Spain a text that reveals the growing interest in questions related to science: the *Vejamen que el poeta Anastasio Pantaleón de Ribera dio en la insigne Academia de Madrid* (Lampoon that the poet Anastasio Pantaleón de Ribera gave in the illustrious Academy of Madrid, 1634).[10] The reading of Pantaleón de Ribera, given at the academy in 1626—the poem was then published five years after his death—reveals from its beginning a notable familiarity with certain proposals by Kepler and Galileo, whom he obviously does not cite by name but from whom he certainly takes the particular construction of the space that he presents. He enters with extraordinary caution, for example, in the then-current debate on the possibility that the moon was inhabited, a debate that had already been mentioned by writers such as Angelo Poliziano, Giovanni Pico della Mirandola, and Justus Lipsius, and which had also been dealt with in Kepler's *Somnium* (Dream, 1634). The verse that opens the text, in fact, begins in the following way: "the surface of the moon is inhabitable" (*que el cuerpo de la Luna es habitable*) (11). I agree with Frederick A. de Armas that Pantaleón de Ribera's *Vejamen* "stands apart from other such incidental works of the period in that it responds in part to a new interest in cosmology, one that grew out of Galileo's and Kepler's discoveries during the first third of the seventeenth century" ("Maculate" 59). The initial allusion to Pythagoras, which appears just afterward, comes veiled behind a joke (*disparate*) that serves to confirm a covert defense of heliocentrism exactly like that defended by Galileo and Kepler, as well as by Diego de Zúñiga in his audacious *In Job commentaria* (Commentary on Job).[11]

Pantaleón de Ribera's moon is not only corrupt and rough but also dotted with cities. To elude censorship, however, he presents all of this as an error, as a moral and civic joke, and to mark out even more distance, Pantaleón de Ribera frames the story within a (Keplerian) dream, a dream that ultimately facilitates free observation and expression. In effect, if in Kepler's *Somnium* the narrator falls asleep reading Central European folklore, Pantaleón de Ribera's alter ego will do so—as a kind of homage to the famous scientist—reading lunar treatises. Such a humorous journey fashions a celestial Parnassus that mixes hilarity and social critique in rapid verbal bursts made possible by the visits that the protagonist makes to different contemporary poets: Alonso de Oviedo (as Don Lucido); Jacinto de Aguilar (as Don Zafiro); José Camerino (as Don Carinemo), "Italian poet and confirmed madman" (*poeta italiano y confirmado loco*) (21) José Pellicer; Diego de Silva; Juan de la Barreda; Alonso de Castillo Solórzano; Pedro Méndez; Gabriel del Corral; Gabriel Bocángel; Nicolás de Prada; and Pantaleón de Ribera himself complete the Parnassus in this sidereal journey in which the portraits are so outlandish that what remains is the sensorial effect of the landscape. As a commentary on the literary field of the time, the *Vejamen* scarcely resembles other Parnassi of the period. As had already occurred in Salas Barbadillo the figure of the misanthrope appears various times; for example, when introducing Don Zafiro "one never saw light in his room, and he would disrobe by memory and sleep in the choir stall" (*jamás se vio luz en su alcoba, y con todo se descalza de retentiva, se desnuda de reminiscencia, y se acuesta de coro*) (19), and Don Gelcambo, "crazy with no equal" (*loco sin igual*) who "by the light of a candle (because it was a bit gloomy) used a lump of coal to paint what he intended to be a human head, although it really seemed to be a work in progress" (*a la luz de un candil (por ser algo lóbrego) pintaba con un carbón una cabeza que quiso ser de hombre, y parecía de proceso*) (37).

What most stands out in the *Vejamen* are its strong Galilean echoes. For example, when referring to the moon, Pantaleón de Ribera indicates that "the stains that mar and darken the splendor of this planet were cities, mountains, and rivers, like those of this inferior world where we live" (*las manchas que afean y oscurecen el esplendor de este Planeta eran ciudades, montes y ríos, como los de este mundo inferior que vivimos*) (13). Ribera is here reproducing passages from Galileo's *Sidereus nuncius*, just as the notary Benito Daza de Valdés had done some years before. He also indicates that the sun is a burning mass and, as such, is corruptible. In one passage, Don Zafiro states: "I stitched together the entire sphere / of Mars, which is corruptible / although physicists may be / of the opposite opinion" (*Yo zurzí la esfera toda / de Marte, que es corruptible / aunque la opinión contraria / físicos tantos afirmen*) (20).

Within the *Vejamen*'s cosmological vision, the telescope plays a fundamental mediating role: the author himself admits, upon awaking from his dream, that he has: "so thoroughly affixed . . . a spyglass to his left eye that to come back without it he had been forced to remove his hand, like a man returning from war" (*pegado [. . .] al ojo izquierdo un anteojo, tan embarazosamente que, por traerle solo, tenía una mano menos, como quien viene de la guerra*) (20). The image brings us once again to the idea that I have discussed earlier in the present book, that is, that the telescope is understood to be an extension of the human body, in this case "affixed" to the left eye as if it were a prosthetic. Pantaleón de Ribera essentially converts the physical incapacity of this figure, a thinly veiled Cervantine echo, into an advantage; and once again a Spanish author plays with the bivalence of the word *anteojo*, which could well refer to a telescope that for the poet allows one to discern the arrival of a new science. Is the *Vejamen* merely an invitation to observe the firmament through a telescope? Is it a vision with *anteojo* or simply the narration of a dream journey? Although the text ends with the phrase "may this be taken in jest" (*esto se haya dicho en burla*) (44), it is certain that the poetic narrative that Pantaleón de Ribera presents remains eclipsed—an apt image—by the weight that cosmology has in the text; a vision of the universe mediated by an *anteojo* that, as I indicated at the beginning of this chapter, plays a kind of game of telling without telling, of making the new known through false condemnation.

I would like to end this survey on fictional dreams and voyages with a piece that has received little scholarly attention but that I consider of utmost importance in the development of this particular genre. Juan Enríquez de Zúñiga's *Amor con vista* (Love with vision, 1625), the subtitle of which reads, *Lleva una sumaria descripción del mundo, así de la elemental como de la etérea* (Containing a summary description of the world, both the elemental and the ethereal), manages to blend the elements explored in the previous pieces with new ones like the pastoral setting.[12] If Fernández de Ribera focused his authorial lens on Seville, Enríquez de Zúñiga is concerned with a Madrid in which "everything is found with money" (*con dineros todo se halla*). *Amor con vista* is not, however, a typical Menippean satire; rather, it is a hybrid text that combines idyllic characters with social critique, joined through the image of the academic shepherd cultivated a century before. Enríquez de Zúñiga discusses here the nature of love and, as the title indicates, its relation to passion and reason: "Love is not blind; rather, it wears a blindfold. And in this way a lover, as enamored as he may be, can take off this blindfold and love will have sight, and its understanding will be free and not subject, as it is the king of our powers, to the will. . . . It is not difficult to love with consideration and discourse" (*No es estar el amor ciego, sino tener vendados los ojos. Y así un enamorado, por más que lo esté, puede quitarse esta venda, con que quedará el Amor con vista, libre digo el entendimiento, no sujeto, pues es rey de las potencias, a la voluntad* [. . .] *No es dificultoso amar con consideración y discurso*). By reflecting on passion and reason, these lines connect with something that Diego de Saavedra Fajardo, as we will see soon, will apply to effect in his advice to the monarch in his emblem (*affectibus crescunt, decrescunt*): the changing nature of emotions that increase and diminish as if they were observed through a telescope.

Amor con vista, which ends up with an act of amorous clairvoyance through reason, begins with the ardent complaints of the shepherd Felicio. He speaks to his friend Albano of the shepherdess Faustina, who has promised to favor him as soon as she can confirm that her beloved Eusebio—a rich local with scientific knowledge—is in love with a mysterious lady who has recently arrived to the community of shepherds. The woman in question is called Potenciana, and she narrates a series of stories to which are added those of another outsider who, coming across the scene somewhat suddenly, recounts another string of apparently unconnected narratives that end up being pivotal for the denouement of the work. Carlos Vaíllo has pointed out very convincingly the total absence of amorous Neoplatonism in these "newly coined" shepherds, and it is true that the most complex and interesting of them is Dionisio who, similar to Cervantes's Grisóstomo, lives between the pastoral village and the University of Alcalá de Henares, where he has studied various scientific disciplines.

Inspired, as with other texts that we have previously analyzed, by Lucian's *Icaromenippus*, Enríquez de Zúñiga's *Amor con vista* enters into oneiric territory on folio 40r. The main character, Dionisio, is snatched up by the eagle of Jupiter, and he ascends through eight of the eleven spheres that make up a cosmic vision still based on Ptolemy. In each stage of the journey, he distinguishes the four natural elements, he measures distances and dimensions, he speaks of meteorological phenomena, and he engages in dialogue with the deity that represents each planet. The sky, as Bartolomé Leonardo de Argensola reminds us in his famous sonnet, "is no longer blue" (*ya no es azul*) (44r), and what is observed from above is that "the entire populated and unpopulated world is full of thieves" (*todo el mundo, poblado y despoblado, está lleno de ladrones*) (75r). So that readers might protect themselves from so many miseries, the text offers a good deal of teaching that, while nothing new, is very detailed. The attacks are, as always, of a varied nature, and the targets include the verbal excess of poets; *culteranista* and impenetrable poetry; and the authors of comedies, although—as in many other cases—the great Lope de Vega emerges unscathed. Upon arriving on the moon, the traveler finds a "sumptuous palace and a majestic fortress" (*Palacio suntuoso y un Alcázar soberbio*). From here, he ascends to the second sphere, where Mercury offers him a panoramic view of the earth that features a very complete and highly satirical review of the period's social types. Here the earth is nothing but "an anthill" (*un hormiguero*) (50r) containing a collection of motley cities, such as the Madrid seen already in Vélez de Guevara.

After Mercury, Enríquez de Zúñiga's celestial explorer arrives at the third sphere (Venus) before going on to the fourth (the sun). At the sun, he encounters Apollo, who shows him the Ptolemaic "trajectory that it follows each day" (*camino que cada día cursa*) (52r). The explorer then visits the fifth sphere, where he encounters the royal fortress of Mars, before going on to the sixth (Saturn). Vaíllo has written very convincingly that Enríquez de Zúñiga is principally concerned with providing his reader a "scientific construction . . . endowed with allegorico-moral value in order to perhaps gain the benevolence of his readers" and relegates aesthetic and conventionally "literary" considerations to the margins (*construcción científica* [. . .] *dotada de valor alegóricomoral, de la que espera tal vez principalmente el reconocimiento de los lectores y que empuja hacia los márgenes los factores literarios del entretenimiento*) (394). Some of *Amor con vista*'s moral judgments are typical of the period, as when the traveler observes that "honor is like glass in that it is pure, clear, and transparent; but it is also fragile, and it will break with the first blow" (*la honra tiene la calidad de vidrio, que es pura, clara y transparente; pero también su fragilidad, pues al primer golpe se quiebra*) (83v). *Amor con vista* is, generally speaking, a text with only relative importance for the Spanish literary canon; however, it is significant insofar as it reveals to us that there are Spanish authors, even as late as the 1630s, who continue to produce the Ptolemaic model in their descriptions of celestial morphology.

Fictional journeys define, to a great extent, Spanish satire from the first half of the seventeenth century. But next to any serious commentary, one will also henceforth find a continuous attack—often through jokes and parody—on judiciary astrology. A good example of such an interweaving of dreams and pseudoscience is the *Universidad del amor y escuelas del interés: Verdades soñadas, o sueño verdadero. Al pedir de las mujeres* (University of love and schools of interest: Dreamed truths, or a true dream. At the request of women, 1636), written by Fray Benito Ruiz (better known as *maestro* Antolínez de Piedrabuena). Ruiz's "university of love" boasts a classroom for jurisprudence, mathematics, arithmetic, and astrology (*sala de la Jurisprudencia, la Matemática, la Aritmética y la Astrología*) where students read a *Repertorio y pronóstico general de los sucesos del Amor, para todo el género de gentes* (Repertoire and general prediction on the happenings of love, for all sorts of people) supposedly put together by Cupid. Other later texts, such as Francisco Santos's *El no importa de España* (Why Spain doesn't matter, 1667) also combine dreams, mirrors, and magical lenses in almost equal measure. It is not surprising, therefore, that in the eighteenth century there would appear a series of pieces in which this curiosity becomes the absolute center of the narrative, such as Torres Villarroel's *Anatomía de todo lo visible e invisible* (Anatomy of all that is visible and invisible, 1738) and *Sueños morales: Visiones y visitas de Torres con don Francisco de Quevedo por la Corte* (The visions and visits of Torres with Don Francisco de Quevedo to the Court, 1743), Lorenzo Hervás y Panduro's *Viaje estático al mundo planetario* (Stationary journey to the planetary world, 1793–1794), or even more important, Manuel Antonio Ramírez de Góngora's *Óptica del cortejo* (Optics of the courtship, 1774). In this last work, a dream brings the author to a mysterious palace in which he finds a machine located in an "optics room" (*sala de óptica*) that allows him to see everything. This hall is a beautiful room containing an enormous mirror before which sits a microscope and, between them, a great number of frames that directly project figures onto the mirror, which are then reflected in the microscope; the result, in theory, is a much more precise view of reality. Both the *Óptica del cortejo* and Alfonso de la Torre's *Visión deleitable* contain an element found in the texts that I consider in the following chapter, namely, the presentation of Understanding (*Entendimiento*) as an allegorical personage, a solid link to sanity in an otherwise wholly fantastic dream landscape.

IV

The refracted muse

✷ 7 ✷
Interventions

Up to this point, I have focused primarily on the tensions that emerged between Florence and Rome during the early modern period as a result of advances in modern science; as part of this focus, I have also examined the figure of Galileo, which was in many ways the epicenter of the debate regarding these advances. I have also argued that the instrumentalization of Galileo, promoted in part by Galileo himself, morphed very early on into a matter of state and then later, with the commodification of the telescope, into a form of transnational innovation. In a sense, Galileo and his telescope became icons of knowledge and were incorporated into the collective imaginary through citations and direct manifestations in early modern art and literature. As we have also seen, some of the greatest European fortunes played a central role in the manufacture, diffusion, and reception of the telescope; but this is true also of small nuclei of intellectuals who knew how to make the most of its many applications. If Galileo's device found itself transformed into an instrument of scientific inquiry, it also became an object of exchange, a product, and a symptom of the new spirit of the time. And even glass itself, along with the mysteries hidden between the lenses of Galileo's optic tube, came to form the center of numerous debates related to progress, the power of vision, the nature of appearances, and the vanity of urban society during the seventeenth century.

Glass would also come to connote new geopolitical meanings on the diplomatic scene. I have already pointed out the important role played by the rivalry between Venice and Madrid at a time when Spanish authors were busily proposing models of ideal government and codes of conduct meant to limit abuses of power by the monarch and his ministers. I have also alluded to the power wielded by the scientific culture of Tuscany, from which emerged numerous innovations and advances in knowledge found in the writings of its leading scholars and fiction writers. From all this brisk activity emerge written testimonies, such as those by Traiano Boccalini and Tomaso Garzoni, that make use of glass as a literary motif to construct a poetics that revolves around a number of pressing issues related to the exercise of power, the limits of the human intellect, and the perplexing traps of vanity, to name but three. These issues gave rise in Spain to written reflections by Francisco de Quevedo and Diego de Saavedra Fajardo (both of whom enjoyed intense and fruitful relations with Italy) that can rightly be counted among the most significant of the period. In the present chapter, I examine specific (and very different) texts by these two moralists, with international affairs and the politicization of the telescope as a unifying theme. At stake in these texts, as we shall see, is a fascinating case of political engagement in which the personal voice of each of these writers, both authors veterans of numerous diplomatic missions, warns the reader of the telescope's great reach—a capability that manifests itself not only on the material plane but also through its immense symbolic fecundity in discussions of both the power of the image and the image of power.[1]

Political intervention I: The transatlantic prism

The appearance of the telescope in the work of Francisco de Quevedo should come as no surprise to modern readers, given the ambitious, almost universal scope of his literary production. Critics such as Alessandro Martinengo and José Julio Tato Puigcerver have in fact demonstrated how Quevedo not only was au courant with respect to the technical advances of the moment but also incorporated into his own texts, whether as literal references or as metaphors, numerous themes and lexical items from the world of science. As a skilled writer deeply engaged with the trends of the time, Quevedo also knew how to make use of the intricate, the precious, or the dynamic—in the scientific sense of the term, with its internal mechanisms—as figures that served to support his analyses of the world around him. One might recall, for example, his famous poem "Érase un hombre a una nariz pegado" (There was once a man stuck to a nose), a masterpiece in which the intimate resonances between form and content are exploited to the maximum, along with the surprising effect of the object's internal functioning and an idea of the body as something composed of disparate elements. It is worth mentioning also that in this poem, as Quevedo presents it, the man's nose launches outward from his face, just as the optic tube propels itself outward when attached to the eye. What emerges from all of this, as we shall see, is that even as the Spanish monarchy—and its loyal servants—maintained a deep enmity for Venice, the latter would yet remain a central figure in all of Quevedo's work.[2]

It is also well known that the sphere of optics, translated by Quevedo into a fictional domain that warns of the limits of external and individual knowledge, does not escape the reach of his pen. Quevedo reflects on the powers of vision in texts such as *España defendida* (Spain defended, ca. 1609–1612) and in passages of "El mundo de por dentro" (The world from within), which forms part of his masterpiece *Sueños y discursos* (Dreams and discourses, 1627). In the latter text, Quevedo takes note of the mediocrity and shallowness of the world "debajo de la cuerda," that is to say, beyond the realm of appearances.[3] His exercises in corrective vision would later find themselves transformed into the thematic scaffolding of a famous political essay, *El lince de Italia u zahorí español* (The lynx of Italy or the Spanish diviner, 1628). Within *El lince de Italia*, Quevedo makes use of a telescope to warn Spanish king Philip IV of the dangers then presented by Charles Emmanuel I, Duke of Savoy. These warnings have their roots—although the 1627 War of the Mantuan Succession would provide the more immediate context—in Charles Emmanuel's forceful retaking of the neighboring marquessate of Montferrat in early 1616, a move that served to renew the war between Savoy and Spain. It is worth pointing out that Venice finds itself deeply implicated in these renewed tensions, for as it happens, a good part of Charles Emmanuel's war chest in fact came from Venice. This was part of a general trend: while it is true that the Republic of Venice chose to confront Spain militarily on only a handful of occasions, it nonetheless offered almost continuous economic support to the Habsburgs' enemies. In early 1616, Venice conceded to the Duke of Savoy a stipend of fifty thousand ducados per month to help with preparations for a war with Philip IV of Spain. This amount grew when hostilities opened up on September 14, 1616, and it eventually reached eighty thousand ducados per month according to Italian sources—Spanish sources put the figure as high as one hundred thousand ducados per month.

It is worth lingering a bit over *El lince de Italia* and Quevedo's deep-seated mistrust of both Venice and Savoy; for while the telescope does not appear in the text as such, Quevedo nonetheless names it as an instrument that permits him to sharpen his finely honed ability to offer useful advice to a monarch somehow unaware of the danger right before his eyes. I have already stressed the central role of the Academy of Lincei in the development and utilization of the figure of the telescope as a means of expounding on the faculty of clear, astute, and profound vision—this much is well known. I do wish to call attention, however, to the manner in which Quevedo makes use of a scientific invention to construct a wide-ranging rhetoric meant to offer an alternative view of the pressing political intrigues and rivalries of the moment. His vision is ultimately a strategic one, and it is at once supported by its immediate historical context and endowed with an innovative perspective that touches all points on the political map. And while Quevedo mentions the telescope only halfway through *El lince de Italia,* it is nonetheless pivotal to take into account for a full understanding of his intentions.

These rivalries defined the balance of power in the following decade. In the early spring of 1628, as part of the War of the Mantuan Succession, Savoy and Spain formed an unlikely alliance to prevent Charles Gonzaga II, Duke of Nevers, from gaining control of Montferrat and the Duchy of Mantua. Spanish reservations regarding the Duke of Nevers stemmed from his French background, and it was feared that he might enter into an alliance with Cardinal Richelieu that would jeopardize Spanish control of the Duchy of Milan, which was situated directly between Montferrat and Mantua. In March 1628, Charles Emmanuel—who would, as it turns out, subsequently ally Savoy with the French—began his invasion of the areas of Montferrat that had been granted to him in 1627 through a treaty with Gonzalo de Córdoba upon the death of Vincenzo II Gonzaga, Duke of Mantua. In *El lince de Italia,* Quevedo's narrator refers to this invasion, asking Philip IV: "If this is so, why is it that the duke, claiming sovereign power and operating outside of your protection, makes war and violently takes land?" (*Pues si esto es así, ¿cómo hoy el duque, siendo imperial y renegado de vuestra protección le hace guerra y le arrebata los lugares?*). He then goes on to offer the ideal instrument for combatting the "cataracts" that compromise the vision of his monarch:

> To this, the Emperor will neither consent nor will he consider it good; nor is Your Majesty unaware of this fact. And Italy is even now in the process of curing itself of the cataracts that made it blind to this aggression. I, my Lord, will put such a telescope in your hands so that from Madrid you should be able to see into the very depths of Turin.
>
> *Esto, ni el emperador lo consentirá, ni él lo disimula bien, ni Vuestra Magestad lo ignora; y Italia se va curando de las cataratas que le hacían no ver este tropezón. Yo, Señor, pondré tal antojo de larga vista en vuestras manos, que desde Madrid se registre en Turín las entrañas.* (88–89)[4]

Quevedo shows here his misgivings with respect to the support that the Duke of Savoy offers Spain, and he makes it clear that these fears extend, above all, to France and Venice. The image that Charles Emmanuel I had forged for himself as Italy's liberator was even then being propagated throughout the Italian Peninsula in texts such as Boccalini's aforementioned *Ragguagli di Parnasso* and *Pietra del paragone politico,* as well as in Valerio Fulvio's *Castigo essemplare de'calunniatori* (Exemplary punishment of the slanderers, 1618), which includes passages that directly attack both Spain and Quevedo himself. In fact, when Quevedo refers to a "touchstone" (*piedra paragón*) in *El lince de Italia,* he is alluding specifically to Boccalini, who for Quevedo is full of "malice and assumptions" (*malicias y suposiciones*) (74). Going even further, the Spanish moralist warns the king that Boccalini "speaks of Your Majesty with insufferable shamelessness" (*habla de Vuestra Magestad con desvergüenza insufrible*) (78). *El lince de Italia* is in many ways a typically Quevedan work, insofar as the Castilian writer and diplomat cautions that the true danger is an internal one. The title's two metaphors (*lince* and *zahorí*) further reveal Quevedo's intention to put into practice both his sharp, lynxlike vision and his ability to "divine" what is hidden below the most immediate reality (Juárez Almendros; Azaustre Galiana 50).

With *El lince de Italia,* we find ourselves, as a good deal of scholarship has pointed out, before a type of ethico-pragmatic historiography set in motion to unmask the dangers that exist in Italy. As Quevedo puts it: "I hold it as healthy state policy to anticipate the malice of powerful external forces. May Your Majesty hear these, my outpourings of focused thought and my warnings based on real fears" (*Tengo por salud de la materia de Estado la malicia anticipada en las cosas de más calificado exterior. Vuestra Magestad oiga estos arrojamientos de mi atención y estas cautelas de mis miedos*) (90). This admonition to Philip IV includes also an ironic wink to the reader, insofar as Quevedo, who was severely myopic and of generally "poor vision" (*de ojos divertidos*), is the person called on to offer the Spanish king the gift of clear sight: "even with my ailing vision and weak eyesight, I have come to see the malice that passed beneath the surface, which was, if poorly founded, even less skillfully covered" (*pues aun de vista enferma y de ojos divertidos se dejó conocer la malicia que iba debajo, si mal fundada, peor cubierta*) (73). Through this perspective, Venice emerges in *El lince de Italia* as the great agitator of European tensions, which ignites (*enciende*) and "blows upon" (*sopla*) or stokes war:

> Venice (which looks for peace with its mouth, and war with its money) will always seek unrest in Your Majesty's kingdoms, and more in Italy than in any other place, because only in this way can it match the power of other

Italian territories and that of your monarchy; and it knows that in foreign countries it is necessary to ignite war and blow on it, and that in Italy this process has no end.

Venecia (que busca la paz con la boca, y la guerra con los dineros) siempre procurará la inquietud de los reinos de Vuestra Magestad, más en Italia que en otra parte, porque solo con eso se contrapesa ella con Italia y con vuestra monarquía, y sabe que en otros países es menester encender la guerra y soplarla, y que en Italia ella se atiza sin fin. (95)

In keeping with a specific connotation of the Spanish term *soplar* that renders it synonymous with *espionage*, the narrator goes on to warn his reader a bit later that Venice

is the world's chief gossip and the quicksilver of princes: it is a Republic that one must neither believe nor forget; it is larger than one would like it to be, and smaller than it claims; it is very powerful in treaties and very decadent in power. . . . Venice is more harmful to its friends than to its enemies; and it undoes the peace of the elements, insofar as it states one thing with one party and then contradicts itself with another; in this way its embrace is a peaceful war.

es chisme del mundo y el azogue de los príncipes: es una república que ni se ha de creer ni se ha de olvidar; es mayor de lo que convenía que fuese, y menor de lo que da a entender; es muy poderosa en tratos y muy descaecida en fuerzas. [. . .] *Es Venecia más dañosa a los amigos que a los enemigos, y es remedio de las paces de los elementos, que con sus contrarios simboliza con una calidad, y se contradice por otra; y así su abrazo es una guerra pacífica.* (101–2)

That the embrace of Venice should be understood as "a peaceful war" is, perhaps, the best possible definition of the dangers that lurk below the smooth surface of the glass that defines this Republic.

Glass in fact serves as a crucial figurative instrument in Quevedo's prodigious satire *La Hora de todos y la Fortuna con seso* (The Hour of all and Fortune with sense, 1650). Composed, it is believed, between 1633 and 1635, it is a masterpiece of wit and political commentary in which he once again makes use of technical objects that come from contemporary cultural debates. *La Hora de todos* is, in fact, one of Quevedo's most accomplished satirical compositions, "a kind of final summary of this great critic of society and Spanish politics," and a "satirical *summa*" that offers a very complete vision of that which for Quevedo was the equilibrium of forces in seventeenth-century Europe.[5] As a result, one finds in this piece not only the pessimism of the author himself but also the profound crisis that challenges Philip IV and the controversial leadership of the Count-Duke of Olivares. The presence of the telescope in chapter 36, which bears the title, “Los holandeses en Chile” (The Dutch in Chile), facilitates a deep analysis of the difficult tensions that exist between the overseas interests of Olivares, the maritime power of the Dutch, and the attitude of legitimate defense on the part of Araucanian people. Beyond these, as a kind of backdrop, one also finds the shadow of Galileo. Quevedo never alludes to him directly; however, the latter’s scientific and commercial instinct flows through and informs much of the cultural conflict that this episode narrates.[6]

This is not the first time that we see in *La Hora de todos* a fascination with the mechanical: just before his commentary on the Dutch, Quevedo had written that politics, especially that which involves international intrigues, functioned like the mechanism of a watch, without being “seen or heard” (*sin ser visto ni oído*) and without “ceasing or turning back” (sin *cesar ni volver atrás*) (268).[7] When speaking once again of Venice, he affirms that “our reason of State is like blown glass in that it is air that gives things form and substance, and of all that we sow in the earth, we employ fire to manufacture ice” (*nuestra razón de Estado es vidriero que con el soplo da las formas y hechuras a las cosas, y de lo que sembramos en la tierra a fuerza de fuego fabricamos hielo*) (283–84). If in other examples we had seen the refracting or reflecting properties of glass, what is valued here is its malleability, and the process by which it is molded from the forge (*fuego*) to the final product (*hielo*) with *sembrar* (to sow) as the supreme colonizing verb. Nevertheless, the use of the word *soplo*, as we shall see, is far from accidental, given that its dissemination points to the idea of gossip and, by extension, to the aforementioned activity of espionage; an activity, it bears mentioning, that was attributed to Quevedo (albeit erroneously) during his Italian embassies in the service of the Duke of Osuna.

At the moment when *La Hora de todos* was published, two of the most important Dutch expeditions of the time—led by, respectively, Joris van Spilbergen (1614–1615) and Jacques l’Hermite (1624)—had failed to establish dominion over the coast of Chile.[8] Prior to the expedition of Hendrik Brouwer (1643), and with it the establishment of Dutch control in Valdivia, it is clear that around the years 1633–1636 the Spanish seemed already to fear an attack supported by the Araucanians, who returned from their overseas distance to the center of Spanish politics, and whose collective imaginary was reborn in 1632 with the reedition of Alonso de Ercilla’s epic *La Araucana* (The Araucaniad). Quevedo participates in the debates revolving around this phenomenon when, in scene 28 of *La Hora de todos,* he criticizes the Dutch contingent and the Count-Duke of Olivares more or less equally (the latter for having signed a treaty with the Duke of Orange), while in scene 36 he describes the encounter between the crew of a Dutch pirate ship and group of indigenous people in a Chilean port.

The scene depicts the arrival of the Dutch in Chile, and presents an "optic tube" (254) at the center of this encounter. Quevedo tells us how the Northern pirates have enthusiastically come to pinch (*pellizcar*) and chew (*roer*) on American cartography, propagating itself dangerously as a "cancer" (257). The Araucanian leadership (*los más principales*) receives the Dutch, who have arrived on the Chilean coast with weapons in their hands. Quevedo tells us that the Araucanians are a "nation very attentive to what is possible and suspicious of what is apparent" (*nación tan atenta a lo posible y tan sospechosa de lo aparente*), underscoring their natural intelligence in the face of the hostile plans of the Dutch, who merely wish—like the astrologers that Quevedo attacks in other texts—to confuse or swindle (*engatusar,* a word also used by Vélez de Guevara) them. Not in vain, Quevedo writes that, in their mistaken strategy, in their misreading, the Dutch see the Araucanians as persons "inclined to toys and curiosities" (*inclinados a juguetes y curiosidades*) (257). But the confusion is mutual: when the Araucanians confuse the Dutch with the Spanish, the captain of the ship, to gain favor, highlights in a short passage their hard-earned freedom and sovereignty vis-à-vis the Iberian yoke.[9] After this diatribe, many of Quevedo's own preoccupations take center stage, as he here utilizes the opponent of Habsburg imperialism as an alter ego to present a moral judgment against the absolutist arrogance of the period.

After the failed gift exchange involving a series of trinkets of little value, the Dutch pirate offers the Araucanians "an optic tube, which they call a telescope" (*tubo óptico, que llaman antojo de larga vista*) (310), which corresponds with the notion that these are the countrymen of Hans Lippershey, Zacharias Janssen, and Jacob Metius (or Jacob Adriaanszoon), the device's first inventors. With the telescope, the Dutch pirate promises his indigenous counterpart, who "among his group held the most authority" (*entre todos tenía mejor lugar*) (311), that he will be able not only to see ships from far away in the distance but also to observe the secrets of the firmament up to that point never seen: new stars, the "eyes" and "mouth" of the moon, and sunspots. The rhetoric associated with the controversial instrument is revealing in the hands of Quevedo, insofar as he embeds within both cosmographic and astronomic insights. As he has it, the Dutch recommend its use to them, and rightly so, saying that with it they would see ships that were still ten to twelve leagues away, and they would know by the clothing of the crew and the standard if they were friend or foe, and the same thing on dry land; they added that with the telescope, the Araucanians would be able to see stars in the sky that had never before been seen and that without it could not be seen; that they be able to make out very distinctly and clearly the spots on the moon's surface that look like eyes and a mouth as well as sunspots. And that these *marvels* were possible because with those two lenses were brought to the eyes things that were infinitely far away.

encareciéndoles su uso, y con razón, diciendo que con él verían las naves que viniesen a diez y doce leguas de distancia, y conocerían por los trajes y banderas si eran de paz o de guerra, y lo propio en la tierra, añadiendo que con él verían en el cielo estrellas que jamás se habían visto, y que sin él no podrían verse; que advertirían distintas y claras las manchas que en la cara de la luna se mienten ojos y boca y en el cerco del sol una mancha negra; y que obraba estas maravillas *porque con aquellos dos vidrios traía a los ojos las cosas que estaban lejos y apartadas en infinita distancia.* (311, emphasis mine)

The citation reproduces some of the promises made by Galileo in a letter sent to Philip III several years earlier and which the monarch, undoubtedly intrigued, shared with none other than the Duke of Osuna: "it makes it possible to determine longitude and so facilitate and make more certain oceanic navigation and . . . it also offers a way for galleys in the Mediterranean to make out enemy ships at ten times the distance of ordinary vision" (*dar el modo para poder graduar la longitud y facilitar y asegurar la navegación del océano y* [. . .] *ofrecía también otra invención para las galeras del Mediterráneo con que se descubrían los bajeles del enemigo diez veces más lejos que con la vista ordinaria*) (311). The second part of the promise to the Araucanians takes up the findings presented in *Sidereus nuncius* with respect to the sun and the moon. The Araucanian puts the tube to his right eye, "directing his vision to some mountains" (*asestándole a unas montañas*), and gives a great shout, frightening the others and saying that he as seen "cattle, birds, and men up to four leagues away, and rocks and groups of trees with such precision and so close up that they seemed, with the hind glass, incomparably larger in size" (*a distancia de cuatro leguas ganados, aves y hombres, y las peñas y matas tan distintamente y tan cerca, que aparecían con el vidrio postrero incomparablemente crecidas*) (311). The rejection is unequivocal: grabbing the telescope with his left hand, he tells his Dutch counterpart:

> An instrument that finds spots on the sun, corrects our mistaken perceptions of the moon, and discovers what the skies obscure, is a subversive one; it is a glass muckraker, and it cannot be esteemed by heaven. To bring to oneself that which rests far away is worrisome for those like us who live far away; with it you should look at yourselves from such a great distance, and with it we have seen the intentions hidden in your offerings. With this artifice you examine the elements and set out to rule over whatever you like: you live underwater without getting wet, and you're tricksters of the sea.
>
> *Instrumento que halla mancha en el sol, y averigua mentiras en la luna, y descubre lo que el cielo esconde, es instrumento revoltoso, es chisme de vidrio, y no puede ser bienquisto del cielo. Traer a sí lo que está lejos es sospechoso para los que estamos lejos; con él debisteis de veros en esta grande distancia, y con él hemos visto nosotros la intención que vosotros retiráis tanto de vuestros ofrecimientos. Con este artificio espulgáis los elementos, metéisos de mogollón a reinar: vosotros vivís enjutos debajo del agua, y sois tramposos del mar.* (311–12)

The Araucanians thus accuse the Dutch of attempting to impinge upon their freedom, of promising false protection from the king of Spain, when what they are actually doing at that moment is merely working to take power over Brazil, which from 1580 to 1640 was under Spanish control:

> The Dutch, inspired by their felicitous betrayals, are now seeking to convert their betrayals into a kingdom, and they strive now to go from being vassals of the King of Spain to being his competitors. They stole their vassalage from him and now seek to usurp from him what is far from them—Brazil and the Indies—and make these lands part of their own, separate dominion.
>
> *Los Holandeses, animados con haber sido traidores dichosos, aspiran a que su traición sea monarquía, y de vasallos rebeldes del gran Rey de España, osan serle competidores. Robáronle lo que tenía en ellos y prosiguen en usurparle lo que tan lejos dellos tiene, como son el Brasil y las Indias, desinando sus conquistas sobre su corona.* (337–38)

But America, the Araucanian reminds us, is a "rich and beautiful whore" (313) (*ramera rica y hermosa*) that is loyal to no one, especially not to a *rufián*, or "pimp." The natives, enlightened by the character of *La Hora de todos*, reject an alliance with the Dutch and expel them from their territory. The Araucanian leader gives the Dutch pirates two hours to leave:

> Since you are inventors, invent a device to take us far away from what is right before our eyes, as we swear to you that with this instrument of yours that brings to the eyes that which is far away, we will still never see your homeland or Spain. And take with you this glass spy, this snooper of the heavens, for with our eyes directly on you we already see more than we'd like, and so it's not necessary. And thank the sun that with this device you managed to find a black spot on its surface, for without this blemish, given the sun's golden color, you'd undoubtedly try to coin it and turn it from a planet into a doubloon.

> *Pues sois invencioneros, inventad instrumento que nos aparte muy lejos lo que tenemos cerca y delante de los ojos, que os damos palabra que con éste que trae a los ojos lo que está lejos, no miraremos jamás a vuestra tierra ni a España. Y llevaos esta espía de vidrio, soplón del firmamento, que, pues con los ojos en vosotros vemos más de lo que quisiéramos, no le hemos menester. Y agradézcale el sol que con él le hallasteis la mancha negra; que si no, por el color, intentárades acuñarle, y de planeta hacerle doblón.* (313)

In this episode, as André Stoll has written: "The art of ideological demolition practiced by Quevedo reaches . . . an unprecedented level of poetic radicality" (36). The moralizing allegorical figure of *La Hora de todos* institutes an order capable of revealing the vanity of the world. Quevedo combines here three of the most important debates then taking place in Europe: that of imperial strategies of expansion against the Dutch enemy, that revolving around the rights of indigenous communities in the Americas, and those related to the advances of post-Copernican science. Modern editors of this piece consistently point out that, more than the Dutch themselves, Quevedo is in fact singling out the Count-Duke of Olivares for having entered into a treaty with them.[10] And, in fact, the Araucanian announces that "we will never see either your homeland or Spain" (*no miraremos jamás a vuestra tierra ni a España*) (313), leaving the Spanish Crown in a disfavored situation. With the presence of the allegorical figure one passes from one situation to another without any apparent improvement.[11] The allegory does not cancel out the effect of scientific experimentation upon its Chilean audience; rather, it deploys

> a series of irritating syllogisms or ingenious and intertwined paradoxes, in whose development the scientific invention of the optic tube demonstrates objectively, even before its situated use is condemned on moral grounds, its contrary utility as a vehicle for the hermeneutic process carried out by the author of the work.

> *una trama de irritantes silogismos o ingeniosas paradojas entrelazadas, en cuyo desarrollo el invento científico del tubo óptico demuestra inopinadamente, aún antes de que se condene moralmente su uso interesado, su utilidad contraria como vehículo del proceso hermenéutico intentado por el propio autor de la obra.* (43)

That is, besides acting as an instrument of the Araucanian leader's "sense of self" (*toma de conciencia*)—although he is precisely the person whom it is intended to trick—the telescope serves also as a metanarrative agent, "scientifically" confirming the existence of "marvels" beyond the optic sort, insofar as these stem from the ludic genius of Quevedo himself, which forms the unmistakable frame of each scene of the text. The Araucanian, however, holds the telescope in the wrong hand, and therefore the benefit to be extracted from it is more accurately moral in nature; he sees himself as others see him—far off—and he objectifies himself and so sees the danger of his new proximity vis-à-vis a master in possession of modern science. In essence, he recognizes through the telescope that he is now at the point of losing the freedom that his distance had previously guaranteed him. If the telescope was presented to him by the Dutch as an equalizing device, as a democratic instrument capable of uniting the distant and the nearby, the Araucanian somewhat paradoxically rejects it precisely on these grounds, in a search for distance and individuality that might leave him beyond the purview of the syncretic expansionism of the empire. All this makes the Araucanian resistant to the entreaties of the Dutch pirate, and he is likewise not tempted to rise up in rebellion against the far-off king of Spain. What he requests, in the end, is to preserve his essence while avoiding any form of foreign occupation; he also states his desire to avoid any sort of displacement, whether it be via sea or sky. In a sense, the telescope does for the Araucanian what the re-edition of Alonso de Ercilla's epic *La Araucana* (1569) would do for its readers: offer up a new interrogation of the collective Hispanic imaginary.

The text, obviously, reveals an open rejection of the Dutch, given that it joins the figure of the pirate to that of the Jew. Quevedo was both profoundly isolationist and anti-Semitic, and he saw the Netherlands as a center of Sephardic Jews and conversos, with Amsterdam as its "Jerusalem of the North"; in this sense, the Dutch threat in Brazil was for him not just a Protestant one but also Jewish in nature.[12] But the episode also contains a second reflection of an epistemological sort that is very typical of the Baroque, as Stoll has pointed out:

> To desire to know—the cognitive appetite—already implies, of course, a grounding in the aesthetic appetite. The first link for sacrilegious original desire resides, for Biblical scripture, in the eye, that same visual organ to which (Neo)platonic philosophy would attribute the faculty of generating and transmitting beauty, that is to say, Love; and with it they would suppress the idea of any original condemnation of the many generations of humans. To condemn the modern equivalent of the organ associated with vision (in aesthetic and epistemological terms) as Quevedo does, by means of the illuminated Araucanian in *La Hora de todos*, implicitly signifies, it follows, the rejection of the inheritance of those Neoplatonic philosophers of the Renaissance, who, like Ficino, Pico della Mirandola, Manetti, or also Juan Luis Vives (*Fabula de homine*, Louvain, 1518) had forced themselves, through spectacular treatises on human dignity (*dignitas hominis*), to explore the consequences of that original telescope for the formation of a new and sovereign humanity. What we have, then, is an Amerindian man capable of articulating his own "illuminating lucidity."

> *Desear conocer—el apetito cognitivo—radica, desde luego, en el apetito estético. El primer vínculo del sacrílego deseo original es, en efecto, para la escritura bíblica, el ojo, ese mismo órgano de la visión al que la filosofía (neo)platónica atribuiría la facultad de generar y transmitir la beldad, es decir, el Amor, suprimiendo con eso la idea de una condena original de las generaciones humanas. Condenar el equivalente moderno del órgano de la visión (estética y epistemológica), como lo hace Quevedo por intermedio del indio iluminado por la Hora, significa implícitamente, por eso, rechazar la herencia de esos filósofos neoplatónicos del Renacimiento, quienes, como Ficino, Picco Della Mirandola, Manetti, o también Juan Luis Vives (*Fabula de homine, *Lovaina, 1518), se habían esforzado, a través de sus espectaculares tratados sobre la* dignitas hominis, *en explorar las consecuencias de aquel antojo original para la formación de una humanidad nueva y soberana. Indio capaz de articular su propia "lucidez desengañadora."* (52)

What most interests me here is to see what Quevedo says with respect to the gains of the "new physics," since it appears evident that he offers no open condemnation of the telescope as an object in itself; rather, he takes aim at the hand in which it falls, at who uses it. The Araucanian's unwillingness to enjoy the telescope's marvels—marvels that, once again, Quevedo never denies but rather celebrates—underscores the failure of the colonial project, the failure to found a heterogeneous colonial society through the mediation of the two lenses of which the text speaks. The

Araucanian's final joke, however, confirms his intelligence rather than his backwardness.

In a highly suggestive essay, Miguel Martínez has spoken of a "polyphonic multiplicity of voices" and of the "presence of a satiric 'I' that is much less defined" (*polifónica multiplicidad de voces* [...] *presencia de un yo satírico mucho menos definido*) in *La Hora de todos* that makes Quevedo's stance much less clear than is the case with his other texts (113). We know that Quevedo was in Rome in the spring of 1617, and that he wrote on some of the events of the moment and on their most important participants. Only a few months had passed since Galileo's trial, and there can be no doubt that its result was still weighing heavily on the collective memory. In her edition of *La Hora de todos,* Luisa López Grigera argues, "Quevedo is not unaware of the debate" (*Quevedo no desconoce el debate*) (28n49), while Pablo Jauralde has affirmed in a more cautious way: "If Quevedo had the time to do more than simply carry out his part-time ambassadorial duties, and if he allowed his eyes take in Rome, it's probable that he became aware... of the defenses of Galileo's doctrine with respect to those of the Church authorities.... If this was indeed the case, however, it has left no perceptible trace in his work" (*Si Quevedo tuvo tiempo de algo más que ejercer sus tareas como embajador ocasional y de pasear sus ojos por la urbe universal, es probable que tuviera noticia* [...] *de la defensa de las doctrinas de Galileo Galilei frente a las autoridades eclesiásticas* [...] *Si así fue, no ha dejado huella perceptible en su obra*) (350). If Quevedo effectively mocks Copernicanism in the prologue to *La Hora de todos* when he speaks of the sun as a "red" (*bermejo*) and moving (*andante*) planet, Schwartz-Lerner in her edition of the text (303n919) and Riandière La Roche Saint-Hilaire (27–30) nevertheless speak of an "ambiguous" and prudent position to which I myself subscribe. They coincide in signaling that the episode with the telescope adheres to an ideology with roots in Stoicism that considered "high culture"—and, in this case, the telescope as an instrument for elites—the exclusive dominion of the learned. This exclusive domain, it is worth mentioning, was defended in particular by the aforementioned Juan de Espina, both when he opened the doors of his home as a place of study to those with a certain degree of formal learning and in his correspondence to Philip IV; it was also a constant element in the work of the giants of the period who, as Cervantes reminds us in the epigraph that opens the present book, praised the virtues of new things only if they had been cultivated by expert hands: "no science is deceptive in itself; the deception lies in the person who doesn't understand it" (*ninguna ciencia en cuanto ciencia engaña; el engaño está en quien no lo sabe*).

Political intervention II: The transalpine prism

Emblems that make use of objects linked to science were rare in early modern Spain, and this is particularly true of instruments related to optics.[13] In the first part of Juan de Borja's *Empresas morales* (Moral advice, 1581), for example, we find one *empresa* (number 46) that makes use of eyeglasses to symbolize the control of one's passions. The commentary goes as follows:

> Although the passions and the affectations that combat us may not seem at first to be large or powerful, we should not for this reason be careless in working to resist and control them. If we allow ourselves to be controlled and vanquished by them, they will not be satisfied with dominion over our will; rather, they will also strive for control over our understanding, blinding it and making it impossible to determine black from white, light from dark, and falsehood from truth; we will move down step by step until we find ourselves among the deepest errors of understanding, which are truly to be feared. This is illustrated in the present "*Empresa* of the Eyeglasses" with the motto SIC ANIMI AFFECTUS, which means: "So is the soul affected by the passions." Because when one looks through glasses, everything appears according to the color of the lenses themselves and all that is big or small likewise appears as the lenses present them. In the same way, the passions and affectations of the soul make it so that everything takes on the tone of the particular passion that rules over the soul at a particular time, putting itself in front of the eyes of reason and clouding its vision. If, for example, it is with love that one looks, everything seems good, beautiful, easy, and pleasant; if with hatred, then that same thing appears bad, ugly, difficult, and unpleasant.

> *Aunque las pasiones y afecciones que nos combaten no nos parezcan al principio grandes ni fuertes, no por eso debemos descuidarnos en resistirlas y sujetarlas; porque si nos dejamos señorear y vencer dellas, no sólo se contentarán con que les rindamos la voluntad, sino también querrán entregarse de nuestro entendimiento, cegándolo y haciéndole juzgar lo negro por blanco, lo claro por oscuro y lo falso por verdadero; bajándonos de escalón en escalón hasta dar con nosotros en el profundo de los errores de entendimiento, que son tanto de temer. Esto se da a entender en esta Empresa de los Antojos con la letra sic animi affectus, que quiere decir: así hacen las pasiones del alma. Porque como el que mira con antojos, todo lo que ve le parece de la color que ellos son, y así le parecen las cosas grandes o pequeñas conforme a la hechura que ellos tienen, de la misma manera las pasiones y afecciones del alma hacen que todo parezca*

conforme a la pasión que la señorea, poniéndose delante de los ojos de la razón y perturbándola, de manera que si es con amor lo que se mira, todo parece bueno, hermoso, fácil y gustoso; si con aborrecimiento, aquello mismo parece luego malo, feo, áspero y dificultoso.

The Spanish scholar Sebastián de Covarrubias develops a similar idea in emblem 18 of the first section of his masterpiece *Emblemas morales* (Moral emblems, 1610). Presenting his reader with the motto "Shadows spring from the truth" (*sombras son de la verdad*), he places below this a *pictura* of a curious pair of eyeglasses with faceted lenses (hanging from a nail) that deceptively multiply the image of a chess pawn that passes through them. This multiplication is especially problematic for Covarrubias, given that for him, the truth "has but one face, is always one, and remains firm" (*no tiene más de una cara, siempre es una, y está firme*). The poetic *subscriptio* below the emblem reads:

Eyeglasses cut into many prisms
turn a single thing into a hundred.
All things, so equally presented
that to grasp them one must fumble:
false opinions disguised as truth,
and without discernment, will confuse us;
they present multiple lies as one abiding truth
and mislead those who don't know better.

Los antojos de lunas cuadreadas
de una sola cosa, hacen ciento.
Todas, tan igualmente pareadas
que echarle mano a de ser a tiento:
las falsas opiniones, disfrazadas
al no advertirlo, sacarán de tiento,
representando por verdad constante
la mentira, que engaña al ignorante.

The poetic tension that Covarrubias opens up between the oneness of truth and multiplicity of lies has much to do with religious ideologies in force during the Counter-Reformation; however, his use of eyeglasses—and especially eyeglasses with prismed lenses—within an emblem that can itself be classified as anamorphic places him in very limited company. One such fellow "moral optician" worth noting is Francisco Núñez de

Cepeda, whose *Idea del buen pastor copiada por los santos doctores representada en empresas sacras* (Idea of the good shepherd copied by the church doctors and represented in pious figures, 1682) contains a crystal prism that breaks the light down into its color spectrum with a corresponding motto that reads: “Don’t put too much stock in your beauty” (*Nimium ne crede colori*).[14]

For all that the previous examples reveal about the links between optics and ethics, it is in the work of Diego de Saavedra Fajardo (1584–1648) that one finds perhaps the most sustained and comprehensive treatment of this material. Saavedra Fajardo’s dense and complex oeuvre, and especially his emblematic work, provides modern readers with an extremely useful and unmistakably serious reading of how the telescope was received in seventeenth-century Europe. His point of view is on the surface a conservative one (the principal readers were, after all, the Spanish monarchy); however, his work is also impregnated with the material and intellectual sources that so profoundly shaped his understanding both of the political sphere and the unstoppable force of the new science. The telescope, given its evident modernity, was central to Saavedra Fajardo’s vision of the perfect monarch as well as to his idea of the model government under the guidance of an astute and cautious leader; however, this link between caution and the telescope also manifests itself, very significantly, in his more explicitly literary production. Saavedra Fajardo’s emblematic work is especially indicative of his stance vis-à-vis science and politics insofar as it traces out a very revealing link between education (with respect to theme) and diplomacy (with respect to form) that without a doubt enriches, by way of contrast, the path taken by the present chapter, and the principal coordinates of its line of analysis.

A fertile link between ethics and vision serves as the foundation for the educational project underlying Saavedra Fajardo’s *Empresas políticas* (Political advice, 1640). In this work, as Mariano Baquero Goyanes pointed out decades ago, “the foregrounding of optics connects with a modality of literary perspectivism that is very typical of the Baroque” (*el predominio de lo óptico se conecta con una modalidad de perspectivismo literario muy típica del Barroco*) (10).[15] For Saavedra Fajardo, a monarch is required not only to train his vision but also to learn to project an image in accordance with his nature so that, in the end, he might realize his maximum ruling potential without having to resort to trickery. For Saavedra Fajardo, this decidedly anti-Machiavellian model of royal supremacy should be articulated,[16] as has been pointed out in recent years, through subtle mechanisms of dissimulation that revolve around an effective use of the visual, as much in the monarch's presence as in his absence: "Who cannot feign cannot see" (*qui nescit fingere nescit videre*); and this presence must be cultivated always for a "people" wholly seduced by pomp and circumstance.[17] If the famous "I wear a mask" (*Larvatus prodeo*) of Descartes had been converted into an insignia for the entire period, important texts such as Torquato Accetto's *Della dissimulazione onesta* (On honest dissimulation, 1641) had raised dissimulation to the level of art, an art that could offer "a certain hiatus to truth" (*cierta tregua a la verdad*) (Beltrán Marí, *Talento y poder* 18).

Saavedra Fajardo's own reading of truth and feigning would cover much of the same territory as that of Acceto. To provide just two brief examples, in *empresa* 43 of *Empresas políticas*, Saavedra Fajardo counsels the prince to show "dissimulation in appearance" (*disimulación en el semblante*) (530); and in *empresa* 39 he recommends the absence of the royal body as an instrument of seduction, given that "what is not seen is venerated more" (*lo que no se ve se venera más*) (500). Regarding the power of the image, and especially with respect to the handsome ruler, Saavedra Fajardo offers an interesting judgment within the history of medieval Iberia that he develops in *Corona gótica* (Gothic crown, 1646), a piece that represents a kind of "putting into practice" of all the theory that he develops in relation to the reason of state in his *Empresas*. Writing on Liuva, the "nineteenth king of the Goths in Spain" (*decimonono rey de los Godos en España*), he writes:

> The handsomeness and good disposition of the monarch tends to win the spirits of his people, because these are more moved by appearances than by the qualities of one's soul; they tend to believe that a presence that is pleasing to the eye always accompanies virtue and goodness, and so they are pleased to obey as king that man who exceeds others in physical grace.
>
> *La hermosura y buena disposición del príncipe suele ganar los ánimos del pueblo, porque se mueve más por las apariencias externas que por las calidades de ánimo; y juzga que a una presencia grata a los ojos acompaña siempre la virtud y la benignidad, complaciéndose de obedecer por rey a quien excede a los demás en las gracias corporales.* (916)[18]

It is not surprising, following certain thematic resources of Menippean satire and other literary genealogies of the period, that Saavedra Fajardo, a diplomat educated in courts and embassies, develops in many of his *empresas* a connection between the human eye and the use and abuse of power through the incorporation of long-established topoi: a rejection of blindness and of those "blind men who seek to guide" (*ciegos guiando*); an appreciation of the virtues of good vision for not letting oneself be fooled and for being able to resolve disputes judiciously, the need for a certain kind of foresight to be beyond what others might see, and so on. And there is nothing better to inculcate these aptitudes in a prince than to do so by means of the tripartite message of motto, *pictura*, and declaration, which converts *Empresas políticas* into a stimulating compendium of political finesse, and one that is likewise eminently modern.[19] *Empresas políticas*, in fact, presents itself today as a useful collection of learning, situated halfway between the pragmatism of its author (an effective political agent, and an authentic doer) and the distractions and hesitations of the Spanish Habsburgs who are, in the end, continuously represented in its pages.

Perhaps precisely because of all the doctrinal padding that underlies and forms much of the structure of Saavedra Fajardo's biography, *Empresas políticas* is an almost archetypically interdisciplinary work, open to a critical rereading in which philosophy, political theory, ethnography, and even the history of science, to name but a few fields, come together. And while it is true that Saavedra Fajardo's use of emblems is anything but novel by 1640, what does merit great interest is his fertile imbrication of traditions—mirrors, windows onto the human heart, eyeglasses to improve vision, telescopes directed at the firmament, water as a reflecting surface—in which the field of optics exercises an active instructional role.[20] It is worth remembering, for example, the famous image of the oar broken by the force of water that opens up his *empresa* 46 (fig. 9).

The dialogue that Saavedra Fajardo undertakes with the field of optics and, by extension, with the controversial theses of the new physics, endows his masterwork with a singularity that has been scarcely studied to the present day. This oscillation between the ancient and the modern, and at the delicate crossroads for Europe in which Saavedra Fajardo lived and operated, make glass a material that will acquire important theoretical dimensions given its capacity to receive and project senses; or, in line with its original purpose, to reflect and refract images saturated with pedagogy, constructing in its reflection severe admonitions (*empresas* 33 and 76, for example) and projecting new images in its refraction (as occurs in *empresa* 46).[21] Much good work has been done on the ethico-political reading that *Empresas políticas* offers with respect to the use of the mirror, a reading well summarized by José María González García:

FIGURE 9. *Empresa 46, Fallimur Opinione*

The prince is a mirror that easily fogs up with anger or pride; or the prince is a mirror in which his subjects see themselves, since in him 'as in a mirror, the people compose their actions." But also, in a game of facing mirrors, the prince must see himself in the mirror of his people if he wishes to find the truth, since this resides far from the court; or the State and its counselors are mirrors of the prince. The past and the future are conceived of as mirrors that reflect the scepter, and which should be consulted by the prince.

el príncipe es un espejo que fácilmente se empaña con la ira o la soberbia; o el príncipe es espejo en que se miran los súbditos, ya que en él, "como en un espejo, compone el pueblo sus acciones". Pero también, en un juego de espejos enfrentados, ha de mirarse el príncipe en el espejo del pueblo si quiere hallar la verdad ya que ésta habita lejos de los palacios; o el Estado y los consejeros son espejos del príncipe. El pasado y el futuro son concebidos como espejos que reflejan el cetro y con los que debe consultar el príncipe. (14)[22]

What can be added to this summary is that the entire book, in fact, should be read as a large mirror in which the reader might study his or her own reflection; in this sense, it is not unlike its predecessors from the previous century.

It is no stretch to claim that there truly is no surface that resists Saavedra Fajardo's didacticism developed through the play of light and its optical effect: when, for example, the book speaks of friendship in *empresa* 91, the reader is reminded that "the broken glass is useless (*inútil queda el cristal rompido*) (594). Here Saavedra Fajardo warns not only of the fragility of human relations but also of the need to develop a critical reading of appearances and of our role in their maintenance, whether we are directly or indirectly involved in this process. There can be no light, after all, without a surface to reflect or refract it.

Apart from the ethico-moral projection of the mirror that one finds in *Empresas políticas*, there are also other, much less discussed pedagogical devices that Saavedra Fajardo employs. For example, the use of the telescope in *empresa* 7, which will form the center of what remains of the present chapter, takes up a very new sort of dialogue with the world around it, the political and philosophical parameters of which have been scarcely discussed. Saavedra Fajardo's use of the telescope acquires new reverberations insofar as it connects with the existing debates revolving around the heretical character of the telescope and its promoters. From an object of wonder possessed by specialists and eccentrics, the telescope would go on to appear in many of the fictional texts of the period and become, in some cases, the very center of the plot. I wish, therefore, to connect the iconography of the Baroque telescope with its symbolic ramifications to offer a more complete reading that might situate this instrument in the very center of the religious and political unrest that so shape the historical background of *Empresas políticas*.

The diplomatic tour that Saavedra Fajardo completes of a Europe so shaken by novelty is, strictly speaking, contemporary to the trajectory taken by Galileo.[23] It is also nearly certain that the former knew the work of the latter, insofar as every good diplomat would have to be able not only to calibrate the weight of politics and history in making decisions but also to possess knowledge of fields such as astronomy. And while it is true that Pope Paul V had shown himself to be reactionary with respect to culture in general, Galileo had been well protected by the Medici family during the former's papacy. If in the two published versions of Saavedra Fajardo's *República literaria* (Literary republic) one finds a discussion of the new discoveries on Mars achieved through the use of *anteojos largos* (i.e., the telescope), it is worth noting that in the second redaction of this text, there is also an allusion to the coetaneous debates over the imperfections of the moon's surface—"one finds all the world's eyes fixed on its workings and imperfections" (*que en sus trabajos y defectos halla fijos todos los ojos del mundo*) (161)—and the (in)corruptibility of the sun, which tore down the classical image of this heavenly body as pure and perfect: "some, who lack the eyesight of an eagle, . . . say that among its light there are dark spots and stains" (*hay quien, sin tener ojos de águila* [. . .] *dice que entre sus luces hay escuridades y manchas*) (266, 273).[24] In fact, as Baquero Goyanes has pointed out when underscoring the "fundamentally optic intentionality that Saavedra assigns to the most common symbols" (*intencionalidad fundamentalmente óptica que asigna Saavedra a los más reiterados símbolos*) (22), his emblematic work indicates that he was familiar with the findings of his contemporary, perhaps even as early as his years as a student in the University of Salamanca, where he studied civil and canon law from 1600 to 1608. I have already pointed out that under the powerful Gaspar Quiroga, always moderate in his inclinations, Copernican science had been allowed to be studied at Salamanca since 1561, as is indicated in the curricular plan contained in the *Estatutos hechos por la muy insigne Universidad de Salamanca* (Statutes adopted by the much celebrated University of Salamanca).[25] More than the presence of astrology in *Empresas políticas,* a fact pointed out already by Abel A. Alves in a highly stimulating essay, what most interests me is the role of astronomy and, in particular, that of the sun as both a cosmic agent and the supreme source of light, given that its use registers a fascinating oscillation between the ancient and the modern very wisely applied by Saavedra Fajardo to his own political and pedagogical doctrine.

As early as *empresa* 4, Saavedra Fajardo condemns astrological divination as Covarrubias had done before in his emblem 97, and he writes that "if our vision perceives things in reverberation with the sun, then it knows them as they are; but if it tries to look directly at the sun's rays, our eyes become so disoriented that they cannot distinguish the form of things" (*si la vista mira las cosas a la reverberación del sol, las conoce como son; pero si pretende mirar derechamente a sus rayos, quedan los ojos tan ofuscados, que no pueden distinguir sus formas*) (226). The solar star represents the truth, and, as the author indicates in *empresa* 12: "and if someone tries to study its rays and penetrate its secrets, he will find in it such deep gulfs and absences of light that they will stun his eyes, and he will have no explanation for what his eyes have seen" (*y si alguno intenta averigualle sus rayos y penetrar sus secretos, halla en él profundos golfos, y escuridades de luz que le deslumbran los ojos, sin que puedan dar razón de lo que vieron*) (288–89).

There can be no doubt that what is at the center of this discussion are the teachings of Galileo on the corruptibility of the sun, something that was up to that moment unthinkable. Moving evidently from the image of the monarch as the supreme source of luminosity, the citation also offers a political reading in which what is recommended is an always oblique view of things that captures only a part of the whole—since looking directly at the sun brings only blindness—but also a view that perhaps gives us a new angle, not before perceived, that might *reveal to us the truth*. In this theorization of sight, *Empresas políticas* represents an interesting reflection on the uses of distance insofar as Saavedra Fajardo advises the prince not to let himself be seen too closely in order not to reveal his imperfections: what remains more distant—adds the author—is more respected. And the sun, like the monarch, is visually incomprehensible in his intermittent presence, in his continuous movement; in this way, *empresa* 49 announces that the sun "traverses" (*tramonta*), and *empresa* 86 adopts an emblem in which the solar star moves from one tropic to the other just as a good monarch visits his territories, always completely present (*rebus adest*) but never entirely visible. In using the sun as an emblem for the good prince who is continually in motion distributing justice—"generating presence" (*generando presencia*)—among his subjects, Saavedra Fajardo adheres still to a Ptolemaic vision of the universe that seems to reject Copernican heliocentrism. These are, it is worth noting, precisely the years during which Galileo and the Jesuit Christoph Scheiner were engaging in disputes over the Aristotelian principle of solar incorruptibility.[26] It is also worth keeping in mind Cardinal Bellarmine's famous *Carta a Foscarini* (Letter to Foscarini, 1615), which sought to demolish all of Galileo's new propositions (his argument being that one might only speak *ex suppositione*; that is, one may advance a technical and astronomical hypothesis, but not a truth), turning to Ecclesiastes 1:5: "Also, the sun rises and the sun sets; and hastening to its place, it rises there again." This geocentric reading is likewise found, with some variants, in *empresas* 24, 94, and 101, and it is also applied to the moon, which comes to be an emblem of the courtier as the shadow, or eclipse, of his ruler (*empresas* 12, 49, and 77).[27] And while it is true, as I have pointed out, that Saavedra Fajardo condemns astrological divination, he consistently makes use of astronomy as his guide when elucidating the comportment of the monarch. In this sense, Alves has spoken of the "astral influences" (82) in Saavedra Fajardo's understanding of geopolitics, defending the latter as a "cautious stargazer" (84) who believes in a certain mixed causality when determining the fortune of the individual, a combination of personal responsibility and celestial design:

> The universe of the *Empresas políticas* is one without complete epistemological closure, but one in which the principle of causality may still play a role as an investigative device. The stars, geography, history, and medicine all have some bearing on the art of maintaining the body politic, but so do human and divine volitions. Saavedra's wise man does not deny causality and reason, but he does stand in awe of the multiplicity of causes at work in an intertwining, organic universe ultimately reliant on God's will and favor. (82–83)

Nevertheless, the telescope of *empresa* 7 leaves no doubt regarding this "epistemological closure," since it is through this that Saavedra Fajardo adheres still to the idea that the new science is but a distorted measure of the cosmos and the individual. The telescope, in offering a view previously prohibited to the human eye, embodies the complete realization of the image and, as a result, it veers off into a cancellation—a saturation, to be precise—of this desire that until that time provoked an intermittence of sight: the human eye should not strive to see that which belongs only to God. This ambition is but a desire that can lead only to error, distortion, and, in the end, heresy, as presented in *empresa* 7 of the first edition of *Empresas políticas*: "Recognize things as they are, without the passions either waxing or waning. It increases and diminishes. (The emotions wax and wane)" (*Reconozca las cosas como son, sin que las acrescienten o mengüen las pasiones. Auget et minuit.* [*Affectibus crescunt, descescunt*]). From this stoic rejection of the affective component of knowledge, Saavedra Fajardo situates himself in an emblematic tradition that predated him and, given the enormous popularity of his *Empresas políticas,* would have numerous followers throughout the seventeenth century (Robbins, *Arts* 16). From *Empresas morales* of Juan de Borja and *Emblemas morales* of Sebastián de Covarrubias to the *Devises et emblemes* (Devices and emblems, 1691) of Daniel de la Feuille, the corrective power of lenses was associated during the Baroque with defects more than with virtues. In the two first cases, the reader is presented with eyeglasses that, in their imperfection—"prismed" (*de cuadrillos*), as Covarrubias puts it—prevent the individual from correctly discerning; in de la Feuille's text, however, the telescope in his emblem, points not so much to ambition but rather to envy (fig. 10).

The motto *auget et minuit,* which precedes Saavedra Fajardo's as well as de la Feuille's emblem, had, by the seventeenth century, possessed a distinguished history as a term associated with the cosmos. It appears embedded within a cosmological digression in Vitruvius's *De architectura* (IX.2.4), in a passage of the book in which the author, as his nineteenth-century commentators pointed out, includes a short treatise on astron-

FIGURE 10. Daniel de la Feuille, *Devises et Emblemes* (1691), *Une Lunette d'aproche regardant un point à la masuë d'Hercules* (A telescope observing a point in the club of Hercules)

omy: "I maintain that the sun, crossing each month the space of a [zodiacal] sign, lengthens and shortens the days and the hours" (*Nunc, ut in singulis mensibus sol signa pervadens auget et minuit dierum et horarum spatia, dicam*).[28] The connection between Vitruvius and Saavedra Fajardo is more than just a hypothesis, since, as both Francisco Calvo Serraller and Aurora Egido have pointed out, Saavedra Fajardo had definitely read the work of his Roman predecessor.[29] In fact, this motto signifies a dramatic shift in that it installs within an *emblema* focused on human passions that which was in effect an argument taken from traditional cosmography, namely, the sun in continuous rotation. Nevertheless, this *auget et minuit* is not solely a Ptolemaic conceit; rather, Copernicus also takes it up in his *De revolutionibus*.[30] Speaking in book 3 of the movements of the earth, Copernicus writes: "There will be one movement which changes the inclination of those circles by moving the poles up and down in proportion to the angle of section. There is another which alternately increases and decreases the solstitial and equinoctial precessions by a movement taking place cross-wise" (*Alius igitur motus erit, qui inclinationem permutat illorum circulorum, polis ita delatis sursum deorsumque circa angulum sectionis. Alius qui solstitiales aequinoctialesque praecessiones auget et minuit, hinc inde per transversum facta commotione*) (134).[31] What Copernicus argues here is completely unacceptable: namely, that the fixed stars are unmoving, and any movement that they might appear to have is explained by the rotation of the earth on its axis over an approximately twenty-four-hour period (postulate 6), and that the annual movement of the sun is in reality the annual movement of the earth around the sun (postulate 6). These ideas would provoke an immediate negative reaction against the text that would quickly translate into a complete prohibition of Copernican thought; in fact, the decree against Copernicus was made public with placards in the streets of Rome, where Saavedra Fajardo was then living.

By the time Saavedra Fajardo wrote his treatise, it was well known that it is the earth and not the sun that moves, in terms of both its own rotation and orbit. With this, Saavedra Fajardo's motto takes on a strange relevance, in that it appears to be associated with a (ultramodern) post-Copernican device such as the telescope. This in turn generates a reading that is simultaneously traditional and innovative, a Galilean approach that can be made out only after having read the declaration that follows it. Nevertheless, what is most significant about this *empresa* is the way in which Saavedra appropriates a motto traditionally associated with cosmography and transforms it, through its imbrication with the chosen picture, into a useful reflection not only on the nature of the monarch and his passions but also on the very act of looking. In its metapolitical constitution, the *empresa* reminds us that "the diversity of judgments and opinions" (*la diversidad de juicios y opinions*) and "the varying estimation of objects" (*la estimación varia de los objetos*) are always the result of "the light to which one puts them" (*la luz a que se los pone*). There is little that is new in this perspectivist argument, as evidenced by the innumerable examples one finds in earlier decades and the elevation of the conceit to the level of philosophy by satirical works as well as the prose of Miguel de Cervantes. What is in fact significant and new about Saavedra Fajardo's *empresa* is the fact that this personal gnoseology, applied to the royal body, is undoubtedly nourished by new discoveries in optics:

> Inevitably, when we look at things through a telescope, things present themselves to us either much larger and substantial or much smaller and reduced. The lenses and the things seen through them are of course different and discrete things. The difference, however, comes from the fact that when the species, that is, the visual rays, pass through one part of the tele-

scope, they move from the center of the lens to the circumference and so spread out and multiply, making the objects seems larger; and when they pass through the other part of the telescope, they move from the circumference to the center, and they arrive smaller. Such difference is produced by looking at things in one way or another.

> *No de otra suerte nos sucede con los afectos que cuando miramos las cosas con los antojos largos, donde por una parte se representan muy crecidas y corpulentas, y por la otra muy disminuidas y pequeñas. Unos mismos son los cristales y unas mismas las cosas, pero está la diferencia en que por la una parte pasan las especies, o los rayos visuales del centro a la circunferencia, con que se van esparciendo y multiplicando, y se antojan mayores los cuerpos, y de la otra pasan de la circunferencia al centro, y llegan disminuidos. Tanta diferencia hay de mirar desta o de aquella manera las cosas.* (244)

This short digression on epistemology and optics announces in itself the familiarity that Saavedra Fajardo had with what was then an essential element of the Baroque *Wunderkammer*, namely, the *antojo largo* or *visorio*. Through his use of the telescope, Saavedra Fajardo manages to construct a "reflection" that the entire book seeks to develop through its concern with Horatian *prodesse et delectare* (instruct and delight).

Within *empresa* 7, there is also an important refraction manifested in a series of less evident motifs that are appreciated, upon a second reading, after probing more deeply into the very equilibrium of forces that Saavedra experiences as a mediator, upon passing through that initial effect of the telescope. This historical substrate is of enormous importance from a diplomatic point of view, since the telescope had been in essence a Venetian agent, associated with vision, espionage, and betrayal. I have already commented on how the concave and convex lenses of the telescope to which Saavedra Fajardo refers had been polished and perfected by Galileo in collaboration with the glassworkers of Murano, and Galileo had presented his *cannocchiale*—as Emanuele Tesauro called it a few decades later—to the doge of Venice with great success, and he was also recognized with honors from the senate for his work. It is not surprising, then, to see Boccalini and his *Ragguagli di Parnaso* (Newsletter from Parnassus) converted into a possible target by Saavedra Fajardo in *empresa* 12, although the latter does not mention the former by name: "What defamatory libels, what false manifestos, what feigned Parnassoi, what malicious placards have not been dispersed against the Spanish monarchy?" (*¿Qué libelos infamatorios, qué manifiestos falsos, qué fingidos Parnasos, qué pasquines maliciosos no se han esparcido contra la monarquía en España?*)

(289).[32] The answer to this question takes on life, as I have pointed out, in the premature death of Boccalini, a death attributed—falsely, as it was later proved—to a group of Spanish thugs financed by Madrid.

The Venetian origin of the telescope, with its affective charge of antimonarchic pride, leads also to another epistemological consideration associated with the idea of equality and democracy. As Timothy Reiss has pointed out, the telescope had already by the seventeenth century come to play a metaphorically revolutionary role to the extent that, given its *versabilità,* was able to unite and divide, to connect and separate, to increase and diminish (Reiss 26–27). It is an equally literal metaphor in that it becomes not only the bridge that unites two realities, or the instrument by which one might appreciate the finer details of an object (for example, the equally sensual and controversial imperfections of the moon's surface), but also the object itself. To put it another way, Saavedra Fajardo transforms the telescope into an instrument accessible to all, one that breaks with the broader spirit of the *Empresas* as much in political as in religious terms. The sun emerges now as corrupt, and the moon, with its famous "face" popularized by Plutarch, is seen to be filled with imperfections and "wrinkles" thanks to the optical effect produced by its mountain ranges, destroying the utility of these celestial bodies as emblems of power, of an inaccessible, uncontestable power. For Saavedra Fajardo, each exercise of authority seems to be constituted by complex folds, by surfaces of action filled with rocky outcroppings. One might note, for example, how the emblem in question presents neither the observer nor the object observed, since the telescope is the question itself regarding the arbitrary nature of language, of the medium, of that bridge, which is, ultimately, a metaphor; the telescope is, in the end, the removal of a mask: for Galileo, Reiss reminds us, "knowledge is a sign-manipulating activity" (33). No other mediating means from the period—compasses, clocks, architectural drawing tools—acquires its own life to the extent that the Galilean telescope does, because none of these is able to place so thoroughly in doubt the incorruptibility of the monarchy. Commenting on Saavedra Fajardo's *empresa* 57, titled "Uni reddatur" (One is given), Otto Mayr reminds us: "Government should work with the harmony of a smoothly running clockwork. The power of the sovereign was to be unlimited and absolute, and government was to be centralized with the king as the source of all power and as the author of all initiative" (103–4). Scientific mechanics thus turns into political mechanics; and Galileo deeply complicates Saavedra Fajardo's work by taking away one of the latter's preferred pedagogical conceits: the sun as an incorruptible body.

In revealing the details of a reality previously hidden from the human eye, the telescope presents itself as a threat to the protective mask that Saavedra Fajardo proposes in his *Empresas políticas.* There also pulses within the telescope, it is worth noting, an antithetical spirit of great magnitude: from a geopolitical point of view, it embodies the commercial activities of guild societies such as Padua and Venice, modern cities in which scientific experimentation could flourish without any of the tight surveillance to which it was routinely subjected in Madrid and Rome. According to Battistini, the telescope possessed no aristocratic prestige, given that artisans and mechanics had developed it ("Telescope" 13).[33] In general, the nobility considered it a frivolous and even childish instrument. With respect to Saavedra Fajardo, who was acutely aware of the steep decline that characterized the decade in which he was working, the telescope was little more than a distraction for the serious spirit—with no time for bagatelles and toys—that runs through *Empresas.*[34] Given that Saavedra Fajardo had forged his career beside the popes in Rome, the figure of Galileo (and, by extension, his telescope) could not logically signify anything for him but a grave threat.

I do not wish to close the door, however, on another possible reading of Saavedra Fajardo's reckoning with Galileo's telescope. This reading concerns a man relatively open to innovation, a man who traveled extensively and saw things that were quite simply hidden from his contemporaries back in Madrid. It also involves a writer who suggests, with great diplomacy, in his *empresa* 29: "Innovations are not always dangerous. Sometimes it is worth adopting them. The world could never be improved if we did not innovate" (*No siempre las novedades son peligrosas. A veces conviene introducillas. No se perficionaría el mundo si no innovase*) (423).[35] It is in this way that the telescope situates itself in Saavedra Fajardo's text: dislocated, uncomfortable, without knowing how to adapt to its circumstances, but tremendously fertile as an object to be read, given its marvelous ambiguity. It is thus likely most useful to understand Saavedra Fajardo to have been a thinker profoundly defined by the geopolitics of his time, who recognizes this important oscillation in his *Empresas,* and, as a result, who presents the Galilean telescope not so much as a heretical instrument but rather as something quite different: as a symbol for the suppression of dissimulation, and as a leveling agent, contrary to the designs and the nature of the goals explicitly sought in *Empresas.* Through the very "laconic brevity" of expression that for Jorge García López (67) links Saavedra Fajardo to other innovators of Galileo's time (including Galileo himself), *Empresas políticas* takes up two forms of inertia of significant interest to the modern reader: on the one hand, the capacity to stitch together a lexicon taken from science with political counsel; on the other hand, the ability to situate the telescope at the very center of the energy produced by the three-way collision between the ancient and the modern, the heretical and the orthodox, and last, the totalitarian and the democratic. For this reason, Saavedra Fajardo's *auget et minuit* stands as a motto that is nothing if not extraordinarily symptomatic of the situation in which its author was living—at the crossroads of the new (in its waxing) and the old (in its waning), and of what was then being made visible by Baroque science and the opacity of a gaze so focused on finding the diminished off in the distance. His entire declaration, in fact, is a probing gloss on the power of mechanics, translated by him to the realm of politics. Because if Saavedra Fajardo's *empresa* seeks to limit one's gaze, Galileo's telescope unbinds it and sets it free. And if Quevedo and Saavedra Fajardo teach us anything through their respective readings of the complex exercise of power—exactly as their contemporary Baltasar Álamos de Barrientos (1555–1640) had demonstrated only a few years before them in his *Tácito español* (Spanish Tacitus, 1614)—it is that one must always understand that politics is a science just as science is always the practice of politics.[36]

✺ 8 ✺

Reverberations

The middle third of the seventeenth century in Spain manifests signs of ambivalence with respect to post-Copernican science. While one may still find during this period more than occasional adherence to geocentric theories of the cosmos, there are also enthusiastically positive descriptions of new optic instruments that take seriously their capacity to provoke doubt or suspend judgment. We find an interesting case of this in Baltasar Gracián's *El criticón* (The critic, 1651), which simultaneously grants the sun a central place in the known universe and describes it as "flawless," the latter a vision that seems to adhere to more traditional theories of the sun's immaculate nature:

> The sun, thought Critilo, is the creation that most obviously demonstrates the majestic greatness of the Creator. It is called *sun* because in its presence all other light sources withdraw. It alone remains. It is situated at the center of the celestial spheres, the heart of illumination and perennial spring of light. It is flawless, that is, unchanging, unique in beauty. It opens up all things to being seen but it cannot be seen, concealing its decorum and hiding its decency; it influences and helps to give being to all things, even man himself. It is affectedly communicative with its light and its happiness, spreading itself everywhere and penetrating even to the very entrails of the earth: it bathes and influences everything, making the earth happy, illuminated, and fertile. It is just, since it shines for everyone; it has no need of anyone below and everyone recognizes their dependence upon it. The sun is, in the end, a creature of ostentation; it is the most brilliant of mirrors, in which divine greatness is represented.
>
> *Es el sol—ponderó Critilo—la criatura que más ostentosamente retrata la majestuosa grandeza del Criador. Llámase sol porque en su presencia todas las*

> *demás lumbreras se retiran; él solo campea. Está en medio de los celestes orbes, como en su centro, corazón del lucimiento y manantial perene de la luz. Es indefectible, siempre el mismo, único en la belleza. Él hace que se vean todas las cosas y no permite ser visto, celando su decoro y recatando su decencia; influye y concurre con las demás causas a dar el ser a todas las cosas, hasta el hombre mismo. Es afectadamente comunicativo de su luz y de su alegría, esparciéndose por todas partes y penetrando hasta las mismas entrañas de la tierra: todo lo baña, alegra e ilustra, fecunda e influye. Es igual, pues nace para todos; a nadie ha menester de sí abajo y todos le reconocen dependencias. Él es, al fin, criatura de ostentación, el más luciente espejo, en quien las divinas grandezas se representan.* (528)

These lines were published in the 1650s. Galileo had died in 1642, and one can discern in many Spanish writers born after 1581 an increasing acceptance of the new and an unquestionable respect for the art of innovation that was at the moment so profoundly modifying the perceived physiognomy of the universe. Italy was not the only source of inspiration, however. The influence of ideas and texts coming from Central and Northern Europe invites us to discuss, even if briefly, a number of pieces that made their way to print in the Iberian Peninsula during this period. A good case with which to begin this cursory survey is the *Cursus philosophicus* (Philosophical course, 1632) of Rodrigo de Arriaga (1592–1667), an Aragonese Jesuit and professor at the University of Prague who argued that arguments supporting the incorruptibility of the sun were at best specious. He joined Diego de Saavedra Fajardo, who advised the readers of his *Empresas políticas* to remain open to change as a catalyst for progress. Described as a technological wonder by another well-traveled writer, Francisco de Quevedo, the telescope allowed fictional writers in Spain to develop a deeply subtle worldview, more open than ever before to the fascination that this instrument provoked. In doing so, these authors sowed seeds that would produce remarkable fruit in the second half of the century. The sun, of course, was at the core of many reflections of the period. The engraving that served as the frontispiece for the 1670 edition of Emanuele Tesauro's *Il cannocchiale aristotelico* (The Aristotelian telescope) presents a feminine figure holding in her left hand a telescope pointed toward the sun, but a sun with obvious stains, a fact that underscores the dissonance between Aristotle, who is helping the figure to aim her telescope, and Galileo, who through his stained and corruptible sun had dismantled an entire tradition (fig. 11).

This is a case in which, as Mercedes Blanco has argued ("El meca-

FIGURE 11. Emanuele Tesauro, *Il cannocchiale aristotelico* (1670)

nismo" 25), "speaking of the Aristotelian telescope is a true oxymoron, a paradox, and even a kind of erudite joke" (*hablar del catalejo aristotélico es pues un verdadero oxímoron, una paradoja, y hasta una especie de chiste erudito*). We know by now that the reverberations of Galileo's intellectual legacy within the sphere of Spanish Baroque fiction give form to works of true technical virtuosity. For a genre like theater, which experiences a phenomenal development in the 1640s and 1650s from a technical point of view, certain kinds of optical play derived from glass are salient when not central features. In some of its best examples, glass is not just a material; it is also a metaphor for power, an identitary allegory, and a vehicle for knowledge. If mirrors elevate the rank of dramatis personae in many *come-*

dias of the period, equally important is the optical illusion produced by new perspectival effects. Keenly aware of the world around her, the muse refracts into new modes of expression and new manners of representing a reality that is always multifaceted; and it is during this time that a new generation of thinkers and authors produce works defined by a deep engagement with scientific innovation. Central among these are Pedro Calderón de la Barca, Bernardino de Rebolledo y Villamizar, Miguel Barrios, Baltasar Gracián, Andrés Dávila y Heredia, and Francisco Santos. These six figures seem to assume with less ambiguity and with greater expository freedom the benefits offered by the political lens, which they readily employ as a powerful literary instrument.

Foreign muses, local verses

The careful elaboration of sophisticated atmospheres marks Baroque Spanish prose during the middle third of the seventeenth century, especially in the construction of inner—and sometimes intimate—spaces by novelists such as María de Zayas and Mariana de Carvajal y Saavedra. This is going to be achieved as well in other genres like *comedia,* as new trends in stage design coming from Italy incorporated more and more complex notions of light and sound on stage. In response to María Alicia Amadei-Pulice's analysis of the particular case of Calderonian theater in her book *Calderón y el barroco,* Francisco Muñoz Marquina has linked the nature of the universe to that of the Baroque theatrical stage, united by their shared ability to trick the eyes:

> It is no coincidence that Baroque theater follows very closely Galileo's precept that "we should learn with all of the certainty that the evidence of our senses affords us." In the wake of the observations carried out by the Tuscan astronomer with his telescope, the space of the cosmos and that of Baroque theater come to have a shared perspectivist point of orientation. Theater, with its foregrounding of the visual and especially visual illusion, duplicates the human error proven by Galileo with respect to the universe: that it is wrong to believe that the sun revolves around the earth.
>
> *No es casualidad que el teatro barroco siga de cerca el precepto de Galileo Galilei: "debemos aprender con toda la certeza que nos brinda la evidencia de nuestros sentidos". Después de las observaciones realizadas por el astrónomo toscano con el telescopio, el espacio cósmico y el teatro barroco tienen un mismo punto de orientación perspectivista. El teatro, con su disposición visual e ilusio-*

> *nista, duplica el equívoco humano comprobado por Galileo en el universo: el que cree ver el Sol girar alrededor de la Tierra se engaña.* (281)

It is true, as Amadei-Pulice has written, that the arrival of Cosimo Lotti in Madrid radically altered both the production and the enjoyment of theater, establishing "the technical bases of a new Baroque genre, conceived as a compendium of all the arts and sciences" (*las bases técnicas de un nuevo género barroco, concebido como compendio de todas las artes y ciencias*) (27) that had been developed up to that time in Italy. For Calderón, she argues, Florentine mannerist technique applied to theatrical practices lends itself very well to the production of effects such as marvel, surprise, and expectation in the audience, an essential part of early modern theater. We know that as of 1626, Lotti begins to focus on Philip IV's gardens, fountains, and royal theater, incorporating in a systematic way the scenic formulas developed in Italy. He arrives with a profound knowledge of the Italian courtly theater of the period, from the theories of Vitruvius to cutting-edge set design, the latter of which included the use of elaborate machinery and perspectival sets. With regard to perspective, Lotti brought with him ideas inspired by Giulio Parigi's work with the Academy of the Arts of Drawing (Accademia delle Arti del Disegno) in Florence. The academy, which served Tuscany in much the same way that the Academy of Mathematics did in Madrid, had by then been converted into a prestigious institution where one could study civil and military architecture, mathematics, Euclidean geometry, and the new science of perspective with applications in military strategy as well as in theatrical design. Galileo himself had studied *disegno,* a term encompassing fine art, and in 1588 had obtained the position of instructor in the academy, teaching perspective and chiaroscuro. Being inspired by the artistic tradition of the city, he had acquired an aesthetic mentality, befriending the Florentine painter Lodovico Cardi (1559–1613)—also known as Cigoli—who included Galileo's lunar observations in one of his paintings. Beyond artists, the Florentine academy had among its members some of the brightest stars of the city's vibrant intellectual scene: Guidobaldo del Monte, Giulio Caccini, Ottavio Rinuccini, Claudio Monteverdi, Bernardo Buontalenti, Francesco Buonamici, and Angelo Ingegneri. The academy was responsible for substantial advances in the study of percussion, voice, optics, and perspective. Galileo's father, Vincenzo Galilei, had been one of its members, and one of singular importance for the development of music theory; his famous *Dialogo di Vincentio Galilei nobile fiorentino della musica antica e della moderna* (Dialogue by Vincenzo Galilei, noble Florentine, on antique and modern music, 1581), in which he brought together his research and theories on Greek music, became very well known in Spain thanks to the work of Jusepe Antonio González de Salas, who transmitted the work of Galilei by applying modern musical theory in his pivotal treatise *Nueva idea de la tragedia antigua* (New idea of ancient tragedy, 1633).[1]

It is important not to assume that the intersection of Baroque theater and the new science in Spain can be reduced to Lotti and González de Salas, as many Tuscan writers and thinkers in fact traveled to Spain to put into practice what they had learned. This was the case with, for example, Giulio Fontana, Francesco Ricci, Bernardo Monnani, and Giulio Rospigliosi (the future Pope Clement IX). Amadei-Pulice reminds us also that other important figures were linked to the Florentine academy, such as the Marquis of Sant'Angiolo, a master field marshal during the wars in Flanders and the siege of Breda; the Marquis Della Stufa, whom the king admitted into the knightly order of Alcántara for his work as a field marshal in the war of Milan; Vicente Carducho, best known, as we saw already, for his *Diálogos de la pintura* (Dialogues on painting, 1633); and Ludovico Incontri (Volterrano), who was an Italian ambassador to Philip IV, mathematician, and disciple of both Giulio Parigi and Galileo. The science of optics, therefore, had innumerable practical applications from which the Spanish were able to extract ample benefits on the military field, and with it constituted something more than a discipline of knowledge: if, as we have seen in the case of Diego de Saavedra Fajardo, it served to construct a certain image of the vigilant monarch, it was also a foundational element of political hegemony for an empire in need of immediate solutions. This was a trajectory that also manifested itself in the opposite direction: Daniel Selcer, in fact, reminds us of the great "theater of the world" (60) and the "theatrical scenes" (62) that organize Galileo's *Dialogue concerning the Two Chief Systems*. If there is an expressive mode that is able to unite technology and language, it is without a doubt courtly theater; but Baroque science, as Selcer reminds us, will also extract what is necessary from theatrical language to achieve an optimal effect in its own representational mechanisms.

This dialogue yields groundbreaking results in Spain. The 1650s brought with them a series of important changes, many of which began emerging in 1649 with the arrival in Madrid of Mariana of Austria, Philip IV's second wife. In 1650, the Royal Coliseum was reconditioned, and the following year Baccio del Bianco, a Florentine engineer, took over for Cosimo Lotti, assuming the role of director, set designer, and choreographer, among many other functions. He and Calderón de la Barca collaborated in a series of classic productions: in 1652, there was the premiere of *La fiera, el rayo y la piedra* (The beast, the ray, and the stone); and in 1653, he produced a spectacular version of *Andrómeda y Perseo* (Andromeda and Perseus) promoted by the princess Maria Theresa to celebrate the restored health of her stepmother. A few years later, before his sudden death in 1657, Baccio del Bianco would stage *El golfo de las sirenas* (The gulf of the sirens) and the light opera *El laurel de Apolo* (The laurel of Apollo) together with Calderón, who, it bears mentioning, did not always approve of the extravagance of his partner.[2] In this six-year period, during which theatrical technology would develop in significant ways, Madrid's theaters made elaborate stage productions their specialty.

In a classic study, Ferruccio Marotti has connected the aforementioned Guidobaldo del Monte with Galileo with respect to a shared conception of space, one that sought to overcome the pictorial two-dimensionality of the Renaissance and replace it with sets that would be three-dimensional, changeable, symmetrical, and concave. Spanish Baroque theater shows us, in fact, how Galileo's *prospettiva del mezzo* would influence a wholly new systematization of space; that is, the world was now imagined in another way, and with sight as the most noble of all the senses. Roberto González Echevarría reminds us that Calderón had studied mathematics in addition to law at the University of Salamanca, stressing the impact of modern science reflected in his dramatic output: "The Baroque ecstasy that Calderón expresses in his theater . . . is the counterpoint to the vertigo caused by the sense of emptiness triggered by the infinite brought about by the new cosmography" (*el éxtasis barroco que expresa Calderón en su teatro* [. . .] *es la contrapartida del vértigo causado por la sensación de vacío que genera el infinito producto por el conocimiento de la nueva cosmografía*) (157). Defined as a "staunch Galilean" (*acérrimo galileista*) (176) by Amadei-Pulice, Calderón documents these epistemic transformations in his little-known treatise "Deposición a favor de los profesores de pintura" (Deposition in favor of the professors of painting, 1677) in which he speaks of these changes using a number of scientific terms like "waxes and wanes" (*aumentos y disminuciones*).[3] In this treatise, Calderón makes it explicit that a goal of Baroque theater is to duplicate the great human error pointed out by Galileo with respect to the universe, consistent with the mistaken belief that the sun revolves around the earth. This has profound epistemological consequences, as William Egginton has indicated: "The Baroque puts the incorruptible truth of the world that underlies all ephemeral and deceptive appearances on center stage, making it the ultimate goal of all inquiry; in the same vein, however, the Baroque makes a theater out of truth, by incessantly demonstrating that truth can only be an effect of the appearances from which we seek to free it" (*Theater* 2). This "theater of truth" essentially magnified invention; and as the century advanced, the by then firmly established sense that nothing was what it seemed, only intensified. Advances in set design played a central part in this "theater": they facilitated the transfer of the audience to marvelous places through continuous visual play, forcing them at the same time to reflect on the journey itself.

With the premiere of *La púrpura de la rosa* (The purple of the rose) on January 18, 1660, the last decade of Philippine rule officially opens up. This is a moment in the history of Iberian culture in which innovations in set design had essentially turned Madrid into a leading center of European theatrical activity.[4] This technological and metatheatrical activity, was not, however, limited to Spain, as we also see innovations in other parts of Europe: a set designer such as Stefano della Bella, who arrived in Paris in 1641, had painted on the wings an image of Galileo showing his famous Medicean stars to three Florentine ladies; Giacomo Torelli, for his part, created magnificent sets for Pierre Corneille in the court of Louis XIV; in England Inigo Jones and Costantino de'Servi made similar innovations; and in Ulm, Joseph Furttenbach made known the theories of Galileo on scenic perspective. By the middle of the seventeenth century, we are in a period of great advances in optics, since after 1660 enormous lenses were constructed thanks to improvements in the techniques of polishing glass. Robert Hooke would publish his *Micrographia* in 1665, and Marcello Malpighi and Anton van Leeuwenhoek would develop his work further with important results. Simple microscopes, very popular toward the end of the century, were being exchanged for compound ones—manufactured primarily by Christiaan Huygens—that possessed three lenses: an objective, an eyepiece, and a field lens.

The many visual mechanisms of the Baroque would generate, in the end, a final game characterized not by the presence of the marvelous but by its very absence. As the century advances, one finds with greater frequency this type of production, which relates only tangentially to the focus of the present book but merits the mention of a small number of important titles: Álvaro Cubillo de Aragón, for example, would explore the theatrical effects of invisibility in *El invisible príncipe del baúl* (The invisible prince of the trunk, 1637); in *El hechizo de Sevilla* (The spell of Seville, 1653), Ambrosio de Arce de los Reyes explores the consequences of enchantment; Francisco Bances Candamo, in *El entremés de las visiones* (The interlude of visions, 1691), develops a similar treatment of this novel inclination; and José de Cañizares, already at the crossroads of two very different moments, would exploit the scenic possibilities of invisibility in *El anillo de Giges, y el mágico rey de Lidia* (The ring of Giges, and the magical king of Lydia, 1764). In all of these one finds reflection, at times implicit, of the marvels of optics: of the consequences of making the visible invisible and the invisible visible, of mixing comic elements with a serious reading of reality, of the ethical foundation of the Baroque framework that defines this historical moment.

These concerns, it is worth pointing out, were not limited to the stage. Spanish poetry from the middle third of the seventeenth century also provides a series of examples that must be considered when attempting to reconstruct the post-Copernican literary scene. Such is the case with the poetry of Bernardino de Rebolledo y Villamizar (1597–1676), which up until recent years had been largely neglected. A highly experienced soldier and diplomat, Rebolledo served as plenipotentiary minister to Philip IV in Denmark between 1648 and 1661. It was there that he composed nearly all of his poetic oeuvre, made up of numerous sonnets, long poems and brief theatrical pieces, didactic works, and biblical translations. The fact that he worked far from the learned societies and literary academies of Madrid endows his work with a strange originality—particularly evident in his sonnets—although one can also detect in its doctrinal content echoes of his contemporaries Francisco de Quevedo and Bartolomé and Lupercio de Argensola (González Cañal).

Rebolledo published his collected poetry in *Ocios* (Leisures, 1650), initially printed by the Officina Plantiniana in Antwerp and reedited in 1660 as part of a collection of his complete poetic works. These include *Selva militar y política* (Military and political forest, 1652) and a poetic genealogy of the royal Danish house titled *Selvas dánicas* (Danish forests, 1655). In various places throughout his poetic work, one finds a preoccupation with new scientific currents, in particular with respect to heliocentric theory. If he admittedly never dared openly to defend the new theories emerging from Galileo's *Dialogue*, one can nevertheless detect a state of doubt in his work, a doubt that crystallizes into what continues to be his most famous maxim: "who doubts nothing knows nothing" (*todo lo ignora quien nada duda*). In the second tercet of *Ocios*, for example, Rebolledo provides a poetic list of the authors who had attributed movement to the earth, mentioning Copernicus and Galileo at the end. The poem provides evidence of a certain familiarity with the new physics, although he presents it through a thick filter of irony. In speaking of Galileo, however, the verdict appears to be clear:

> Copernicus, so close to our own time,
> wishes to send the earth around the sun,
> against human and even divine sensibility.
> Galileo, follows and prefers him.
> He lit this obstinate flame in the moderns
> such that no one expects to see it extinguished.
> But I, along with Owen, would judge
> that he had just dined or was sailing somewhere
> when the earth appeared to him to be moving.
>
> *Copérnico, a estos tiempos ya vecino,*
> *alrededor del sol traerla quiere*
> *contra el sentir humano y aun divino.*
> *Galilei, que le sigue y le prefiere,*
> *encendió en los modernos la porfía*
> *tanto que no hay quien apagarla espere.*
> *Pero yo con Oveno juzgaría*
> *que acabó de cenar o navegaba*
> *cuando le pareció que se movía.* (ll. 329–37)[5]

With that fascinating "lighting of the flame," Rebolledo suggest that there is no turning back, even if it may seem that what Galileo had seen is merely the product of altered faculties. In the end, it is not surprising that Rebolledo chooses a traditional path; however, beyond his humorous swipe at Galileo, what remains clear for him is that the obstinate "modern flame" of heliocentrism was not expected to be put out, given that Galileo's ideas were by then largely unquestioned by many of his contemporaries. Rebolledo would take up these ideas once again in his *Selvas dánicas,* where he exhibits his knowledge of and concerns regarding the scientific questions that were likely of great import in Denmark (manuscript copies of Tycho Brahe's work were by that time actively circulated and recopied) but did not convince him completely.

In the *Selva militar y política,* Rebolledo states that "all here that moves / remains constant in divine essence" (*todo cuanto aquí tiene movimiento / está constante en la divina esencia*), and he advises the prince that he should not be concerned with these matters:

The monarch who loses sleep
worrying about the revolution of the heavens
when he should be correcting those of the earth,
and occupied by carefree study,
neglects the government of the state.

El monarca que aplica sus desvelos
a las revoluciones de los cielos,
debiendo corregir las de la tierra,
y en ociosos estudios ocupado,
descuida del gobierno del estado.[6]

These verses call to mind Saavedra Fajardo's admonitions to the king in his *Empresas políticas*, as we saw in previous pages, in which he counseled the monarch on how to lead effectively through emblems that exploited the metaphoric possibilities of glass and the mechanics of different instruments of calculation. The "revolutions" that are mentioned in the poem bring to mind a Copernicus that should not become a distraction when there is so much to be concerned about with the wars—the other revolutions mentioned in the poem—taking place in Europe. This was clearly a concern shared by other Spaniards abroad, and in particular by those with expertise on technical matters: similar reverberations would be felt in another important piece of the period, *Breve descripción del mundo* (Brief description of the world, 1688), which was penned by the engineer and professor in the Military Academy of the Spanish Army in Flanders, Sebastián Fernández de Medrano, and which also warned against placing the sun at the center of the universe.

However, these last decades of the century also witnessed a number of fascinating conflations of the scientific with the purely fictional. We might take, for example, the case of the crypto-Jewish poet Miguel Barrios (*né* Daniel Leví, 1635–1701). Barrios spent his youth in Portugal and Algeria, later going to Italy and still later to Nice, where he lived for a time. In Brussels, he joined the Spanish army in Flanders, and soon earned the rank of captain. There he wrote his *Flor de Apolo* (Flower of Apollo), a handful of dramatic pieces, and his *Coro de las musas* (Chorus of the muses), with panegyric compositions devoted to notable European princes. In 1672, he moved to Amsterdam, where he lived until his death. From 1674 onward he focused much of his energy on business, but he also composed poetry and a history of the Spanish language, both the product almost certainly of the freedom that he enjoyed in the Netherlands. Barrios founded a literary academy in 1676, the Academia de los Sitibundos, of which he was president. At the beginning in 1685 he was named president of another academy, the Academia de los Floridos, whose thirty-eight members were devoted to literature and represented the most elite sector of the Jewish community of Amsterdam. In Barrios, we find an entire life dedicated to intellectual pursuits in the context of a Netherlands that was, according to what Simon Schama and others have argued, in the middle of its own "golden age," excelling in the arts thanks to artists like Hals, Vermeer, and Rembrandt.

As Inmaculada García Gavilán has pointed out, Barrios includes within the first forty verses of his little-known *Fábula de Prometeo y Pandora* (The story of Prometheus and Pandora) an interesting *geographia mundi* that proves his knowledge of Copernicus's *De revolutionibus*. He concludes this cosmography with an ethical discourse filled with motifs that remind one of the condemnatory biblical tradition of heaven and hell, and in which he expresses his pessimism regarding the spiritual decline that was consuming his contemporaries (ll. 41–72). Barrios's celestial cosmography, based on the *Astronomica* (attributed to Marcus Manilius), the *Catasterismi* (pseudepigraphically attributed to Eratosthenes of Cyrene), and Andreas Cellarius's *Harmonia macrocosmica* (1660), provides an excellent example of the dexterous handling of which he was capable not only with respect to mythical allusion or catasteristic mythology, but also to the astral cartography of his time. There can be no doubt that his crypto-Jewish upbringing, which was enriched by his travels through Europe, as well as his participation in literary academies and the contacts he established in places such as Brussels and Amsterdam—where most of his work was written—provided him with an environment that was much more open to innovation. He constituted, along with Rebolledo, one of the most important lyrical voices in a cultural scene that, after the death of Quevedo in 1645, produced very few talented poets. These contacts between the local and the foreign created networks of information that established, it bears mentioning, new circuits of knowledge while also imbuing the poetic lexicon with new and highly subtle links between the erudite and the practical. And if optics and astronomy remained relevant in genres like poetry and theater, the novel would not fall short in the use of glass as a powerful narrative tool.

Strained vision: The eyewear shop

One finds a literary scene in late seventeenth-century Spain that is both deeply engaging and almost wholly dependent on earlier sources. The eyeglass shop, for example, once again appears as a motif, and it serves principally to denigrate, through caricature, the Baroque subject's surfeit of vanity and, by extension, the collection of ills that weighed down Spanish society during the last third of the century. It is important to remember that the eyeglasses of this later period are primarily associated with artisans and fairs, as well as with older, respectable people; on other occasions, however, they were the target of ridicule and convert their users into the object of frequent jokes. They are largely uncomfortable to wear, and in some cases, they are even considered harmful to one's health. It is well known that in *El criticón* (1651) the reference that Gracián makes in his prologue to Trajano Boccalini—"the crises of Boccalini" (*las crisis de Boquelino*)—as well as his description of the Parnassus store in the seventh *crisi*, in which one could buy, among other things, eyeglasses and gloves. Gracián introduces, in fact, a good deal of Lucian and Boccalini into his text, which allows him to "manipulate" the properties of glass and describe certain optical effects. These include the "universal fair" (*feria universal*) in *crisi* 13, as well as the "window to the human chest" (*ventanilla del pecho humano*), which Salas Barbadillo had cultivated with excellent (satirical) results in *La estafeta del Dios Momo* (The god Momo's post office, 1627). For Salas Barbadillo, such a window is wholly unnecessary for the person who knows how to see correctly, and the figure of the seer, so profitably used before by Quevedo, returns once again:

> Momo truly aimed low when he asked for a window in the human chest; this was no case of censorship but rather confusion, since (it should be mentioned) the seers of hearts . . . do not require even the smallest crack in order to penetrate into the most private interior. Transparent glass is of no use to one who sees through long-distance lenses.

> *Muy a lo vulgar discurrió Momo cuando deseó la ventanilla en el pecho humano; no fue censura, sino deslumbramiento, pues debiera advertir que los zahoríes de corazones* [. . .] *no necesitan ni aún de resquicios para penetrar al más reservado interior. Ociosa fuera la transparente vidriera para quien mira con cristales de larga vista.*[7]

Gracián's diverse and excessive *feria universal*, at once cacophonous and marvelous, leads us to two of the texts with which I wish to end my survey. These are *Tienda de anteojos políticos* (Store of political eyeglasses, 1673) by Andrés Dávila y Heredia, and *El sastre del Campillo* (The tailor of El Campillo, 1685) by Francisco Santos.

Dávila y Heredia (1627–1686) writes from a singular perspective that serves to make his prose somewhat strange. Having studied in some of the most prestigious universities in Europe while serving as an imperial soldier, he actively defended university astronomy against the local practices of the necromancers, erasing in this way the borders between science and literature at a time when the royal body of Charles II symbolized, in its weakness, the contemporary social and cultural body (Martín Vega 125–42). His most important work is the novel *Tienda de anteojos políticos*, which begins with a devil opening up a business in the royal court with Lucifer at its helm. The exchanges that follow present an entire gallery of courtiers who are severely ridiculed according to the eyeglasses that they request. In each case, Lucifer sends his customer home after offering a stern tongue-lashing. This is one of the most relevant parts of the book, insofar as the author deploys substantial erudition with respect to historical and literary themes while denouncing the social ills of his time. The novelty with respect to earlier novels resides in the fact that the lenses that Dávila y Heredia presents (and Lucifer sells) do not transform one reality into another, as in Fernández de Ribera, nor do they lay bare the object under scrutiny, as in Vélez de Guevara. In Lucifer's eyeglass shop, the client is invited to try on different eyeglasses in order to see, as if through a kaleidoscope, a kind of cinematic allegory that only exists within the invention. Dávila y Heredia thus offers his reader something akin to a telescope, which corresponds more with astronomical discoveries at the end of the century achieved through powerful lenses used not so much as filters but rather as means to access new realities. Lucifer occasionally receives the requests of his customers with interest, but there is also anger, as is the case of the page at whom the devil hurls a flask for having asked for eyeglasses that might improve his forecasts. Having launched the flask, Lucifer then tells him: "such eyeglasses do not exist, nor have they been made, nor can they be made" and "the privilege of seeing the future remains ours alone" (*no hay semejantes antojos, ni se han fabricado, ni se pueden fabricar* and *este privilegio de arbitrios se queda para nosotros*) (148, 156).

Throughout the text, which lacks any discernible plot and ends abruptly, the reader sees different possible uses for lenses in an urban society that was progressively converting itself into what Norbert Elias has referred to as "the civilizing process"—or here, in the hands of Dávila y Heredia, perhaps more a kind of "impoverishing process." He substitutes the virtuosi of Mount Parnassus with vain and insane men who pass through a disorienting urban landscape that, without being as grotesque as that of its predecessors, is equally disheartening. In *Tienda de anteojos políticos*, Lucifer complains bitterly about the vanity of the court, claiming at one point that "here in the court eyeglasses are not sold to discover truths, but to cover them up" because "this entire age is one of ignorance and malice" (*aquí en la Corte no se venden anteojos de descubrir verdades, sino de encubrirlas* [*porque*] *toda la era de hoy es ignorancia y malicia*) (100–101). In many cases, clients ask Lucifer for eyeglasses simply to seem more clever, more distinguished, more honest, and even to give a good first impression. The text is filled with fascinating revelations, such as when a group of courtiers enters into the shop looking for eyeglasses to "make political predictions" (*pronosticar en lo político*) provoking the anger of the shopkeeper "because such predictions have more to do with politics than with astrology, since their goal is to destroy spirits, and to wreck hopes" (*porque semejantes pronósticos más tienen de política que de astrología, siendo su mira a destruir los ánimos, y a variar las esperanzas*) (118–19). The swan song of a satirical tradition, *Tienda de anteojos políticos* not only confirms the legacy of Boccalini six decades after his death but also ends a long controversy revolving around the possibility (or not) of rectifying the moral and intellectual nearsightedness of Baroque Spain—or, more modestly, the possibility of writing on the act of seeing and of "being seen seeing."

The act of double vision in *Tienda de anteojos politicos* situates the reader at the very center of a new phenomenology of perception, based on the belief that it is through the effect of perspective that the subject creates the object. In this way, we witness a progressive change of paradigm from the earthbound observation of the Babylonians and high vantage points to celestial observation in *El diablo Cojuelo*; and in the end, we are returned to earth in *Tienda de los anteojos*. Each of these texts is without a doubt conditioned by the position of the observer and the place of the observed objects. From a narrative point of view, the use of political lenses as a means to access the everyday is double: on the one hand, it allows authors to take a metacritical turn and reflect not so much on their surrounding mediocrity as on how to register it in writing; on the other hand, it converts the genre of satire into the object (rather than just a tool) of commentary, through a focused reflection on language, truth, and the moral perspective of the writer. In a solipsism very typical of the Baroque, the trompe l'oeil aesthetic in these pieces becomes in fact an aesthetic of disheartened relativism each time the "internal eyes of the mind" scrutinize reality from new angles that are, as we have seen, as much physical as literary. But in approaching the object from an anamorphic perspective, the surface of things becomes a powerful reminder that nothing is what it seems. As Fernando R. de la Flor has argued in his classic book *Pasiones frías*: "it forces a particular ethics and the aesthetics of creation, one that knows how to penetrate through winding paths into the multifaceted reality of the world" (*ello fuerza una ética y una estética del ingenio, justamente aquella que sabe penetrar por caminos desviados en la realidad plurifacetada del mundo*) (285). From this new perspective, we might add, vision becomes exemplary while its objects do not.

Francisco Santos (1623–1698) constructs in the fifth *puntada* of his *El sastre del Campillo* (1685) a strikingly disenchanted portrait of the period through the vision afforded him through several sets of eyeglasses. Little is known of Santos's life except for what he has left us in his writing. He was born in 1623 near the Campillo de la Manuela in the Lavapiés district of Madrid, and during his life he lived in some of Spain's most important cities, such as Seville (in 1666) and Toledo. His work as a soldier seems to have given him an informed understanding of the local underworld, the various sordid environments, and the "itinerant" Spain constituted by active soldiers and retired combatants. He devotes numerous pages to these military figures, and we may take as an example the soldier that he describes as a "shattered mass" (*bulto despedazado*) (117r) in *El sastre del Campillo*. He was, as he states in the front matter to his *El no importa de España* (Why Spain doesn't matter, 1667), "raised by his Majesty in the old Spanish Royal guard" (*criado de su Majestad en la Real Guarda Vieja Española*) (Hammond, *Francisco Santos' Indebtedness* 4). In Madrid, he lived on the Calle del Olivar (in the houses of Juan Martínez), attended the theaters, and in 1647 he married María Múñoz and had nine children, one of whom followed in his father's professional footsteps albeit with less success. We know, for example, that Santos was a witness in 1672 to the burning of the Casa de la Panadería in Madrid's Plaza Mayor, to which he devoted his pseudojournalistic work, *Madrid llorando* (Madrid in tears, 1690). He was, according to his own words, a family man attached to his neighborhood and to his neighbors. Seriously affected by gout and impoverished in the final years of his life, he died in 1698.

It must be admitted that the written work of Francisco Santos continues to enjoy a secondary status in relation to some of his contemporaries and immediate predecessors; the weight of other seventeenth-century figures has made Santos a lesser-known figure even for specialists in spite of the notable success that his work enjoyed during his lifetime and immediately afterward. Let us remember, for example, the esteem that Diego de Torres Villarroel held for him, or the influence that Santos's *Periquillo el de las gallineras* (The little parrot of the henhouses, 1668) had upon the Mexican writer José Joaquín de Lizardi when the latter composed his masterpiece *El periquillo sarniento* (The mangy parrot, 1831). The constant similarities with Juan de Zabaleta, Francisco de Quevedo, and other earlier prose writers, such as Luis Vélez de Guevara, have diminished somewhat the prestige and the perceived originality of a written corpus that remains poorly understood but nevertheless contains numerous elements of interest.[8] Even if we limit our survey only to works such as *Día y noche de Madrid* (Day and night in Madrid, 1663) and the aforementioned *Periquillo*, as Phyllis Eloys Czyzewski has pointed out, Santos merits critical attention.[9]

The interest that Santos had in science is difficult to miss. In his *El arca de Noé* (Noah's Ark, 1697), one finds several passages dealing with medicine, astrology, alchemy, and magic, with numerous allusions to cures, plants, and recipes. The same occurs in his novel *La tarasca de parto en el Mesón del Infierno* (The tarasque in labor in the Inn of Hell, 1671), in the section devoted to the eve of the feast of San Juan. Superstition and magic are principal themes this piece—as well as in other narratives such as *La verdad en el potro* (The truth in the colt, 1686)—and Santos often uses the term "physicians of the soul" (*médicos del alma*) rewriting Gracián's famous "physicians of heaven" (*médicos del cielo*). His novels, in fact, reveal extensive cultural baggage and many possible readings: next to the echoes of the medieval poet Jorge Manrique (ca. 1440–1479) in his numerous litanies for a glorious past, one detects the influence of Quevedo, Gonzalo de Céspedes y Meneses, Pedro Liñán de Riaza, Suárez de Figueroa, Saavedra Fajardo, and above all Gracián, whom he plagiarized repeatedly. Santos employs a form of language that is a cross between a highly satirical Salas Barbadillo and a highly sententious Juan de Zabaleta, with short and forceful phrases, full of license and registers readily recognizable as Baroque.[10] Modern scholars have tended to disapprove of what they consider his excessive moralism and the rigidity of his opinions, with two of his most important modern critics associating him with "an outrageous conception of the present" (*una delirante concepción del presente*) (Rodríguez-Puértolas 3:419), and of a "pestering, obsessive morality" (*moralidad atosigante, obsesiva*) (Navarro Pérez xxvi). Aware of the general environment of decline and corruption in which he lived during the reign of Charles II, Santos is sensitive to the political, social, cultural, and economic moment in which he lives and, as a consequence, his complaints—not lacking in sporadic proposals and practical solutions—tend to be more sanctimonious than didactic. With this, however, a reading of his work nonetheless offers valuable insight, stylistic originality, engaging characters, and a good deal of information on the festive and marginal Madrid of the late seventeenth century, which remains poorly understood. We may take as an example a passage from *El sastre del Campillo* that describes (and decries) the customary rise in meat prices after Lent—a phenomenon about which Santos would complain in more than one text—that manages to provide the modern reader with a rich and surprisingly nuanced idea of the social repercussions of such rituals celebrations.[11] As a result, nobody comes out worse in Santos's merciless portraits than the residents of Madrid. Inheriting earlier tastes, the body submits itself in Santos to the tribunals of the urban landscape, and the flesh converts itself into a motif toward which the great part of his narrative gravitates: bodies that consume and are consumed by others, bodies that generate and are degenerated. The lacerated body is exploited over and over with material and symbolic ends to create surreal scenes that remind one of the landscapes of Hieronymus Bosch—whom Santos openly praises—or the fantastic excesses of François Rabelais. The "dangers" of the flesh are, for Santos, at once a social threat and a literary motif, and in a great majority of his texts the festival occupies the central allegorical space.

One of the last notable pieces of the century, *El sastre del Campillo* consists of a series of complaints—at once picaresque and costumbrista in tone—directed at a Madrid languishing under Charles II's rule (Navarro Pérez ix–lxxiii; Arizpe). The central character of the piece is a tailor from Campillo de la Manuela, and he serves as the focal point for a disillusioned and horrendous parade of characters, each one more extreme than the one before. The influence of Boccalini and Gracián is evident, for example, in this *feria universal* as well as in the warnings of a blind man who walks along with "those that have good eyesight, since those with poor eyesight don't share my outlook" (*aquellos que tienen buena vista, que los de la mala no son de mi escuela*) (120v). Sight is one of the central themes in Santos's portrait of Madrid: in this tribunal which in certain form reminds one of Sancho Panza's Barataria, the question of the correct interpretation of reality runs through the text from beginning to finish, whether in reflections on appearances or in the very observation of surrounding miseries. There is a judgment on offer, and the judgment is always symptomatic.

In *El sastre del Campillo* we are faced with numerous complaints about the present in the form of attacks on the powerful as well as on the man on the street. The thematic framework of the piece takes form through the motif of the visit. The tailor, as Sancho Panza had done in his brief stint as governor, goes out to meet his neighbors and visitors who come from far off, each one with complaints or laments. One after the other, they test his good judgment and common sense. In the fifth *puntada*, for example, we meet a foreign merchant who sells *antojos* and *anteojos* in his shop. He is not the only who appears in the text—the tenth *puntada* opens with a merchant who enters escorted by Truth and Justice—however, he is the most important one given the nature of what he sells. What in some earlier texts appeared as a more or less specific, even technical term, is now turned back in Santos with surprising results, since the *antojo*, or vain madness, is not here the *anteojo* used to see far away. The emissary who announces to the tailor who is coming initially speaks of a shop—a movable one, it is understood—of "pregnancies or miscarriages" (*preñeces o abortos*) (51r). This lexicon reproduces the aesthetic device of the "monstrous birth" commonly found in his novels *Las tarascas de Madrid* (The tarasques of Madrid, 1665), *Los gigantones en Madrid por de fuera* (The giants in Madrid go outside, 1666), and the above-mentioned *La tarasca de parto en el Mesón del Infierno*, which made use of this motif to denounce the celebratory excesses of Madrid. The allusion, in any case, marks the tone of what is to be presented, which generally results from the host's displeasure. As the merchant states quite adamantly, what he brings is not an *antojo*—madness or vanity—but rather exactly the opposite, a varied selection of eyeglasses that will allow the tailor to gain a new perspective on the world and also see himself: "What I have to sell, oh, great tailor of Lavapiés, aren't *antojos* but rather *anteojos* for the sight of humans, useful for their conservation and largesse, and for all ages; and so that you might experience them firsthand and find out what they can do for you, take these eyeglasses, and look at yourself" (*Lo que yo traigo, ¡oh, gran sastre de Lavapiés!, no son antojos, son anteojos para la vista humana, para su conservación, y largueza, y todas sus edades; y para que lo experimentes, y te hagas capaz, toma estos anteojos, y mírate a ti propio*) (52r). Here eyeglasses are used principally for introspection, as if this moment at the end of the century were the ideal time to reflect on how such a state of misery and spiritual emptiness had been reached. We are invited to reflect, in the end, on what Santos terms "the nothing of vanity" (*el nada de la vanidad*) (52v).

Santos's tailor puts on a pair of eyeglasses and utters words such as *desire*, *shame*, and *vanity*. The eyeglasses permit him to see a reality in which pride and hypocrisy run wild, causing a kind of stupor. The merchant then confirms that these are a popular set of lenses that sells well but morally breaks its wearer. It significant that Santos, as Saavedra Fajardo had done earlier, associates glass with the desire to speak of "low desire" (*vil deseo*) and the "danger of desire" (*peligro del deseo*) (52v), as these are lenses that change people: they suspend their wearer's judgment and compel him or her to follow the pull of sensuality. As had occurred in Dávila y Heredia, the shop effectively censors itself and the lens, which had served in other texts to help people perceive a reality lying behind the veil of illusion, has become the enemy of its wearer.

As the plot advances, more lenses are offered. The tailor then puts on another, completely different set of eyeglasses, with which he can see nothing. This blindness connects the tailor to the common figure of the blind seer, and these new eyeglasses end up being, because of their inappropriate graduation, the most benevolent of all, since they prevent the observer from having to take in the depressing landscape that surrounds him, a landscape that curiously includes a Cervantine reference meant to underscore the state of delirium in which the tailor lives, his moral blindness, and the madness of his neighbors, all urban Quixotes. As Santos tells us, "There is more happiness at the moment in not seeing at all" (*hay más dicha como no ver en el tiempo presente*) (54r). The visit of the merchant brings together a condemnation of the present with a critique of the custom of wearing eyeglasses as a social marker, a critique that had also been made a half century before by Daza de Valdés. Taken as a whole, *El sastre del Campillo* is a text in which many of the disenchanted facets of the Baroque urbanite come together, closely following a pattern that we can also detect in his contemporary Juan de Zabaleta and his *Día de fiesta por la mañana y por la tarde en Madrid* (Day of celebration in the morning and in the afternoon in Madrid, 1666). It is, like all of Santos's work, a powerful witness to the exhaustion of the last third of the seventeenth century, and it significantly makes extensive use of eyeglasses as a means to narrate the miseries of Madrid—consistently presented as an infirm social body, spent and contaminated. It is a situation so severe that it goes beyond the earth. In the first *discurso* of his *Periquillo*, Santos even goes so far as to anthropomorphize the moon, presenting it as a corrupt mirror of the depraved world that it occasionally illuminates: "the light from that celestial body is misleading. . . . The moon, in the end, is a reflection of the small world, that of humans, that is so similar in its imperfections, since it waxes and wanes, is born, dies, is now something, is now nothing" (*Equívoca la luz de aquel lucero* [. . .], *luna, en fin, retrato del pequeño mundo, digo del hombre, tan parecida en sus humanas imperfecciones, pues ya crece, ya mengua, nace, muere, ya es algo, ya es nada . . . Equívoca, digo, la luz de este retrato de la criatura humana*) (54r).[12] This rugged, imperfect moon is more Galilean than Aristotelian and embodies in its humanity the sunset of an era, as Oscar Barrero Pérez has eloquently written:

> Santos lives in the last moments of the Baroque, a time when the increasingly evident decadence of Spain exacerbated in the mind of the Spanish artist the confusion between ideal and reality; reality now is not, in ideological or existential terms, monolithic and inalterable . . . , and these

traces of exterior reality shape, in some measure, the structural decomposition of narrative art in the seventeenth century.

> *Santos vive ya las postrimerías del Barroco, un tiempo en que la cada vez más ostensible decadencia de la Patria exacerbaba en la mente del artista español la confusión entre ideal y realidad; la realidad ya no es, ni ideológica ni existencialmente, monolítica e inalterable,* [. . .] *esos rasgos de la realidad exterior rigen, en alguna medida, la descomposición estructural del arte narrativo del siglo XVII.* (37–38)

If Santos's vision is a disenchanted one, it can be said also that it is an exhausted one, lacking in solutions beyond turning to God for final salvation. Placed next to the scientific and innovative spirit of the turn of the century, Santos's legacy is one that has unjustly been forgotten and which helps us reflect on new forms of approaching literary expression as a network of multiple tensions, among which science begins to take an increasingly evident role.

Santos helps us connect with a series of contemporaries that deserve mention. In the last third of the seventeenth century the works of Juan de Caramuel y Lobkowitz (1606–1682), Vicente Mut (1614–1687), and José de Zaragoza y Vilanova (1627–1679), and then later Juan Bautista Corachán (1661–1741) offer interesting astronomical proposals. In Caramuel's *Enciclopedia* (1670), for example, Spanish readers could follow the controversies surrounding the isochronism of the pendulum, the validity of Galileo's law of free-falling objects, or the satellites of Jupiter discovered by the German Capuchin Anton Maria Schyrlaeus de Rheita, which would be added to those previously discovered by Galileo. With respect to the world system, Caramuel showed his preferences for the Tychonic system, although he also pointed out that the physical and astronomical arguments against heliocentric theory were weak and that astrological theories were completely wrong. Mut, for his part, defended as valid Kepler's ellipses, rejecting Aristotelian physics, and his student José de Zaragoza y Vilanova wrote in his *Esphera en común, celeste, terráquea* (The common, celestial, and terraqueous sphere, 1675) on the theories of Copernicus, Brahe, Kepler, Galileo, Descartes, Gassendi, and Clavius, refuting the notion of the incorruptibility of the heavens and the celestial spheres, and suggesting a heliocentric vision of the cosmos.

The seventeenth century closed with one more notice: following the steps of Fernando Pérez de Sousa, the crypto-Copernican mathematician, physician, and astronomer Juan Bautista Corachán left in manuscript form his *Avisos de Parnaso* (Notices from Parnassus, 1690), a very personal creation that Gregorio Mayans y Siscar would end up editing in 1747, with one more edition appearing in 1754, and many partial translations found in various manuscripts held in different European libraries. It reproduced in Spanish various aspects of the new scientific currents and included a fragment of Descartes's *Discourse on Method*, the first translation it is believed, of this philosopher in Spain. The figure of Corachán is nothing short of fascinating: he participated in numerous academies of the period, such as the Academy of Mathematics, which Tomás Vicente Tosca (1651–1723), the author of *Compendio mathematico* (Mathematic compendium, 1707–1715), also attended (Navarro Brotons, "La renovación" 367–70). Tosca was, without a doubt, one of the most significant exponents of Galileo in Spain at the turn of the eighteenth century, and he was part of an important group of *novatores* working in Valencia, work that culminated with a study of the compiled work of Galileo as the founder of modern science produced by Juan Andrés (1740–1817), a Valencian Jesuit and university professor. Andrés, for his part, is widely considered the founder of historiography specifically focused on Galileo and a scholar with a fascinating trajectory: with the expulsion of the Jesuits from Spain in 1767, Andrés went to Italy, where he spent the rest of his life, and where he wrote his monumental *Dell'origine, progressi e stato attuale d'ogni letteratura* (On the origins, developments, and actual state of each literature, 1782–1799), an ambitious history of culture in seven volumes that would be reedited twelve times in complete form and five times in incomplete form until 1844, in Italian, Castilian, and French. The work's volumes dedicated to the history of philosophical and scientific literature constituted the first history of the sciences written by Spanish author. In 1776 Andrés joined the Academy of Sciences in Mantua, where he had already presented an extensive work on fluid dynamics, in large measure given the success of his *Saggio della filosofia del Galileo* (Essay on Galileo's philosophy, 1776), in which he examined distinct facets of Galileo's legacy in mechanics, aesthetics, astronomy and cosmology, sea tides, meteorology, music, optics, and magnetism (Navarro Brotons, "Galileo y España" 825).

If science makes important contributions during these final years of the seventeenth century, the impact of its dialogue with didactic prose fiction is no less relevant. In the last third of the century, figures such as Benito Jerónimo Feijóo (1676–1764), Martín Martínez (1684–1734), and Diego de Torres Villarroel (1693–1770) were born, and it is during this period that the term *telescopio* was incorporated into the Spanish lexicon. As Torres Villarroel would write in his *Sueños morales* (Moral dreams, 1786): "We have come to know the entire state of the heavens" (*hemos llegado a saber todo el estado del cielo*) (4:379), and it is true that during the last decades of the century there were many accounts of its uses in scholarly work and in private use. As Luis Miguel Vicente García has pointed out, Torres Villarroel disseminated that part of natural astrology that was then accepted by the church, and in the most serious and economical way ("Torres Villarroel"). He was important more as a transmitter of ideas than as an innovator, although he clearly possessed his own intuitions regarding the transmission of that which complemented his experience. He would borrow with great prudence from judiciary astrology, horoscopes in particular, and mix them with natural astrology. A master of humor, Torres Villarroel would also make metaphoric use of the telescope in a comical way, speaking at one point in his *Sueños morales* of informers in an image that reminds one of Quevedo's famous poem "Érase un hombre a una nariz pegado" (There was once a man stuck to a nose): "You stick yourself in telescopes, through which the clerks and the sheriffs register the most hidden of crimes" (*Métense a telescopios, por los cuales los escribanos y los alguaciles registran los delitos más ocultos*) (2:243). From a much more serious angle, Feijóo would write in his *Teatro crítico universal* (Universal critical theater, 1726–1740) of the "faulty astronomical knowledge with origins both in mistaken application and the lack of a telescope" (*impericia astronómica originada ya del defecto de aplicación, ya de la falta del telescopio*) (vol. 3, discourse 7, para. 14), and that "since the invention of the telescope we have discovered so many stars, both fixed and otherwise, that they exceed in number those that earlier astrologers observed" (*desde que se inventaron los telescopios se han descubierto tantas estrellas, ya fijas, ya errantes, que exceden en número a las que observaban los astrólogos anteriores*) (vol. 1, discourse 8, para. 29). In his *Cartas eruditas* he then goes on to follow Tycho Brahe, who considered the Ptolemaic system unsustainable: "It should be confessed that the popular system, or Ptolemaic is absolutely indefensible, and it is only dominant in Spain due to the great ignorance of our schools in astronomical matters; but this model can be abandoned together with the Copernican one, embracing that of Tycho Brahe, in which celestial phenomena are adequately explained" (*Debe confesarse, que el Sistema vulgar, o Ptolemaico es absolutamente indefensable, y sólo domina en España por la grande ignorancia de nuestras Escuelas en las cosas Astronómicas; pero puede abandonarse éste juntamente con el Copernicano, abrazando el de Tyco Brahe, en el cual se explican bastantemente los Fenómenos Celestes*) (vol. 3, letter 20, para. 27). Likewise, the physician and philosopher Martín Martínez commented in the fictional dialogue that makes up his *Filosofia escéptica* (Skeptic philosophy, 1730), "Those who see through Aristotelian eyeglasses, see everything as formality, abstraction, reduplication, and virtuality" (*Los que ven por anteojos aristotélicos, todo lo ven con formalidades, abstracciones, reduplicaciones y virtualidades*) (11). We have now entered fully into a new epoch in which there is no longer any doubt; rather, there is an increasing belief in the arguments derived from empirical science—although at times, as we see in Feijóo, these are taken only halfway. These reflections from the first half of the eighteenth century reveal to us a familiarity with the telescope, which in this century would find their way into the homes of many scientists and learned people in general. Of no less importance would be several contributions made during the second half of the century, with texts of great interest that make extensive use of the telescope. We may take as an excellent example of this trend in the anonymous *Anteojo de larga vista para todas las edades, y aviso de contrayentes de futuro matrimonio* (Long-distance lens for all the ages, and advice for those looking to marry, 1796), written in satiric courtly verse and with an unmistakably misogynistic message. Gaspar Melchor de Jovellanos likewise wrote in his diaries about purchasing a telescope for his personal enjoyment, as Manuel Álvarez-Valdés y Valdés has pointed out (277–78).

These final contributions bring to a close a slow and cautious change of paradigm, even if they do not fully close off interest in the intersection of the literary and scientific spheres that after the eighteenth century would enter into much closer contact and, for some, coexist much more harmoniously. The conclusion to this book will serve as a reflection on possible future avenues of study for the phenomena pointed out in these chapters: the projection of Galileo, the social life of the telescope, the unequal adoption of corrective lenses, and curiosity, in general, to see that which is new in the Spanish Baroque. I make no attempt, however, to close any debate; rather, my goal is to maintain these doors open for further study. What I have expressed here constitutes only an invitation to dig more deeply and in a much more detailed way into this century of curious figures and many curiosities.

Conclusions

The mystery of the world is the visible, not the invisible.

OSCAR WILDE

Lights, shadows, eclipses

Galileo's telescope has cast a long shadow throughout time, through to our present, with the controversial "revision" carried out by Pope John Paul II that I alluded to in previous pages. The scholarship on his legacy is still abundant, not only in the fields of history and philosophy of science and technology but also in disciplines such as art history and literary history. Even the publishing industry has benefited from the appeal of specific biographical events, as evidenced by Dava Sobel's 2000 book *Galileo's Daughter: A Historical Memoir of Science, Faith, and Love,* which quickly became a worldwide best seller. These rereadings of Galileo have not eclipsed, by any means, the different lines of academic work that have sought to illuminate things in a less fictional way, including those—like Alice Dreger's provocative study *Galileo's Middle Finger*—which have been faithful in their research to his vision in the face of adversity. Recent scholarly contributions, such as the essays gathered by Jürgen Rehn in his useful compilation *Galileo in Context,* provide a number of informative assessments similar to what I have attempted to delineate here more modestly by focusing on a hundred years of epistemological changes in the Iberian Peninsula—and my attempt, after all, has not been an isolated one considering the existing research on his impact in countries like France and England. In any case, where my itinerary ends, a new one necessarily begins, especially taking into account Galileo's long shadow in twentieth-century Europe as evinced in masterpieces like Bertolt Brecht's famous play *Life of Galileo*—a play that, one must not forget, also had a great impact in Spain.[1]

Galileo's biography and his scholarly work were, in fact, deeply influential for some of the best-known Spanish intellectuals of the past century, starting with the philosopher José Ortega y Gasset (1883–1955) and his *En torno a Galileo* (On Galileo, 1933),[2] whose reflections on the sad state of affairs in Spanish culture still resonate. And Galileo's presence was equally felt, with no small dose of doubt and anguish, in the poetry of Jorge Guillén (1893–1984), who wrote in awe of the achievements of modern science in his poem "La astronomía" (Astronomy):

> Of all the sciences, the flower: astronomy.
> Fantastic its discoveries
> of that enormous world with all its galaxies.
> Enormous! It feels so infinite to us:
> a very solid base for our humility.
> And just one Engine leading the universe,
> interested in one specific dot
> of the Earth, a planet without splendor?
> Or maybe the center of all galaxies?
> Enormous, this universe—and I, so minuscule.
> I confess: minuscule. Minuscule!
>
> *Entre todas las ciencias, la flor: la astronomía.*
> *Extraordinarios los descubrimientos*
> *En ese mundo enorme y sus galaxias.*
> *¡Enorme! Nos parece ya infinito.*
> *Es la base muy sólida de la humildad humana.*
> *¿Y un solo Móvil guía el universo,*
> *Interesado por un solo punto*
> *De la Tierra, planeta sin realce?*
> *¿O tal vez capital de las galaxias?*
> *Enorme el universo. Y yo, minúsculo.*
> *Lo confieso: minúsculo. ¡Minúsculo!* (156)

These are, of course, only two of the many examples that could be offered in these final pages. There is still much to be done on Galileo's presence in modern Spanish thought, as early as the first systematic commentaries on his work in the eighteenth century, and touching on all those fields of knowledge—religion, politics, philosophy—that so deeply connected, for many of his heirs in the Iberian Peninsula, with his life and thought. But there is also more that can be said on early modern Spain. As I announced at the beginning of this book, I have delved into only a few paradigmatic examples that reveal adhesions, fears, and doubts regarding certain scientific phenomena. The decades I have covered in this book are particularly tumultuous, and they coincide with a series of books that appeared in rapid succession in England, Italy, and France, and which laid waste to the Aristotelian natural philosophy taught in universities while building the foundation for "a new home," as Ortega y Gasset eloquently (and poetically) put it: "By 1650, with the death of Descartes, the new house has been built, the building of culture according to the *new mode*. This sense of self, of being in a *new mode* as opposed to the old and traditional one is what was expressed with the word *modern*" (*Hacia 1650, cuando muere Descartes, puede decirse que está ya hecha la nueva casa, el edificio de cultura según el nuevo modo. Esta conciencia de ser de un nuevo modo frente a otro vetusto y tradicional es la que se expresó con la palabra "moderno"*) (28). The authors, of course, were Francis Bacon (*Novum organum*), Galileo Galilei, and René Descartes (*Discourse on Method*). No name of this stature, it is true, can be found in Spain's philosophical arena during the 1600s, but if anything can come out of this analysis, it is the impact of these first thirty years, of this particular "scientific revolution" that brought the kind of knowledge acquired from practical experience together with the kind achieved through reason. This revolution, as we discovered reading classics like Karl Popper and Thomas Kuhn, and as we continue to learn today from historians of science like William Eamon, Peter Dear, and Stephen Gaukroger, was a directionless, fragmented event that involved many different areas of knowledge. The pieces examined in this book point to the evidence of a literary field that was highly permeable to external influences, and highly playful when it came to negotiating the limits of the permissible. The Spanish muse did not reject the light that came from abroad; she refracted it, creating new plots and characters from a different angle.

But my exploration has been, of course, a literary one, and therefore it has tackled only part of the problem. The theoretical question—or questions—are, perhaps, larger in scope and are closely related to the domain of philosophy of science, and in particular to processes of induction, explanation, and scientific change. I have argued that for these early modern writers, the act of seeing was a question of degree in the transition from the naked eye to the privilege of the powerful lens, and that the Spain of Lope de Vega, Francisco de Quevedo, and Calderón de la Barca imposed a movable barrier that separated what could and couldn't be seen. This concern went all the way back to the question of what nature was: if, for Cardinal Bellarmine, much like others before him—such as Giovanni Filopono (490–570), Thomas Aquinas (1225–1274), and even Andreas Osiander (1498–1552), who added his unsigned preface to Copernicus's *De revolutionibus*—astronomy could be equated with mathematics, Galileo, for his part, agreed with Bruno and Kepler in comparing astronomy with philosophy, thus advocating for a physical, and not hypothetical, description of the universe. Therefore, as we step outside this constellation of fictional accounts and gain some distance, we feel compelled to ask ourselves something that, I believe, ties all of these texts together: if something can be seen only with the help of a sophisticated scientific instrument, does it count as observable? In other words, when do the perceived objects stop being a matter of belief, or pillars of a tradition, and become "real"? This point is inevitably tied to a philosophical question that has been central to modern debates about science and religion, namely the issue of realism.[3] If, for the realist, science is in the business of proving accurate descriptions of entities, the antirealist is to remain doubtful about the accuracy of such descriptions and to hold that science is in the business only of providing accurate predictions of observable phenomena. All this, of course, takes us back to the sheer nature of the Baroque: as Thomas Dixon has written, the realist intuition is that our sense impressions are caused by an external world that exists and has properties independently of human observers, so that it is reasonable to try to discover what those properties are, whether the entities in question are directly observable by us or not; the antirealist intuition is that all we ever discover, either individually or collectively, is how the world *appears* to us. We have no knowledge of the world beyond the impression it makes on us, and so we should remain agnostic about the hidden structures that scientists hypothesize about in their attempts to explain those impressions. This is, in a way the sort of "agnosticism" that Rome imposed on Galileo when it forced him to present his findings as a hypothesis, but never as a fact—and never, of course, as a subject to be taught. And this is the tension that his friend Bartolomé Leonardo de Argensola (*Rimas*) revealed in a poem that, by being shared with his contemporaries—we see it in Calderón de la Barca and others[4]—is symptomatic of an era that announced the arrival of the new cosmos, one in which "the philosophical stance that identifies truth with natural beauty has been destroyed by the methods of observation that have revealed Nature's 'deceit,' Nature's other nature" (Wescott 59–60):

> I first want to confess, my friend Don Juan,
> that Elvira's white and colorful skin
> is anything but hers, if you look closely,

for she must have surely paid for it.
I must also confess to you,
how polished is her trickery,
for true and real beauty, in vain,
would dare to compete with her.
So, who cares if I am troubled
by such a lie, if we all know
that nature tricks us thus?
For that blue sky that we all see
is neither sky nor blue. How sad
that beauty is not true either!

Yo os quiero confesar, don Juan, primero:
que aquel blanco y color de doña Elvira
no tiene de ella más, si bien se mira,
que el haberle costado su dinero.
Pero tras eso confesaros quiero
que es tanta la beldad de su mentira,
que en vano a competir con ella aspira
belleza igual de rostro verdadero.
Mas, ¿qué mucho que yo perdido ande
por un engaño tal, pues que sabemos
que nos engaña así naturaleza?
Porque ese cielo azul que todos vemos
ni es cielo ni es azul. ¡Lástima grande
que no sea verdad tanta belleza! (49–51)

After the two quatrains in which the poet complains to his friend Don Juan about the artificial nature of a woman's beauty, the poet's voice moves on to equate this effect with that of nature, equally deceiving, and where this striking depiction of the sky points to a new understanding of astronomical notions of movement, space, and light. Howard Wescott has argued that this is a poem in which Galileo's telescope determines the "speaker's underlying unease, in spite of his languid tone that intends to show his insouciant unconcern with the suddenly unrecognizable infinity of the universe and the unintelligible nature of Nature" (61).

Much like the many passages I have examined in this study, Argensola's extraordinary poem reveals a sort of scientific agnosticism that was shared by some of the most religious men in Spanish history. As we have seen in piece after piece, this doubt is forceful and humorous, but it is a sustained doubt nevertheless, and it is always there. It is always there even if it is through the words of a marginal or a radically peripheral character like Sancho, a limping devil flying over Madrid, or an Araucanian in Chile—three literary types endowed by their creators with a healthy dose of irony, sagacity, and doubt. By their troubling presence and their telling frequency, these pieces suggest that there was not in this century a frontal rejection of the new, if only because material innovation was always welcome in a society shaped by the opening of new markets and the creation of new networks. What ultimately transpires from all these satires and poems, I would conclude, is a fascination with the mechanics of the object, with its internal and hidden powers when they are paired with the naked eye. The telescope, after all, has been considered a liberating measuring tool much like others whose inner mechanisms fascinated collectors, scholars, and princes: the clock, the compass, the astrolabe.

As a result—and as we distance ourselves one final step—what we learn from these decades of upheaval is that the printed word should never be an island of meaning, but rather the piece of a larger puzzle in which its components inevitably touch one another. We have created comfortable categories to separate science from fiction, and to separate science from religion. But the past tells a different story: Galileo, for one, thought that science and religion could coexist in harmony, and, as a true believer, he based this compatibility on what we know as the principle of accommodation, that is, that the Bible can be read in several—and often contradictory—ways according to the cultural competency of its different interpretive communities. Biblical references sometimes used to attack heliocentrism included Psalms 93:1 96:10, and 1 Chronicles 16:30: "the world is firmly established, it cannot be moved." In the same manner, Psalm 104:5 says, "The Lord set the earth on its foundations; it can never be moved." Further, Ecclesiastes 1:5 states, "And the sun rises and sets and returns to its place." What Spanish writers thus experienced when fictionalizing the telescope was not so much the fear of punishment—and we saw that Vélez de Guevara drops Galileo's name without fear—but rather a number of justified doubts about its role as a game changer. This openness toward novelty was also noticeable in Galileo's *Dialogue concerning the Two Chief Systems* through the character Sagredo, who asks Salviati—the mask of a Copernican Galileo—and is progressively convinced about this new paradigm as the dialogue advances, and as all the opinions of the Simplicius of the world are left behind. This is a piece that, in a sense, symbolizes in its development the slow and bumpy transformation that took place in Spain, as ancient ideas were progressively replaced by new ones in a conversation that involved many different social agents.

Likewise, similar care should be taken when divorcing science from fiction. That vague entity we call scientific discourse is even harder to define when placed in dialogue with what we understand as literature, since both feed from neighboring impulses and borrow from a similar rhetoric. In fact, the study of the material creation and transmission of early modern literature, like the history of the book and its different components—ink, parchment, glue—is nothing but the study of technology at work. In this regard, a telling example will suffice: Galileo was a great admirer of the Italian writers Torquato Tasso (1544–1595) and Ludovico Ariosto (1474–1533), and his final major work, the aforementioned *Dialogue concerning the Two Chief Systems*, with its Socratic model and its fictional characters, resembled some of the dialogues—like those of Spanish humanists Juan de Valdés (1500–1541) and Alfonso de Valdés (1490–1532)—that so significantly shaped the Spanish Renaissance. Conversely, another line of inquiry can help us see this transmission of knowledge: the accounts by writers like Juan de Piña and Vicente Carducho of Juan de Espina's magnificent collection were encyclopedic in nature, resembling the botanic catalogs of imperial travelers a century later, and they could very well be read as science disguised as fiction—or vice versa. Categories were not, in the end, so monolithic: even Pope Urban VIII, when he was still Cardinal Maffeo Barberini, wrote a poem with a flirtatious title, *Adulatio perniciosa* (1620), expressing his admiration for Galileo's telescopic discoveries, thus redrawing the boundaries of religion, science, and fiction.

When it comes to Spain, early modern scholarship still is, in my view, far too divorced from the history of medicine and astronomy, the two scientific disciplines with the largest number of published titles at the time—of the more than the 1,800 editions of works of scientific nature, close to 800 belonged to medicine, while nearly 300 dealt with astronomy, astrology, and chronology, or repertoires (*repertorios*). In a recent book on the intersection of optics and fiction in nineteenth-century France, Andrea Goulet reminds us:

> In one of the most important interdisciplinary initiatives of recent years, scholars of art history, philosophy, and history of science have problematized a static conception of the human seeing subject by calling attention to the changing ways in which vision is imagined, defined, and articulated across various ages and cultures. Contemporary scholars have replaced notions of sight as a biologically constituted, universal faculty with the culturally shaped concept of "visuality," ever-shifting according to different historical circumstances and philosophical frameworks. (3)

These are precisely the kinds of dialogues between established areas of knowledge, of which we still know so little, that can open up new fields of inquiry for both cultural scholars and historians of science. The field is ripe for the creation of new disciplinary networks; it is just a matter of pointing the lens in the right direction.

Notes

Preface

1. See Gal and Chen-Morris, who provide a thought-provoking discussion on the current state of the discipline.

2. See, for example, González Echevarría's assessment of Calderón's *La vida es sueño* (*Life Is a Dream*, originally published 1635), in which he argues that Basilio represents Ptolemy's obsolete cosmography; Basilio's anguish, he argues, is also that of Calderón, who seeks to please, through a sort of "double truth" (145), both the church and his audience.

Introduction

1. I do not attempt in this book to discuss the validity of or the current state of research devoted to this term, *Scientific Revolution*, which has been debated a great deal over the past several decades. For a general understanding of this issue, see, for example, Blumenberg, *Genesis*; Cohen; Dear, *Discipline*; Shapin.

2. The *occhiali politici* motif, as we will see, also receives different names according to the genre or author in question: *anteojos de mejor vista, anteojos de larga vista, anteojo de allende, anteojos políticos*, and so on. Already in Counter-Reformation Rome we find the terms *cañón* and *occhiale* in a 1611 letter written by the cardinal Robert Bellarmine in which he asks the Roman College for more information regarding the new optical inventions about which he had so many doubts. For an initial approach to this phenomenon, see García Santo-Tomás, "Fortunes." The Royal Spanish Academy's eighteenth-century *Diccionario de autoridades* (Dictionary of authorities) defines *anteojo de larga vista*, or "long-distance eyeglass," as an "instrument to see far away with facility, which consists of two, three, or four glasses or lenses, each placed one after the other at a distance in line with their focus and placed within one or more cardboard, wood, or metal tubes; the principal benefit of which is to bring to one's field of vision that which is observed through it and to make large and visible those objects which are very distant or remote. It is also referred to as a *Longemira* or Telescope" (*instrumento para ver con facilidad desde lejos, que consta de dos, de tres u de cuatro vidrios o lentes, puesta una después de otra a distancia del encuentro de sus focos y colocadas en uno o más cañutos o cañones de cartón, madera o metal, con cuyo beneficio se acercan a la vista del que mira por ellos y se agrandan las especies de los objetos muy distantes o remotos. Llámase también Longemira o Telescopio*). Although the term *telesco-*

pio did not make its way into the Castilian lexicon until the eighteenth century, *anteojo* continues to be present with the same meaning well into the nineteenth century.

3. An excellent survey of what has been published on the history of science is López Piñero's introduction to his *Historia*, and his "Galileo." A thought-provoking intervention can be found in Slater and Prieto, in a special issue, "History and Representation of Hispanic Science." With respect to the Anglo-American context, perhaps the most comprehensive bibliographic review of the recent state of the field can be found in Smith; although Smith's article curiously contains no reflection on the case of Spain, it does present—albeit in cursory form—an account of science in Spain's American colonies.

4. The different theories of vision employed in this book stem from the works of Lindberg, Jay, and Harries, the latter of whom focuses on Nicholas of Cusa and Leon Battista Alberti. For seventeenth-century Spain, see R. de la Flor, *Pasiones*, especially chapter 4.

5. On the Spanish presence in Italy during this period, especially in politics, see Dandelet; Dandelet and Marino; Signorotto; and D'Amico. In terms of diplomacy, see Levin. For a highly suggestive analysis of the confluence of literature and diplomacy, see Hampton.

6. Perhaps the most renowned author of "the literature of optics" during the period is Sor Juana Inés de la Cruz. Sor Juana's interest in optics is well known, and it finds expression in works such as the praise (*loa*) in the *auto sacramental* "El mártir del sacramento" (The martyr of the sacrament), in which she makes use of Athanasius Kircher's recently developed camera obscura. See, for example, Bergmann, "Amor" and "Sor Juana," as well as Valbuena Briones. On Sor Juana's exploration of geometrical concepts widely used later on, see Vélez Sainz.

7. I quote from page 61 of the 1796 edition of *Sueños morales: Visiones y visitas de Torres con Don Francisco de Quevedo*. On Torres's literary treatment of astrology, see Durán López 45–58.

8. Kirsten Kramer has referred to this "emancipation of the divine," writing that "in Baroque techniques of catoptric illusion, an order of readership and of the visible takes shape that is specific to early modernity. This order not only relativizes the Neoplatonic hierarchy between the original image and the reproduction, between spirit and matter; it also simultaneously moves beyond—and in equal measure—the Christian notion of God as personal creator, suspending with it the entire theological and cosmological framework within which poetic reflection on images had seemed to find its theoretical consolidation" (78) (*en las técnicas de ilusión de la magia óptica catóptrica del barroco se perfila un orden del espectador y de lo visible propio de la temprana Edad Moderna, el cual no sólo relativiza la jerarquización neoplatónica entre imagen original e imagen reproducida, entre espíritu y materia, sino que al mismo tiempo supera igualmente la noción cristiana de un Dios creador personal, suspendiendo con ello el entero marco teológico-cosmológico en el que la reflexión poética sobre las imágenes parecía encontrar su consolidación teórica*).

9. For Albumazar's presence in early modern Spanish letters, see de Armas, "Saturn" and "*De magnis*"; more recently, see Robinson.

10. For an excellent overview of science during these crucial years, see Mosley, who reminds us, for example, that Jerónimo Muñoz's *Libro de nuevo cometa* (Book of the new comet, 1572) was of great interest to Tycho Brahe. The figure of Galileo, however, scarcely appears in the most important studies on this topic. In his classic study

The Spanish Inquisition: A Historical Revision, Henry Kamen devotes a very brief mention to him (135); and in their books on the Inquisition, both Cecil Roth and Joseph Pérez omit all mention of him. Pardo Tomás, however, deals extensively with the matter in *Ciencia y censura* 151–85.

11. For a brief overview of the isolationist politics of Philip II, such as prohibiting his Spanish subjects to study abroad (except in Rome, Coimbra, Naples, and Bologna), see Elliott, *Imperial Spain* 225–31; and Kamen, *Spanish Inquisition* 106–21. According to Kamen, the 1559 index was "the beginning of an epoch of repression in Spanish culture" (111).

12. For a more recent account of scientific publications in sixteenth-century Spain, see Wilkinson. As I have pointed out, these constituted only a small percentage (1–4 percent) of total publications, which were dominated by religious texts (roughly 50 percent), such as bulls and breviaries, a fact that provides an idea of the challenge writers of science faced with respect to what Pimentel has characterized, once again, as an "impossibly heavy and obsolete edifice" (55) (*un edificio demasiado pesado*).

13. The bibliography associated with Quiroga is vast. For a very useful overview of earlier research, see Lynn.

14. Of the extremely large bibliography associated with this topic, see Kuhn's classic study, *Structure of Scientific Revolutions*. For Spain, useful introductions include Green, *Spain*, 2:42–64; Maravall, *Antiguos* 557–67; Rico; Beltrán Marí, *Revolución*.

15. For a more detailed analysis of this topic, see Ruiz Morales.

16. For a comprehensive account of the Salamanca circle, see Fernández Álvarez; Rey Pastor, especially the chapter "El sistema de Copérnico en España." Also of great interest is Salcedo y Ruiz 2:150. See also García Gibert 265–72.

17. I quote from Francisco Rico's 2001 edition of *Don Quijote de la Mancha*, 1:129.

18. One should keep in mind that there were, in fact, very few people who openly presented themselves as followers of Copernicus; Rheticus of course did so at the middle of the sixteenth century, and at the end of the century Thomas Digges, Giordano Bruno, Christopher Rothmann, and Fray Diego de Zúñiga likewise did so. And at the beginning of the seventeenth century we can include Thomas Harriot, Juan Cedillo Díaz, Simon Stevin in Holland, and Michael Maestlin and Johannes Kepler in Germany.

19. On Rodríguez, see Trabulse, *Los orígenes* and "Un científico."

20. See Navarro Brotons, "Reception." Zúñiga's texts are reproduced in López Piñero, et al., 87–91.

21. For more on this, see Beltrán Marí, *Talento y poder*.

22. Navarro Brotons and Eamon also add: "Although Zúñiga eventually changed his mind about the Copernican theory, his decision cannot be attributed to an Inquisitorial condemnation or censure. The condemnation of Copernicus did not occur until the decree of the Roman Inquisition in 1616, in which Zúñiga's work is explicitly cited for expurgation" (32n31). Also of interest to this debate is the 1999 reedition of Picatoste's classic study, which touches on the subject.

23. This temporality has also been pointed out by Badiou: "When Galileo announced the principle of inertia, he was still separated from the truth of the new physics by all the chance encounters that are named in subjects such as Descartes or Newton. How could he, with the names he fabricated and displaced (because they were at hand—'movement,' 'equal proportion,' etc.), have supposed the veracity of his principle for the situation to-come that was the establishment of modern science; that

is, the supplementation of his situation with the indiscernible and unfinishable part that one has to name 'rational physics'?" (401).

24. On Milton and Galileo, see Boesky; Brady; McAdam; Ulreich; Wood; and Fletcher. On the literary portrayals of Mars, see Crossley. For a more general approach, see Campbell. On Galileo and Cavendish, see Spiller. On Giambattista Marino, see Mussio. Hitchcock's imagined dialogue is recommended. On the French reception of Galileo, see Lewis. On Spinoza and Galileo, see Rudavsky. On the presence of natural philosophy in Baroque Germany, see the study by Wagman. On Galileo in English literature, see Drake; Dick reminds us that the same day *Sidereus* came out, it was deemed both groundbreaking and disturbing by the writer and diplomat Sir Henry Wottom, whereas the playwright Thomas Tomkis ridiculed it in his *Albumazar* (1615).

25. These realities were not exclusive of the novel or of poetry, as de Armas's work on the presence of astrology in early modern Spanish literature reveals; on his vast production, see, for instance, de Armas, *Return*.

26. I later return to the literary motif of the watchtower, popularized with extraordinary success by Mateo Alemán in his *Guzmán de Alfarache* (1599 and 1604). For an interesting conversation on the political valences of this device, see Negredo del Cerro; and Tobar Quintanar.

27. See Henchman's stimulating discussion on authors such as Thomas De Quincey and Alfred, Lord Tennyson. For a similar approach in Spanish letters, and in particular to the figure of Leopoldo Alas, known also as "Clarín," see Balbino.

28. On the relations between the Scientific Revolution and disciplines like natural philosophy, theology, and metaphysics, see Gaukroger's monumental study.

29. On this discussion, see Dear, *Literary Structure*; Aït Touati; Marchitello, *Machine*, in particular chapter 4, "Galileo's Telescope," 86–115. Neither Aït Touati nor Marchitello pays attention to the Iberian Peninsula. See also Battistini, "Antagonistic Affair"; on Galileo's "literary elements," see Elías.

30. The term *third dimension* comes from Díaz de Urma's study on the Renaissance mirror.

31. I examine this issue in "Ruptured." On the uses of military weaponry in early modern Spanish theater (primarily Lope de Vega and Cervantes), see Reed.

32. On the specific case of the watch, see Mayr 3–138. The watch triggers in Spain a number of meditations on the uses and abuses of power: Quevedo uses it to refer to European diplomacy in *La Hora de todos*, and Suárez de Figueroa, in discourse 77 of his *Plaza universal de todas ciencias y artes*, reproduces Garzoni's reflections on the watchmaker: "This occupation is extremely honorable and useful, as convenient as it is to know the time and the activities that it determines" (*Esta ocupación es por extremo honrosa y útil, por la gran comodidad de la ciencia de la hora, y de los tiempos para sus ejercicios*) (583). Fernando R. de la Flor ("El cetro" 61) has defined the monarchy as "absolutely invisible in its watchmaking mechanisms" (*absolutamente invisible en sus mecanismos relojeros*) and depicted the king as the "hand of a hidden movement" (*manilla de un movimiento oculto*) that "makes fulminant decisions" (*toma decisiones fulminantes*), thus making the faculty of vision "the guiding principle of action" (*el principio máximo de actuar*).

33. For the study of this figure in early modern Spanish fiction, see Vilar Berrogain's classic study.

34. Turriano was a personal friend of Charles V and Philip II, and highly admired by writers such as Cervantes, Ambrosio de Morales, and Sebastián de Covarrubias.

For an assessment of his work, see Frago Gracia and García-Diego. On Juan de Espina, see Cotarelo y Mori; Bouza Álvarez, "Coleccionistas"; and Susan Paun de García's critical edition of José de Cañizares's *Don Juan de Espina en su patria* and *Don Juan de Espina en Milán*, which provides a comprehensive overview of his persona.

35. His contemporary Matías de los Reyes also uses this motif in *El curial del Parnaso* (1624), according to Williams 52.

36. For a comprehensive study of this phenomenon, see Findlen.

37. On October 31, 1992, just over 350 years after Galileo's death, Pope John Paul II solemnly rehabilitated the Pisan astronomer and criticized the errors of the theologians that led to his condemnation; importantly, the pope did so without expressly disqualifying the tribunal that sentenced Galileo. In a thirteen-page speech, pronounced in the Sala Regia of the Apostolic Palace, John Paul II described Galileo as an "ingenious physicist" and "sincere believer" who showed himself to be more perspicacious in the interpretation of Holy Scripture than his theological adversaries. In light of John Paul II's 1992 rehabilitation of Galileo, the year 2009—during the Year of Astronomy (convened by the International Astronomical Union, and supported by UNESCO), which commemorated the four hundredth anniversary of the first astronomical study and corresponding international congress on Galileo—represented for the Holy See an important occasion to deepen knowledge and engage in dialogue on astronomy and Galileo himself. Archbishop Gianfranco Ravasi, then head of the Pontifical Council for Culture, stated: "Galileo was the first man to look with a telescope toward the sky, feeling a sense of new wonder. He opened up for humanity a world that was up to then little understood, expanding the limits of our knowledge and forcing us to reread the book of nature with new eyes. The Church wishes to honor, therefore, the figure of Galileo, an ingenious innovator and son of the Church." The archbishop added that "the times are now ripe" for a revision of the image of Galileo and Galileo's case itself, and he pointed out that the Second Vatican Council had already made reference to Galileo, deploring "certain mental attitudes, which at times are found even among Christians, derived from not taking sufficiently into account the legitimate autonomy of science." Ravasi recalled how in 1981 John Paul II created the commission to examine the Galileo case and underscored the commission's "courage in recognizing the errors of Galileo's judges," who, incapable of separating faith from a millenarian cosmology, believed that accepting the Copernican revolution would weaken the Catholic tradition and that it was therefore their duty to prohibit those teachings. The prelate added that for that "subjective error of judgment," Galileo had been forced to suffer a great deal. He concluded: "Today, in a more serene climate, we can finally look to the figure of Galileo and recognize the believer who attempted, in his time, to reconcile the results of his scientific investigations with the content of the Christian faith. For this reason, Galileo deserves today our great appreciation and gratitude" (all quotes from Ravasi appear in the January 29, 2009, edition of the newspaper *El Mundo*). In this context, the Vatican reedited the proceedings of Galileo's trial to remind people, as Ravasi has pointed out, that Pope Urban VIII never signed the Inquisition's condemnation of Galileo. Among the initiatives, the February 26 conference in the Pontifical Lateran University of Rome titled "1609–2009: 400 Years since Galileo's *Starry Messenger*" stands out. Later, in Florence, from May 26–30, the Jesuit Niels Stensen Cultural Foundation organized an international scholarly conference on Galileo's trial. On June 21–26 there was a seminar (organized by the Vatican Observatory), and throughout the entire month

of October, one could visit an exhibition at the Vatican titled *Galileo 2009, Fascination and Anguish over a New Vision of the World: 400 Years since the First Observation with a Telescope*. From mid-October 2009 to early January 2010, one could also see the Vatican Museums' exhibition *Astrum 2009: Astronomy and Instruments—The Historical Patrimony of Italian Astronomy Four Hundred Years after Galileo*, which included books, archives, and instruments from the Vatican Observatory and the Vatican Museums, as well as Galileo's *Sidereus nuncius* manuscript, held in the National Central Library in Florence.

38. Recent discussions on John Paul II's 1992 rehabilitation of Galileo are those by Doncel; Beltrán Marí, "'Una reflexión"; and Kerr.

Chapter One

1. On Harriot's work on the moon between 1610 and 1613, see Chapman.

2. For a recent discussion on the topic, see, for example, Reeves, *Galileo's Glassworks*; and Enoch. On the origins and commercialization of the first telescopes, see Ilardi, "Eyeglasses."

3. See Simón de Guilleuma; and López Piñero, *Ciencia y técnica* 190–91, where he provides valuable information on this family of artisans.

4. For an in-depth analysis of this piece, see González-Cano, "Un poema" 40–43.

5. For a nuanced discussion on what Galileo's telescopes could actually achieve, see Doble Gutiérrez.

6. On the relations and connections between Galileo and the architecture of the time, see Hersey's study, particularly chapter 3, "The Light of Unseen Worlds," 52–77.

7. On the development of the Academy of Lincei, see Freedberg, who analyzes a collection of sketches found in Windsor Castle that reproduce fauna and flora gathered by the members of the academy.

8. R. de la Flor ("El *Quijote* espectral") reminds us of the Jesuit practice of wearing a type of lens known as *rayada* (scratched) or *ahumada* (smoky), "with the intent of not seeing the world, thus manifesting their complete disdain for sublunar reality" (*con el objeto precisamente de no alcanzar a ver el mundo, manifestando así su desprecio por la realidad sublunar*).

9. For a more comprehensive discussion on this phenomenon, see García Tapia's studies, *Ingeniería y arquitectura* and *Técnica y poder*; Vicente Maroto and Piñeiro; Navarro Brotons, "La ciencia"; and López Piñero, *Ciencia y técnica*.

10. See Sánchez Navarro 75. The quote refers to Curzio Picchena (1556–1626); the Tuscan ambassador in Spain, Conde Orso D'Elci; and the famous poet Bartolomé Leonardo de Argensola. For a more in-depth view of this issue, see Bedini 7–16; González; and, more recently, Kimmel 316–17.

11. It was the Mexican presbyter Joaquín Furlong who discovered this document in the Archivo General de Indias.

12. See Mele, who spoke "nel negozio il nome del poeta spagnuolo" (259). See also Green, "The Literary Court"; and Wescott.

13. As quoted by Riandière La Roche Saint-Hilaire 28. The letter can be found in the Archivo General de Simancas in Valladolid, Spain, Consejo de Estado, 1883/29.

14. These rivalries were not uncommon in early modern Europe. For the case of Elizabethan London, see Harkness's excellent study.

15. For an informative analysis of Labaña's report, see Floristán Imízcoz.

16. Mancho Duque speaks of an "overcoming of the Latin-vernacular tension in scientific works" (*superación de la tensión latín-romance en las obras científicas*) and of a "democratization of knowledge in the 1500s" (*democratización de los saberes en el quinientos*) (258–9). On the phenomenon of linguistic ambivalence in the period, see Woolard and Genovese.

17. See Bustos Tovar. On the reception of Copernicus from Juan de Pineda (1598) to the controversies at the University of Salamanca, see Navarro Brotons, "Galileo y España" 817.

18. See Navarro Brotons, "El copernicanismo." Piñeiro and Gómez Crespo argue that this is perhaps the first translation of its kind in Europe.

19. These texts are cataloged under Mss. 6150, 9091, 9092, and 9093, as well as in two other manuscripts with the reference numbers Mss. 8934 and 8896. Ms. 9091 includes *Ydea astronomica de la fabrica del Mundo y movimiento de los cuerpos celestiales* (Astronomical Idea of the Fabric of the World and Movement of Celestial Bodies), with the subtitle "Ydea y Cosmología," which is an incomplete translation of Copernicus's *De revolutionibus*; it comprises a small, perfectly preserved notebook with 79 leaves (138 pages) that might be an original, intended for future distribution in print. This text, in the form of a prologue, figures in folios 180r–181r. In Ms. 9092 there is a Spanish translation of Galileo's *Dialogue on the Ebb and Flow of the Sea*, in which tides are explained as a consequence of the movement of the Earth, and in Ms. 9093 there is a short treatise, "On Spheroids," directed toward astronomy—pieces that were never published but that are conserved in manuscript form.

20. On the development of these two institutions, see Simón Díaz; Valverde and Piñeiro; and Piñeiro, "La Casa de la Contratación" 35–52.

21. It is the letter dated March 2, 1611, included in Sliwa's excellent study.

Chapter Two

1. For more on astrology during this period, see Vicente García, *Estrellas*. On astrology in medieval manuscripts, see Page.

2. There is an ample bibliography on this issue since the 1980s. The most recent studies worth mentioning are those by Lara Alberola; and García Soormally.

3. See, for example, Juan Bautista de Arroyo y Velasco's *Entremés del astrólogo burlado* (Interlude of the ridiculed astrologer); Francisco Antonio de Bances Candamo's *El astrólogo tunante* (The rogue astrologer) and *El esclavo en grillos de oro* (The slave in golden shackles); Calderón de la Barca's *El astrólogo fingido* (The fake astrologer) and *El laurel de Apolo* (Apollo's laurel); Antonio de Nanclares's *La hechicera del cielo* (Heaven's witch); and the ever-prolific Lope de Vega's *El ausente en su lugar* (The absent one in his place), *El caballero del sacramento* (The knight of the sacrament), *La niñez del Padre Rojas* (The childhood of Father Rojas), *Quien más no puede* (Who can no more), *Roma abrasada* (Rome burned), *Sembrar en buena tierra* (Sow in good earth), *Servir a buenos* (To serve the good), *Sin secreto no hay amor* (No love without secrets), *El sufrimiento del honor* (The suffering of honor), and *Ya anda la de Mazagatos* (There she goes again!). For a comprehensive overview of this phenomenon, see Lima 15–27; and de Armas's useful special issue *The Occult Arts in the Golden Age*, in the journal *Crítica hispánica*.

4. See Castillo, *(A)wry Views*; R. de la Flor, *Barroco*.

5. A pivotal source like the Real Academia Española's database *Corpus Diacrónico del Español* (http://corpus.rae.es/cordenet.html) yields the following results for the period 1500–1700: 13 uses of *antoxos* in 11 documents, 14 uses of *anteojos* in 4 documents, and 849 uses of *antojos* in 164 documents. On the uses of *anteojo*, see Álvarez de Miranda, "El doblete" and "Algo más."

6. The complete poem is reproduced in García Santo-Tomás, *La musa* 110–14, as well as in Arnaud, 3:797–804.

7. On the dissemination of this text, see Thorndike. For the case of early modern Spain, see Hurtado Torres. Portuondo (39n50) reminds us of eleven different commentaries on the text (eight of them in printed form), including by Francisco Junctino (1582) and Christoph Clavius (1595), both of which appeared in Fray Luis de Miranda's *Exposición de la esfera de Juan de Sacrobosco* (Exposition of Johannes de Sacrobosco's Sphere), published in Salamanca in 1629.

8. For an initial approach to Cervantes's understanding of Renaissance science, see Becedas González, et al.; Domínguez's illuminating essay; Sánchez Ron 7–12; and Vallés Belenguer.

9. Half a century later, the novelist Francisco Santos referred to this adherence to Ptolemy in his piece *El sastre del Campillo* (The tailor of El Campillo, 1685): "you'd better laugh at Ptolemy's Cosmography, Aristotle's Mathematics, and Euclid's Astronomy; throw those books away, and study the buffooneries of the Court" (*ríase de la Cosmografía de Tolomeo, de la Matemática de Aristóteles, y la Astronomía de Euclides; arrójelos a todos, y solo estudie en las bufonadas de la Corte*) (117v). I have modernized the orthography and punctuation in this and subsequent citations to this novel from those of the existing seventeenth-century original.

10. See also de Armas, "Heretical Stars."

11. See Vosters, "Lope de Vega," "Dos adiciones," and "Levinus Lemnius"; Ptolemy's influence is treated in Morby, "Franz Titelmans" and "Two Notes." See also Halstead, "Attitude of Lope de Vega"; and Dixon.

12. It is worth pointing out here how in plays such as *Lo que está determinado* (That which is determined) and *El servir con mal estrella* (To serve under a bad star), Lope makes light of astrology. In *Las burlas de amor* (The trifling of love) he speaks of it as "false judiciary" (*la falsa judiciaria*) and states that the art of palm reading is "uncertain" (*incierta*); he speaks of "useless astrology (*vana astrología*) in *La difunta pleiteada* (The dead woman who was sued) and finds it runs counter to "moral excellence" (*la moral excelencia*), so drops it in *Roma abrasada* (Rome burned). Going even further, he defines astrology as a "useless chimera" (*vana quimera*) in *Los hidalgos del aldea* (The noblemen of the village), and as "insult" (*vituperio*) in *El enemigo engañado* (The deceived enemy).

13. On Tirso's understanding of astrology, see also Halstead's fascinating "Optics of Love."

14. On this phenomenon, see Pérez Magallón; and Vega, *Ciencia*.

15. See also Whitby.

16. Despite its longevity, the best analysis on this phenomenon remains Moll's groundbreaking study.

Chapter Three

1. See Gelabert's influential study *Castilla convulsa*, from which I borrow the phrase.

2. See also the pivotal work by Gómez Moreno, although the pieces analyzed in his book, as he is careful to admit (324–5), are still pre-Copernican.

3. For the impact of *Ragguagli* in England, see Thomas; for its influence in Italy, see Firpo. Williams (9) mentions translations into Latin, German, and French.

4. See Blanco, "Del infierno" 170–75. In *El Lince de Italia* Quevedo writes, as we will see, of the dangers that Philip IV must face in dealing with France, Venice, and the aforementioned Duke of Savoy, according to Azaustre Galiana.

5. For a more in-depth analysis of this and other poetic compositions in which Lope criticizes Boccalini, see Arellano.

6. It is no coincidence that Fernando R. de la Flor ("El cetro" 75) has highlighted this dialectic in Gracián. On vision in the works of the Aragonese moralist, see Cañas; and M. T. Cacho.

7. The debates on this phenomenon were already numerous in the previous century; see, for example, Etreros; Gómez-Centurión Jiménez; Schwartz Lerner, "Golden Age"; Pérez Lasheras; and R. Cacho. For a useful update on this debate, see the recent work by Vaíllo and Valdés.

8. On literary academies in early modern Spain, see Sánchez, *Academias*; on Anastasio Pantaleón de Ribera, see Kenneth Brown; of pivotal interest is Egido, *Fronteras*, especially the chapter "Poesía de justas y academias"; more recently, see Close.

9. See Sánchez, "Nombres."

10. Boccalini returns to this matter in I.89, II.71, and II.89. When quoting from *Ragguagli*, the first number corresponds to *parte* (*centuria* in the original), and the second to *aviso* (*ragguaglio*).

11. On the invention of eyeglasses, see Rosen; on concave lenses, see Ilardi, *Renaissance* and "Eyeglasses". In "Eyeglasses" Ilardi argues that the first known mention of them is from a manuscript dated to 1289 in which one of the members of the Popozo family complains, "I am so debilitated by age that without the glasses known as spectacles, I would no longer be able to read or write" (360).

12. As indicated, for example, by Verene 72. The relation among gaze, light, and truth is examined by Blumenberg, in "Light."

13. See Kemp. On the representation of eyeglasses in portraiture, see Riccini. Recent studies on the "Baroque gaze" are those by Harbison; R. de la Flor, *Barroco*; and Careri.

14. The coinage "exaltation of the gaze," referring to El Greco's portrait, is from R. de la Flor, "El cetro" 78. On optics and Spanish art, see Reeves's excellent studies, *Galileo's Glassworks* and *Painting*.

15. On the pivotal phenomenon of "Baroque curiosity," see Lezama Lima; and Raimondi.

16. For a panoramic view of the development of this "century of satire," see Romero-González.

17. On the composition and the aims of the text, see Cherchi; and Arce Menéndez.

18. The mention of Brahe comes from the twenty-third discourse, where he pays homage to the likes of Scilace Cariandeno, Euclid, Hippias of Elis, Eratosthenes of Cyrene, Theon of Alexandria, Nicephorus Gregoras, Boethius, Leodamonte Tasio,

Eupompus of Macedonia, Francesco Sansovino, Christoph Clavius, David Origanus, Antonio Magini, Pedro Núñez, Juan Bautista Labaña, Giuliano Ferrofino, Juan Arias de Loyola, Diego Pérez de Mesa, and Jerónimo Muñoz, among others (209). The list includes, as one can see, a number of scholars of his time who were at the forefront of scientific exploration.

Chapter Four

1. I use the term *scientist* in a very broad sense, as I am fully aware of the fact that the category of "the scientific" did not yet exist in the seventeenth century.

2. See, for example, Skal; and Haynes. On madness in early modern Spanish literature, and specifically regarding the authors covered in this book, see Atienza.

3. On the arrival of artificial light in early modern Madrid, see Herrero García, *El alumbrado*. For its development in Central Europe, see Koslofsky.

4. The complete citation, which comes from his masterpiece *Discurso de todos los diablos o infierno emendado*, reads: "Los tabacanos, como luteranos, si le toman en humo, haciendo noviciado para el infierno; si en polvo, para el romadizo" (The smokers, like Lutherans, if they smoke it, sending the novitiate toward hell; if they sniff it, toward [an excess in] mucus).

5. For a recent reassessment of the "social life" of tobacco and chocolate in early modern Spain, see Norton's excellent overview; see also García Santo-Tomás, *Espacio urbano*, chapter 5.

6. For a nuanced and thought-provoking reflection on the subject, see Stewart.

7. On the performance of domesticity in early modern Spain, see García Santo-Tomás, "Fragmentos"; and Cirnigliaro. On the specific case of Salas Barbadillo, see García Santo-Tomás, *Modernidad*. On games of chance and homosocial relations, see García Santo-Tomás, "Outside Bets."

8. On the life of Juan de Espina, see Cotarelo y Mori; and San Emeterio Cobo. Caro Baroja provides important information on some of Espina's practices. More recently, see Reula Baquero's reflection on the cinematic nature of Espina's visual culture.

9. On the figure of Ayanz, see García Tapia, *Un inventor*; on Turriano, see Frago García and García-Diego; García Tapia, *Técnica y poder* 265–92; and Aracil 79–90, 312–15.

10. For the circumstances surrounding the acquisition of these documents, see García Tapia, "Los codices"; Helmstutler di Dio; and Sánchez Cantón. The first codex is devoted to mechanics and statics; the second one to geometry and fortification.

11. On the life and deeds of Rodrigo Calderón, see Martínez Hernández's excellent study.

12. On the issue of collecting, see Bleichmar and Mancall; Morán Turina and Checa; and Pimentel and Marcaida, who write, "Among the cast of actors who contributed to the development of modern scientific practices—a cast traditionally dominated by natural philosophers, mathematicians, medical practitioners, botanists, and apothecaries—a place must be found for patrons of the arts, artists, collectors, merchants, and dealers" (114). Marcaida López's study, and in particular its first chapter, is pivotal to this discussion.

13. Bouza Álvarez reminds us of a similar effect produced by the Duchess of Alba's collection on St. Teresa of Ávila: "As she was leaving, recovered from the initial shock

that she had felt during the visit, the saint confessed not remembering anything in particular, but only the sensation caused by the *whole collection*" (*A la salida, recuperada del espanto inicial que le había causado la visita, confiesa la santa que no conseguía recordar, en particular, la hechura de ningún objeto, pero sí la sensación que le había producido* el conjunto) ("Coleccionistas" 248, my italics).

14. On this type of "sociability of strange facts," on its capacity to spark social relations out of wonder and curiosity, see Daston and Park, especially chapter 6; on the connection between magic and secrecy, see Eamon, *Science*.

15. The amount of food in Espina's party was not by any means a unique feat in Baroque Madrid. Díez Borque, for example, reminds us of a famous banquet in 1657 at the Ermita de San Antonio in El Buen Retiro, in which five hundred dishes were served, accompanied by thirty different wines (49).

16. The text is stored at the Biblioteca Nacional de Portugal, Lisbon, H. 6.38. It is partially studied in Cotarelo y Mori 31–7, and is fully reproduced in Asenjo Barbieri 188–201, from which I quote.

17. For a detailed study of the text, see Piccus.

18. For more on this event, see Caturla's two articles.

19. On this phenomenon, see Aracil 297–339. Aracil argues that Espina borrows the concept of the *autómata* from the *hombre de palo* (woodman) perfected by his predecessor Juanelo Turriano.

20. On this event, see Brotton; Robbins, "Spanish"; see also Brown and Elliott.

21. Vicente Maroto and Piñeiro, for example, have defined science in Spain as "interest of the King, indifference of his people" (*interés de un monarca, indiferencia de un pueblo*) (493).

22. The numbers provided after each citation correspond to the lines of the poem in the 1625 edition.

23. Pantaleón de Ribera's sonnet reads: "Curioso (o peregrino) te desea / de este culto edificio la hermosura / en cuya argumentosa arquitectura / feliz el arte mejoró la idea. / Lo que así la atención te lisonjea / fama después venerará futura / que en bronces firme, en pórfidos segura / o sea admiración o envidia sea. / Tesoro es rico de curioso dueño / cuanto estudió naturaleza, y cuanto / obró imitando artífice ingenioso; / la admiración es corto desempeño, / Peregrino, si a objeto tan hermoso / el éxtasis te niegas del espanto" (199).

24. On the different connotations of the term *prodigious*, so pivotal in Espina's public standing and the epoch at large, see Castillo, *Baroque* 79.

25. See García Tapia, *Un inventor* 182–84.

26. There is a vast bibliography on the subject; see, for instance, Beltrán Marí's *Talento y poder*, in which he offers one of the most thorough and balanced accounts of the issue.

27. See, for example, Marchitello's convincing reassessment of disciplinary boundaries in early modern Europe (*The Machine*); see also Aït-Touati.

28. On the artistic interests of Galileo, see Peterson's groundbreaking study; Galileo, in fact, considered the development of theater as one of the most attractive aspects of the Madrid he was never able to visit, despite several attempts in the 1620s and 1630s.

29. On the figure of Lastanosa, see Rey Bueno and López Pérez.

30. This separation is discussed with stimulating results in Gaukroger, chapter 1, particularly note 40, which provides additional bibliography.

Chapter Five

1. In this regard, see Alaminos López; and Carbajo Isla.

2. More recently, see Haney; and Glenn.

3. This name has a long tradition in the literary depictions of marginal characters, according to Étienvre 47n53.

4. On literary depictions of sodomy, see Martin's excellent study, particularly her chapter "Homosexuality and Satire" (43–78).

5. This half-baked construction of the main characters of the novel is also noticeable in other pieces of the genre, as we will see in Antonio Enríquez Gómez, on whom Matthew D. Warshawsky has written: "This description helps account for the general absence of character development in Enríquez Gómez's three dream texts and also in 'The Marquis of the Bottle.' In these works, little is known about the speakers themselves; more important is the content of their message" (*Longing* 68).

6. On anamorphosis in early modern Spanish literature, see Castillo, *(A)wry Views*.

7. Recent research on this topic includes Vinge; and Classen.

8. See Rose; Oelman; Dille; Kramer-Hellinx, especially 1–30. See also Warshawsky, "Spanish Converso's"; Vaíllo, "Afinidades."

9. I quote from the 1992 edition edited by Rose and Kerkhof.

Chapter Six

1. See Gómez Trueba 162–3, who considers this to be the most important piece in the tradition of the fictional dream in early modern Spanish literature. For a more general overview, see Palley.

2. Schwartz-Lerner ("Golden Age") has written that Menippean satire was very much unknown in the Middle Ages, and that it was popularized in the Renaissance in Lucianesque dialogues and with Lipsius's *Satira Mennipea. Somnium. Lusus in nostri aevi critici* (1581). See also Valdés, "Rasgos"; R. Cacho; Smet; and Romero-González. For a more comprehensive take on the subject, see Guillén.

3. On the narrative device of the voyage in Kepler and Galileo, see Piers Brown; Marchitello, "Telescopic Voyages."

4. For a detailed analysis on the development of political satire in early modern Spain, see Egido, *Sátiras*; Avilés Fernández; Etreros; Gómez-Centurión Jiménez; and Pérez Lasheras, especially 61–106.

5. See Assaf; Deffis de Calvo, Levisi, "Los aspectos"; Peale; and García Santo-Tomás, "Artes."

6. On Quevedo's work, see Schwartz-Lerner, *Metáfora*; Valdés, *Los "Sueños."*

7. Gómez Trueba also adds: "We must determine to what extent the presentation of a fiction in the form of a dream modifies or conditions the composition and meaning of a text" (*Se trata de saber en qué medida la presentación de una ficción en forma de sueño modifica o condiciona la composición y significado de la obra*) (20). Pieces like the anonymous *El sueño de la ciudad en ruinas* (The dream of the city in ruins, 1588), which reflects on the 1588 disaster of the Spanish Armada, remain neglected by scholars. On the different theories of the literary revery in early modern Spain, see Gómez Trueba 172–206; Egido, *Cervantes*.

8. On Lucian of Samosata in Spain, see Vives Coll; and Zappala.

9. See the collective volume coordinated by Jean-Pierre Etienvre titled *Las uto-*

pías en el mundo hispánico, which provides a fascinating array of examples on this phenomenon.

10. Brown reproduces this text in *Anastasio Pantaleón de Ribera* 283–303. The manuscript offers a number of variants with respect to the one offered by Balbín Lucas in his 1944 edition of Pantaleón de Ribera's *Obras de Anastasio Pantaleón de Ribera* 2:11–44, from which I quote. A stimulating reading is offered in Lacadena y Calero 99–100. For a more general overview, see Layna Ranz 27–56, especially note 3; also, Madroñal Durán.

11. Still relevant to this discussion is Heninger's classic study.

12. The scholarship on this piece is uneven; see González de Amezúa 1:280–306; Solé-Leris 131–32; Vaíllo, "Un libro"; see also Hafter's stimulating discussion on this phenomenon ("Toward"). I quote from the second edition, published in 1634, of Enríquez de Zúñiga's *Amor con vista* (Cuenca: Julián de la Iglesia, 1634), Biblioteca Nacional de Madrid, signature R-173727.

Chapter Seven

1. Of particular interest on this phenomenon is Carrió-Invernizzi.

2. On the portrayal of Venice in Quevedo, see, for example, Crosby; Roncero López; Juárez Almendros; and Clamurro, "*La Hora*."

3. For a more general context, see the article published in four parts by Tato-Puigcerver in the Spanish journal *La perinola*. Also see Tato-Puigcerver, "Una nota." López-Grigera's 1975 edition of *La Hora de todos* contains useful information regarding Quevedo's possible anti-Copernicanism; Schwartz's 2009 edition of the same, which is even more comprehensive, also offers an excellent critical overview.

4. All citations to *Lince de Italia* are from Pérez Ibáñez's 2002 edition.

5. The first citation is from Clamurro, "*La Hora*" 841; the second comes from Martínez, "'Quien me entendiere" 118. Clamurro's study focuses on Quevedo's critique of the Venetians; Martínez's essay examines the legal and political discourses that constitute the background for the episode, along with the internal tensions that consume Quevedo. With respect to the text's political vision, see also Iffland, "Apocalipsis"; and Kent. All citations here from *La Hora de todos* are from the 1987 edition by Bourg, Dupont, and Geneste.

6. For a thorough and thought-provoking reading of these pages, see Stoll; Riandière La Roche Saint-Hilaire also alludes to the episode in her ambitious book.

7. In fact—as it also occurs in *El diablo Cojuelo*'s seventh episode (*tranco*)—the watch also symbolizes the work of the favorite, or *privado*, always secret(ive) in his actions, much like the internal mechanisms of a clock.

8. See, in this regard, Schmidt's timely piece.

9. A thorough overview of the historical context is offered, for example, in Israel, *Dutch*.

10. Criticism of the Dutch in early modern Spanish literature was, in fact, fairly common. For a representative sample, see Herrero García, *Ideas* 437.

11. The tone in these passages responds, as the editors of the text remind us, to both Quevedo's temperament and his philosophical background: "In these scenes, Quevedo, loyal to the rules of the satirical genre, presents himself as aggressive, passionate, moved, excessive, and, at the same time, profoundly marked by the dialectic of contraries inherited from the Stoics" (*En los cuadros, Quevedo, fiel a las leyes*

del género satírico, se muestra agresivo, apasionado, arrebatado, excesivo, y, al mismo tiempo, profundamente marcado por la dialéctica de los contrarios heredada de los estoicos) (*La Hora de todos*, ed. Bourg, et al., 52).

12. See Böhm; and Israel, *Empires*.

13. Of great interest in this context is Daly.

14. The ultimate source for this motto is Virgil, *Eclogues* 2.7; however, it is perhaps more likely that Núñez de Cepeda took it from Filippo Picinelli's *Mondo simbolico* (Symbolic world, 1635).

15. See also Maravall, "Saavedra"; R. de la Flor, *Pasiones* 183–92; and González García.

16. There is a vast bibliography on this topic. For its European frame, see Bireley, who discusses Saavedra Fajardo in relation to Botero and Lipsius. With respect to the influence of Lipsius over Saavedra Fajardo, see López Poza's introduction to her critical edition of Saavedra Fajardo's *Empresas políticas*; see also López Poza's article "La *Política*." For a more general approach with respect to the other political and philosophical tendencies of the period—particularly the neo-Stoicism and skepticism of the 1640s and 1650s—see the passages on Saavedra Fajardo in Robbins, *Arts* 82–94.

17. For more on this question, see Grande Yáñez 190; and R. de la Flor, *Pasiones* 123–28. For more in-depth information on relations between dissimulation and religious oppression in a broader context, see Zagorin's classic study. On the connections between dissimulation and civic virtue, see Snyder.

18. The citation comes from the sixteenth chapter of Saavedra Fajardo's *Obras completas*. I quote from the 1946 edition by González Palencia.

19. An authoritative analysis of the intersection of writing and the visual in early modern Spanish literature, as well as a comprehensive presentation of modern criticism on the emblematic tradition, can be found in Egido, *De la mano*.

20. See Egido, "La historia" 49–90; Joucla-Ruau; Romanoski; and Fernández-Santamaría. For a more general treatment, see Tuck, especially 123–4, where he situates Saavedra Fajardo in a broader European context.

21. I here follow the numeration used in the second print edition of the *Empresas políticas*, as presented in López Poza's 1999 edition. See also Luciani, especially his fourth chapter ("The 'I' Glass"), which refers to *empresa* 46 on 136–37. See also Blanchard 105–6; and Battistini, "The Telescope."

22. Fernández-Carvajal, in his 1984 edition of *Empresas políticas*, has similarly written that "Saavedra merges the emblematic subgenre with another epideictic subgenre, that of the 'mirror of princes,' which has its roots in classical rhetoric and has its model in Pliny the Younger's 'Trajan's Panegyric,' cited in *Empresas*" (*Saavedra funde el subgénero emblemático con otro subgénero epidíctico, el de los "espejos de príncipes", que se remonta a la retórica clásica y tiene su modelo en el "Panegírico de Trajano", de Plinio el Joven, citado en las* Empresas) (xxxiii).

23. For more on the European context in which Saavedra Fajardo operated, see Aldea Vaquero's monumental study. The third volume of his work, which focuses on Saavedra Fajardo's correspondence with Philip IV's younger brother, Cardinal-Infante Ferdinand of Austria, between 1633 and 1634, is of particular value.

24. I quote from Jorge García López's 2001 edition of *Novelas ejemplares*. It is fascinating to compare these images with Boccalini's allusion to Cornelius Tacitus and the telescope, which also become quite common during the period.

25. For more on Saavedra Fajardo's studies at Salamanca, see Goodman 20–21.

26. On these polemics, for example, Shea, "Galileo the Copernican"; Topper. According to Beltrán Marí (*Talento y poder* 152), the Dutch scholar Johan Fabricius was the first to speak of sunspots, in a small text that had almost no diffusion. For a more general approach to this issue, see Kelter.

27. According to García López, "Quevedo y Saavedra," and Blecua, "Las *Repúblicas*" 86n34, the definition of the seven climates that Saavedra Fajardo offers in *empresa* 81, in opposition to that forwarded at the University of Salamanca, represents a rejection of heliocentrism—an opinion that would not change up to the publication of his *Corona gótica* (Gothic crown, 1648).

28. As Charles Louis Maufras argues in his 1847 edition of Vitruvius's *De architectura* (On architecture): "These first two chapters and the three subsequent ones contain a small treatise on astronomy that is extremely interesting . . . ; we find there the principles of the Ptolemaic system, and we find also that these were known in Rome long before Ptolemy had published his treatise" (1:233).

29. Calvo Serraller argues that "Saavedra Fajardo has perfect knowledge" of Vitruvius's work (450), and Egido maintains, for her part, that Saavedra Fajardo gained direct access to Vitruvius's works through the library of his friend and well-known book collector Vincencio Juan de Lastanosa. She goes on to argue that Vitruvius's treatment of the "window into the human heart" also served as the inspiration for Saavedra when he speaks of the "chest of glass" (*pecho de cristal*) that conditions the purity of the prince's opinions (Egido, "La historia" 58–59 and n29). See also Gállego; and García Melero 23–78.

30. See Kuhn, *Copernican*; Elena.

31. I quote from Hawking's edition.

32. This, however, is not the only tone adopted by Saavedra Fajardo. García López ("Quevedo y Saavedra") reminds us: "Saavedra Fajardo took from the *Ragguagli* their burlesque, festive, and playful tone; likewise, the dynamic swirl of images scattered throughout the work ceaselessly reminds us of Boccalini's work" (*nuestro autor tomó de los* Ragguagli *su tono burlesco, de chanza festiva y ligereza jocosa, así como un hervidero de imágenes desparramadas a lo largo de la obra y que sin cesar recuerdan las páginas del italiano* [. . .] *de quien extrae el alegre gracejo y la gracia festiva a que somete a los autores*) (32, 36). García López goes on to argue that Saavedra Fajardo extracts from Boccalini: "the happy wit and the festive grace to which he submits other authors" (36). As Boadas Cabarrocas has argued, this type of attack—and this defense of the nation, already in grave crisis—emerges also in another of Saavedra Fajardo's great political works, the little-known *Locuras de Europa* (Insanities of Europe, c. 1635).

33. For more on the relations between artisanal activity and the political theory of the period, see Corteguera's excellent study.

34. García López has written in his edition of the *República literaria* that for Saavedra Fajardo, all the sciences "have a speculative value, and those that practice them are inept being when it comes to the resolutions of life" (*tienen un valor especulativo, y los que las cultivan son seres ineptos para las resoluciones de la vida*) (l). García López further maintains that Saavedra Fajardo saw politics in a very different light, namely, that it was a practice designed for the happiness of the republic, given that "those theorists who characterize it as merely a vague philosophical system deserve only disdain" (*no merecían sino desdén los teorizantes que la exponían como un vago sistema filosófico*) (li).

35. See also R. de la Flor, "*La ciencia*" for a stimulating reading of this issue.

36. Álamos de Barrientos, *Tácito Español* 12. For a fascinating reflection on the "deciphering" of nature and human passions in Galileo, Gracián, and Álamos de Barrientos, see Checa.

Chapter Eight

1. On González de Salas's musical background, see the introductory pages by Luis Sánchez Laílla in his 2003 edition of *Nueva idea de la tragedia antigua*, and in particular 129–31.

2. Of the vast bibliography on this partnership, see, for instance, Ferrer, "*El golfo*"; and Greer.

3. See Wilson; Patterson; and Curtius 559–70.

4. See, for example, Cruickshank; and Stein.

5. I quote from González Cañal's 1997 edition of Rebolledo's *Ocios*, 2:355–56.

6. Rebolledo, *Selva militar y política* 175.

7. *El Discreto*, XIX, "Hombre juicioso y *notante*" 127. On the different ocular uses in Gracián, see M. T. Cacho; and Cañas.

8. See Hammond, "Francisco Santos," "References," and "Plagiarium"; and Hafter, "Saavedra Fajardo."

9. In spite of the great popularity that Santos enjoyed in the eighteenth century, there are few modern editions of his works. The only known collection is a volume of his complete works, published in 1723. And with the exception of a few pieces by Milagros Navarro Pérez, Julio Rodríguez-Puértolas, and Víctor Arizpe, as well as a handful of unpublished dissertations, Santos has essentially been ignored over the past thirty years.

10. See, for example, Alfaro, "La anti-picaresca"; and Melero Jiménez.

11. On this, see Simón Palmer.

12. On these lines, see the stimulating comment by Rico 167–69.

Conclusions

1. On this, see Eamon, "Brecht"; Ladra; Suvin; and González Molina.

2. There are several critical editions of *En torno a Galileo*. I have consulted Hernández Sánchez's 2012 edition and Abellán's 1996 study, from which I quote.

3. On this debate, see, for example, Devitt; and Psillos. Coming from the opposite angle, see Van Fraassen's classic study. I draw in the following pages from the superb synthesis carried out by Okasha 58–94.

4. See Green, "*Ni es cielo*."

Works cited

Primary sources

Álvarez de Barrientos, Baltasar. *Tácito español illustrado con aforismos*. Madrid: Luis Sánchez and Juan Hasrey, 1614.

Anteojo de larga vista para todas las edades, y aviso de contrayentes de futuro matrimonio. Madrid: Imprenta de Josef López, 1796.

Argensola, Bartolomé, and Lupercio Leonardo. *Rimas*. José Manuel Blecua, ed. Zaragoza: Institución Fernando el Católico (CSIC) de la Excma. Diputación Provincial de Zaragoza, 1950–1951, 2 vols.

Boccalini, Traiano. *Discursos politicos, y avisos del Parnaso*. Fernando Perez de Sousa, trans. Madrid: María de Quiñones, 1634.

———. *Avisos de Parnaso de Traiano Bocalini; primera, y segvnda centvria; Tradvcidos de lengva Toscana en Española por Fernando Perez de Sovsa*. Madrid: Diego Díaz de la Carrera, 1653.

Cañizares, José de. *Don Juan de Espina en su patria: Don Juan de Espina en Milán*. Susan Paun de García, ed. Madrid: Castalia and Comunidad de Madrid, 1987.

Carducho, Vicente. *Diálogos de la pintura*. Francisco Calvo Serraller, ed. Madrid: Turner, 1979.

Castillo Solórzano, Alonso de. *Donaires del Parnaso*. Luciano López Gutiérrez, ed. Madrid: Universidad Complutense, 2005.

Cervantes, Miguel de. *Don Quijote de la Mancha*. Francisco Rico, ed. Barcelona: Crítica, 2001.

———. *Novelas ejemplares*. Jorge García López, ed. Barcelona: Crítica, 2001.

———. *Los trabajos de Persiles y Sigismunda*. Juan Bautista Avalle-Arce, ed. Madrid: Castalia, 2001.

Coello, Antonio. *"Vejamen que se dio en el certamen del Buen Retiro": Sales españolas, o agudezas del ingenio nacional*. Antonio Paz y Meliá, ed. Madrid: Atlas, Biblioteca de Autores Españoles, 1965, 176:315–21.

Copernicus, Nicolaus. *On the Revolutions of Heavenly Spheres*. Stephen Hawking, ed. Philadelphia: Running Press, 2002.

Dávila y Heredia, Andrés. *Tienda de anteojos políticos*. Valencia: Jerónimo de Villa, 1673.

Daza de Valdés, Benito. *El libro del uso de los antojos*. Manuel Márquez, ed. Biblioteca Clásica de la Medicina Española 4. Madrid: Imprenta de Cosano, 1923.

Dreger, Alice. *Galileo's Middle Finger: Heretics, Activists, and the Search for Justice in Science*. New York: Penguin, 2015.

Enríquez de Zúñiga, Juan. *El amor con vista*. Cuenca: Julián de la Iglesia, 1634.

Enríquez Gómez, Antonio. *Inquisición de Lucifer*. Constance H. Rose and Maxime P. A. M. Kerkhof, eds. Amsterdam: Rodopi, 1992.

———. *La torre de Babilonia*. Teresa de Santos Borreguero, ed. Madrid: Universidad Autónoma de Madrid, 1989.

Espina, Juan de. "Memorial que Don Juan de Espina envió a Felipe IV." *Biografías y documentos sobre música y músicos españoles*, by Francisco Asenjo Barbieri. Emilio Casares, ed. Madrid: Fundación Banco Exterior, 1986, 188–201.

Feijóo, Benito Jerónimo. *Teatro crítico universal*. Madrid: Imprenta de Ayguals de Izco Hermanos, 1852.

———. *Sueños morales*. Madrid: Viuda de Ibarra, 1794, vol. IV.

———. *Cartas eruditas, y curiosas*. Madrid: Viuda de Ibarra, 1774, vol. 3.

Fernández de Ribera, Rodrigo. *Los anteojos de mejor vista. El mesón del mundo*. Víctor Infantes de Miguel, ed. Madrid: Legasa, 1979.

Galilei, Galileo. *La gaceta sideral*, and Johannes Kepler, *Conversaciones con el mensajero sideral*. Carlos Solís, ed. Madrid: Alianza Editorial, 2007.

———. *Diálogo sobre los dos máximos sistemas del mundo ptolemaico y copernicano*. Antonio Beltrán Marí, ed. Madrid: Alianza Editorial, 1994.

———. *Le opere di Galileo Galilei*. Antonio Favaro, ed. and comp. Florence: Tipologia di G. Barbèra, 1909, vol. 5, 419–23.

Góngora, Luis de. *Sonetos completos*. Birutė Ciplijauskaitė, ed. Madrid: Castalia, 1969.

González de Salas, José Antonio. *Nueva idea de la tragedia antigua*. Luis Sánchez Laílla, ed. Kassel: Reichenberger, 2003.

Gracián, Baltasar. *Obras completas*. Arturo del Hoyo, ed. Madrid: Aguilar, 1960.

Guillén, Jorge. *Final*. Antonio Piedra, ed. Madrid: Castalia, 1987.

Hurtado de Mendoza, Antonio. *Obras líricas y cómicas, divinas y humanas . . . de D. Antonio Hurtado de Mendoza*. Francisco Medel del Castillo, ed. Madrid: Imprenta de Juan de Zúñiga, 1738.

Martínez, Martín. *Filosofía escéptica*. Madrid: Francisco López, 1730.

Mendoza, Fray Íñigo de. *Coplas de Vita Christi*. Julio Rodríguez-Puértolas, ed. Madrid: Espasa-Calpe, 1968.

Miranda, Luis de. *Exposición de la esfera de Juan de Sacrobosco doctor parisiense. Traducida del latín en lengua vulgar, aumentada y enriquecida, con lo que de ella dijeron Francisco Juntino, Elias Veneto y Christoforo Clavio* . . . Salamanca: Jacinto Taberniel, 1629.

Nieremberg, Juan Eusebio. *Curiosa y oculta filosofía: primera, y segunda parte de las maravillas de la naturaleza examinadas en varias cuestiones naturales*. Alcalá: Imprenta de María Fernández, 1649.

Ortega y Gasset, José. *En torno a Galileo*. Domingo Hernández Sánchez, ed. Madrid: Tecnos, 2012.

———. *En torno a Galileo*. José Luis Abellán, ed. Madrid: Espasa Calpe, 1996.

Piña, Juan de. *Casos prodigiosos y Cueva encantada*. Madrid: Librería de la Viuda de Rico, 1907.

Quevedo, Francisco de. *La Hora de todos y la Fortuna con seso*. Lía Schwartz, ed. Madrid: Castalia, 2009.

———. *Lince de Italia u zahorí español*. Ignacio Pérez Ibáñez, ed. Pamplona: Eunsa, 2002.

———. *El buscón*. Fernando Cabo Aseguinolaza, ed. Barcelona: Crítica, 1993.
———. *La Hora de todos y la Fortuna con seso*. Jean Bourg, Pierre Dupont, and Pierre Geneste, eds. Madrid: Cátedra, 1987.
———. *La Hora de todos y la Fortuna con seso*. Luisa López Grigera, ed. Madrid: Castalia, 1975.
———. *Grandes anales de quince días: Obras de don Francisco de Quevedo y Villegas*. Aureliano Fernández-Guerra y Orbe, ed. Colección Biblioteca de Autores Españoles. Madrid: Rivadeneyra, 1852, vol. 23, 193–220.
Real Academia Española. *Diccionario de autoridades*. Madrid: Gredos, 1990.
Rebolledo, Bernardino de (Conde de). *Ocios*. Rafael González Cañal, ed. Cuenca: Ediciones de la Universidad de Castilla-La Mancha, 1997.
Relación de la fiesta que hizo D. Juan de Espina, Domingo en la noche, último día de febrero. Año 1627. This poem is included in the book *Relaciones poéticas sobre las fiestas de toros y cañas*. Antonio Pérez y Gómez, ed. Cieza, Murcia: Artes Gráficas Soler, 1973. Col. El aire de la almena. Vol. 46, *tomo* 7, facs., fols. 112r–116r.
Ribera, Anastasio Pantaleón de. *Vejamen que el poeta Anastasio Pantaleón de Ribera dio en la insigne Academia de Madrid*. In Kenneth Brown, *Anastasio Pantaleón de Ribera (1600–1629): Ingenioso miembro de la república literaria española*. Madrid: Porrúa, 1980, 283–303.
———. *Obras de Anastasio Pantaleón de Ribera*, vol. 2. Rafael de Balbín Lucas, ed. Madrid: Consejo Superior de Investigaciones Científicas, Instituto "Nicolás Antonio," 1944.
Saavedra Fajardo, Diego de. *República literaria*. Jorge García López, ed. Barcelona: Crítica, 2006.
———. *Empresas políticas*. Sagrario López Poza, ed. Madrid: Cátedra, 1999.
———. *Empresas políticas*. Rodrigo Fernández-Carvajal, ed. Murcia: Real Academia de Alfonso X el Sabio, 1984.
———. *República literaria*. Madrid: Espasa-Calpe, 1956.
———. *Obras completas*. Ángel González Palencia, ed. Madrid: Aguilar, 1946.
Salas Barbadillo, Alonso de. *Fiestas de la boda de la incasable malcasada*. Madrid: Viuda de Cosme Delgado, 1622.
———. *El necio bien afortunado*. In *Dos novelas de D. Alonso Jerónimo de Salas Barbadillo. El cortesano descortés. El necio bien afortunado*. Francisco Rafael de Uhagón y Guardamino, ed. Madrid: Imprenta de la Viuda e Hijos de M. Tello, 1894.
———. *Patrona de Madrid restituida*. Madrid: Alonso Martín, 1609.
Santos, Francisco. *El sastre del Campillo*. Madrid: Lorenzo García, 1685.
Sirtori, Girolamo. *Telescopium, sive Ars perficiendi novum illud Galilaei visorium instrumentum ad sydera*. Francoforte, 1618.
Sobel, Dava. *Galileo's Daughter: A Historical Memoir of Science, Faith, and Love*. New York: Penguin, 2000.
Suárez de Figueroa, Cristóbal. *Plaza universal de todas ciencias y artes*. Enrique Suárez Figaredo, ed. Barcelona, 2004. http://users.ipfw.edu/jehle/CERVANTE/othertxts/Suarez_Figaredo_PlazaUniversal.pdf.
Téllez, Fray Gabriel (Tirso de Molina). *Comedias del Maestro Tirso de Molina*. Juan Eugenio Hartzenbusch, ed. Colección Biblioteca de Autores Españoles. Madrid: Imprenta Rivadeneyra, 1885.
Torres Villarroel, Diego de. *Sueños morales, visiones y visitas con don Francisco de Quevedo*. Madrid: Joseph Doblado, 1796, vol. 2.

Vega, Lope de. *Arcadia, prosas y versos*. Antonio Sánchez Jiménez, ed. Madrid: Cátedra, 2012.

———. *La Dorotea*. José Manuel Blecua, ed. Madrid: Cátedra, 1996.

———. *La doncella Teodor*. Julián González Barrera, ed. Kassel: Reichenberger, 2008.

———. *Rimas humanas y otros versos*. Antonio Carreño, ed. Barcelona: Crítica, 1998.

———. *El laurel de Apolo*. Antonio Carreño, ed. Madrid: Cátedra, 2007.

Vélez de Guevara, Luis. *El diablo Cojuelo*. Ramón Valdés, ed. Barcelona: Crítica, 1999.

Vitruvius Pollio, Marcus. *De architectura*. Ch.-L. Maufras, ed. Paris: C. L. F. Panckoucke, 1847.

Secondary sources

Aït Touati, Frédérique. *Fictions of the Cosmos: Science and Literature in the Seventeenth Century*. Chicago: University of Chicago Press, 2011.

Alaminos López, Eduardo. "Espacio público: Madrid Villa y Corte." *Calderón y la España del Barroco (Catálogo de la Sala de Exposiciones de la Biblioteca Nacional, 16 de junio-15 de agosto de 2000)*. Madrid: Sociedad Estatal "España Nuevo Milenio," 2000, 93–107.

Aldea Vaquero, Quintín. *España en Europa en el siglo XVII: Correspondencia de Saavedra Fajardo*. Vol. 3, *El Cardenal Infante en el imposible camino de Flandes, 1633–1634*. Madrid: Consejo Superior de Investigaciones Científicas, Real Academia de la Historia, 2008.

Alfaro, Gustavo. "La anti-picaresca en *El Periquillo* de Francisco Santos." *Kentucky Romance Quarterly* 14 (1967): 321–27.

Álvarez de Miranda, Pedro. "Algo más sobre anteojo/antojo." *Boletín de la Real Academia Española* 72, no. 255 (1992): 63–66.

———. "El doblete 'antojo' / 'anteojo': Cronología de una recomposición etimológica." *Separata del Boletín de la Real Academica Española*. Madrid: Imprenta Aguirre, 1991, notebook 253, vol. 71, 221–44.

Álvarez-Valdés y Valdés, Manuel. *Jovellanos: Vida y pensamiento*. Oviedo: Nobel, 2012.

Alves, Abel A. "Complicated Cosmos: Astrology and Anti-Machiavellianism in Saavedra's *Empresas Políticas*." *Sixteenth Century Journal* 25, no. 1 (1994): 67–84.

Amadei-Pulice, María Alicia. *Calderón y el Barroco: Exaltación y engaño de los sentidos*. Amsterdam: John Benjamins, 1990.

Aracil, Alfredo. *Juego y artificio: Autómatas y otras ficciones en la cultura del Renacimiento a la Ilustración*. Madrid: Cátedra, 1998.

Arce Menéndez, Ángeles. "Suárez de Figueroa y su versión de *La Piazza universale* de Garzoni: Entre texto y paratexto." *Cuadernos de Filología Italiana* 15 (2008): 93–124.

Arellano, Ignacio. "Lope y Boccalini: Tres sonetos de Tomé de Burguillos." *Revista de Literatura* 74, no. 148 (2012): 387–400.

Arizpe, Víctor. "Francisco Santos: Aclaraciones crítico-bibliográficas a las *Obras* en prosa y en verso." *Hispania: A Journal Devoted to the Teaching of Spanish and Portuguese* 74, no. 2 (1991): 457–58.

Armas, Frederick A. de. "Heretical Stars: The Politics of Astrology in Cervantes' *La gitanilla* and *La española inglesa*." *Material and Symbolic Circulation between England and Spain, 1554–1604*. Anne J. Cruz, ed. Aldershot: Ashgate, 2008, 89–100.

———. "*De magnis coniunctionibus*: Albumasar, Lope de Vega y Calderón." *Mélanges Luce López-Baralt*. Abdeljelil Temimi, ed. Zaghouan: Fondation Temimi pour la Recherche Scientifique et l'Information, 2001, vol. 1, 229–38.

———. "The Maculate Moon: Galileo, Kepler, and Pantaleón de Ribera's *Vexamen de la Luna*." *Calíope: Journal of the Society for Renaissance & Baroque Hispanic Poetry* 5, no. 1 (1999): 59–71.

———. "Saturn in Conjunction: From Albumasar to Lope de Vega." *Saturn from Antiquity to the Renaissance*. Massimo Ciavolella and Amilcare A. Ianucci, eds. University of Toronto Italian Studies 8. Ottawa: Dovehouse, 1992, 151–72.

———. *The Return of Astraea: An Astral-Imperial Myth in Calderón*. Lexington: University Press of Kentucky, 1986.

———, ed. *The Occult Arts in the Golden Age*. Special issue of *Crítica Hispánica* 15 (1993).

Arnaud, Emile. *La vie et l'oeuvre de Alonso Jerónimo de Salas Barbadillo: Contribution a l'étude du roman en Espagne au debut du XVIIe siècle*. Toulouse: Université de Toulouse Le Mirail, 3 vols., 1977.

Asenjo Barbieri, Francisco. *Biografías y documentos sobre música y músicos españoles*. Emilio Casares, ed. Madrid: Fundación Banco Exterior, 1986.

Assaf, Francis. "Aspects picaresques du *Diablo Cojuelo*." *Revista canadiense de estudios hispánicos* 8 (1984): 405–12.

Atienza, Belén. *El loco en el espejo: Locura y melancolía en la España de Lope de Vega*. Amsterdam: Rodopi, 2009.

Avilés Fernández, Miguel. *Sueños ficticios y lucha ideológica en el Siglo de Oro*. Madrid: Editora Nacional, 1981.

Azaustre Galiana, Antonio. "Estructura y argumentación del 'Lince de Italia u Zahorí español' de Quevedo." *La perinola: Revista de investigación quevediana* 8 (2004): 49–76.

Badiou, Alain. *Being and Event*. Oliver Feltham, trans. New York: Continuum, 2007.

Balbino, Marcos. "El catalejo del Magistral en Vetusta." *Letras de Deusto* 15, no. 32 (1985): 69–86.

Baquero Goyanes, Mariano. "Visualidad y perspectivismo en las *Empresas* de Saavedra Fajardo." *Murgetana* 31 (1969): 5–37.

Barrero Pérez, Óscar. "La decadencia de la novela en el siglo XVII: El ejemplo de Francisco Santos." *Anuario de estudios filológicos* 13 (1990): 27–38.

Barthes, Roland. *A Lover's Discourse: Fragments*. New York: Hill & Wang, 1978.

Battistini, Andrea. "The Antagonistic Affair between Literature and Science." *Interfacing Science, Literature, and the Humanities/ACUME 2*. Paola Spinozzi and Brian Hurwitz, eds. Göttingen: Vandenhoeck & Ruprecht, 2011, 61–71.

———. "The Telescope in the Baroque Imagination." *Reason and Its Others: Italy, Spain, and the New World*. David Castillo and Massimo Lollini, eds. Nashville: Vanderbilt University Press, 2006, 3–38.

Baumgartner, Frederick J. "Galileo's French Correspondents." *Annals of Science* 45 (1988): 169–82.

Bayarri, María. "Galileu Galilei en la literatura del segle XVII." *Mètode: Revista de difusió de la investigació de la Universitat de Valencia* 64 (2010): 56–57.

———. "Universos poéticos femeninos: Las amigas de Galileo Galilei." *Revista de la Sociedad Española de Italianistas* 2 (2004): 19–27.

Becedas González, Margarita, Cirilo Flórez, and María Jesús Mancho Duque, eds. *La ciencia y la técnica en la época de Cervantes.* Salamanca: Universidad de Salamanca, 2005.

Bedini, Silvio A. *The Pulse of Time: Galileo Galilei, the Determination of Longitude, and the Pendulum Clock.* Florence: Olschki, 1991.

Beltrán Marí, Antonio. *Talento y poder: Historia de las relaciones entre Galileo y la Iglesia católica.* Pamplona: Laetoli, 2006.

———. "'Una reflexión serena y objetiva': Galileo y el intento de autorrehabilitación de la Iglesia católica." *Arbor* 160, no. 629 (1998): 69–108.

———. *Revolución científica, Renacimiento e historia de la ciencia.* Madrid: Siglo XXI, 1995.

Bergmann, Emilie L. "Amor, óptica y sabiduría en Sor Juana." *Nictimene . . . sacrílega: Estudios coloniales en homenaje a Georgina Sabat-Rivers.* Mabel Moraña and Yolanda Martínez-San Miguel, eds. Mexico City: Universidad del Claustro de Sor Juana, 2003, 267–81.

———. "Sor Juana, Góngora and Ideologies of Perception." *Calíope: Journal of the Society for Renaissance and Baroque Hispanic Poetry* 18, no. 2 (2013): 116–38.

Biagioli, Mario. *Galileo's Instruments of Credit: Telescopes, Images, Secrecy.* Chicago: University of Chicago Press, 2006.

Bireley, Robert. *The Counter-Reformation Prince: Antimachiavellianism or Catholic Statescraft in Early Modern Europe.* Chapel Hill: University of North Carolina Press, 1990.

Blanchard, Jean Vicent. *L'optique du discourse au XVIIe siècle: De la rhétorique des jésuites au style de la raison moderne (Descartes, Pascal).* Saint-Nicolas: Presses de l'Université Laval, 2005.

Blanco, Mercedes. "Del infierno al Parnaso: Escepticismo y sátira política en Quevedo y Trajano Boccalini." *La perinola: Revista de investigación quevediana* 2 (1998): 155–94.

———. "El mecanismo de la ocultación: Análisis de un ejemplo de agudeza." *Criticón* 43 (1988): 13–36.

Blecua, Alberto. "Las *Repúblicas literarias* y Saavedra Fajardo." *El Crotalón: Anuario de filología española* 1 (1985): 67–97.

Blecua Perdices, José Manuel. *La poesía aragonesa del barroco.* Nueva Biblioteca de Autores Aragoneses. Zaragoza: Guara, 1980.

Bleichmar, Daniela, and Peter C. Mancall, eds. *Collecting Across Cultures: Material Exchanges in the Early Modern Atlantic World.* Philadelphia: University of Pennsylvania Press, 2011.

Blumenberg, Hans. "Light as a Metaphor for Truth." *Modernity and the Hegemony of Vision.* David M. Levin, ed. Berkeley: University of California Press, 1993, 30–62.

———. *The Genesis of the Copernican World.* Robert M. Wallace, trans. Cambridge: MIT Press, 1987.

Boadas Cabarrocas, Sònia. "Guerras panfletarias del siglo XVII: *Locuras de Europa* y sus fuentes." *Criticón* 109 (2010): 145–65.

Böhm, Günter. *Los sefardíes en los dominios holandeses de América del Sur y del Caribe, 1630–1750*. Frankfurt: Vervuert, 1992.

Boesky, Amy. "Milton, Galileo, and Sunspots: Optics and Certainty in *Paradise Lost*." *Milton Studies* 34 (1997): 23–43.

Bourland, Caroline. *Boccaccio and the "Decameron" in Castilian and Catalan Literature: Extrait de la "Revue hispanique,"* vol. 12. Paris: Macon, Protat Frères, Imprimeurs, 1905.

Bouza Álvarez, Fernando. *Imagen y propaganda: Capítulos de historia cultural del reinado de Felipe II*. Madrid: Akal, 1998.

———. "Coleccionistas y lectores: La enciclopedia de las paradojas." *La vida cotidiana en la España de Velázquez*. José Alcalá-Zamora, ed. Madrid: Temas de Hoy, 1995, 235–54.

Brady, Maura. "Galileo in Action: The 'Telescope' in *Paradise Lost*." *Milton Studies* 44 (2005): 129–52.

Briesemeister, Dietrich. "Antecedentes de la polémica en torno a la ciencia española." *Dos culturas en diálogo: Historia cultural de la naturaleza, la técnica y las ciencias naturales en España y América Latina*. Norbert Rehrmann and Laura Ramírez Sainz, eds. Frankfurt: Vervuert; Madrid: Iberoamericana, 2007, 39–57.

Brotton, Jerry. "Buying the Renaissance: Prince Charles's Art Purchases in Madrid, 1623." *The Spanish Match: Prince Charles's Journey to Madrid, 1623*. Alexander Samson, ed. Aldershot: Ashgate, 2006, 9–26.

Brown, Jonathan, and John Elliott, eds. *The Sale of the Century: Artistic Relations between Spain and Great Britain, 1604–1655*. New Haven: Yale University Press, 2002.

Brown, Kenneth. *Anastasio Pantaleón de Ribera (1600–1629): Ingenioso miembro de la república literaria española*. Madrid: Porrúa, 1980.

Brown, Piers. "'That Full-Sail Voyage': Travel Narratives and Astronomical Discovery in Kepler and Galileo." *The Invention of Discovery, 1500–1700*. James Dougal Fleming, ed. Aldershot: Ashgate, 2011, 15–28.

Buci-Glucksmann, Christine. *La folie du voir: De l'esthétique baroque*. Paris: Galilée, 1996.

Bustos Tovar, Eugenio. "La introducción de las teorías de Copérnico en la Universidad de Salamanca." *Real Academia de Ciencias Exactas, Físicas y Naturales* 67–68 (1973): 236–52.

Cacho, María Teresa. "Ver como vivir: El ojo en la obra de Gracián." Various eds., *Gracián y su época: Actas de la I reunión de filólogos aragoneses: ponencias y comunicaciones*. Zaragoza: Instituto Fernando el Católico, 1986, 117–135.

Cacho, Rodrigo. "La sátira en el Siglo de Oro: Notas sobre un concepto controvertido." *Neophilologus* 88, no. 1 (2004): 61–72.

Calvo Serraller, Francisco. *La teoría de la pintura en el Siglo de Oro*. Madrid: Cátedra, 2007.

Cámara, Alicia. "Madrid en el espejo de la corte." *Madrid, ciencia y Corte*. Antonio Lafuente y Javier Moscoso, eds. Madrid: Comunidad de Madrid, 1999, 63–74.

Campbell, Mary Baine. *Wonder and Science: Imagining Worlds in Early Modern Europe*. Ithaca: Cornell University Press, 1999.

Cañas, Dionisio. "El arte de bien mirar: Gracián." *Cuadernos Hispanoamericanos* 127 (1982): 127–39.

Carbajo Isla, María F. *La población de la Villa de Madrid desde finales del siglo XVI hasta mediados del siglo XIX*. Mexico City: Siglo XXI, 1987.

Cardenal Iracheta, Manuel. "Galileo y España." *Comentarios y recuerdos*. Madrid: Eds. de la Revista de Occidente, 1972, 15–24.

Careri, Giovanni. *Baroques*. Princeton: Princeton University Press, 2003.

Caro Baroja, Julio. *Vidas mágicas e Inquisición*. Madrid: Istmo, 1992, vol. 1.

Carrió-Invernizzi, Diana. *El gobierno de las imágenes: Ceremonial y mecenazgo en la Italia española del siglo XVII*. Frankfurt: Vervuert; Madrid: Iberoamericana, 2008.

Castillo, David R. *Baroque Horrors: Roots of the Fantastic in the Age of Curiosities*. Ann Arbor: University of Michigan Press, 2010.

———. *(A)wry Views: Anamorphosis, Cervantes and the Early Picaresque*. West Lafayette: Purdue University Press, 2001.

Caturla, María Luisa. "Documentos en torno a D. Juan de Espina, raro coleccionista madrileño, El testamento de 1624." *Arte español* 26 (1968–69): 5–8.

———. "Documentos en torno a D. Juan de Espina, raro coleccionista madrileño." *Arte español* 24 (1963–66): 1–10.

Chapman, Allan. "A New Perceived Reality: Thomas Harriot's Moon Maps." *Astronomy & Geophysics* 50, no. 1 (2009): 1.27–1.33.

Checa, Jorge. "Gracián and the Ciphers of the World." *Rhetoric and Politics: Baltasar Gracián and the New World Order*. Nicholas Spadaccini and Jenaro Talens, eds. Minneapolis: University of Minnesota Press, 1992, 170–87.

Cherchi, Paolo. "Suárez de Figueroa e la traduzione spagnola della *Piazza universale* di Garzoni." *Studi ispanici* 1 (1997–98): 75–84.

Cirnigliaro, Noelia. *Domus: Ficción y mundo doméstico en el Barroco español*. Woodbridge: Tamesis Books, 2015.

Clamurro, William. "Quevedo y la lectura política." *La perinola* 5 (2001): 95–105.

———. "*La Hora de todos* y la geografía política de Quevedo." *Actas del X Congreso de la Asociación Internacional de Hispanistas: Barcelona, 21 a 26 de agosto de 1989*. Antonio Vilanova, ed. Barcelona: Prensas y Publicaciones Universitarias, 1992, 841–48.

Clark, Stuart. *Vanities of the Eye: Vision in Early Modern European Culture*. Oxford: Oxford University Press, 2009.

Classen, Constance. *Worlds of Sense: Exploring the Senses in History and Across Cultures*. London: Routledge, 1993.

Close, Anthony J. *Cervantes and the Comic Mind of his Age*. Oxford: Oxford University Press, 2000.

Cohen, H. Floris. *The Scientific Revolution: A Historiographical Inquiry*. Chicago: University of Chicago Press, 1994.

Corteguera, Luis R. "Artisans and the New Science of Politics in Early Modern Europe." *Journal of Medieval and Early Modern Studies* 43, no. 3 (2013): 599–621.

Cotarelo y Mori, Emilio. *Don Juan de Espina; noticias de este célebre y enigmático personaje*. Madrid: Imprenta de la Revista de Archivos, 1908.

Crosby, James O. "Quevedo's Alleged Participation in the Conspiracy of Venice." *Hispanic Review* 23, no. 4 (1955): 259–73.

Crossley, Robert. *Imagining Mars: A Literary History*. Middletown: Wesleyan University Press, 2011.

Cruickshank, Don. *Don Pedro Calderón*. Cambridge: Cambridge University Press, 2009.

Curtius, Ernest. *European Literature and the Latin Middle Ages*. Princeton: Princeton University Press, 2013.

Czyzewski, Phyllis Eloys. *Picaresque and "Costumbrista" Elements in the Prose Works of Francisco Santos*. Doctoral dissertation, University of Illinois, 1975.

Daly, Peter M. "Emblems through the Magnifying Glass or Telescope." *Emblematica: An Interdisciplinary Journal of Emblem Studies* 18 (2010): 315–37.

D'Amico, Stefano. *Spanish Milan: A City within the Empire, 1535–1716*. Basingstoke: Palgrave Macmillan, 2012.

Dandelet, Thomas J. *La Roma española, 1500–1700*. Lara Vilà, trans. Barcelona: Crítica, 2002.

———, and John A. Marino, eds. *Spain in Italy: Politics, Society and Religion, 1500–1700*. Leiden: Brill, 2006.

Daston, Lorraine J., and Katharine Park. *Wonders and the Order of Nature, 1150–1750*. London: Zone, 1998.

Dear, Peter. *Revolutionizing the Sciences: European Knowledge and Its Ambitions, 1500–1700*. Princeton: Princeton University Press, 2001.

———. *Discipline and Experience: The Mathematical Way in the Scientific Revolution*. Chicago: University of Chicago Press, 1995.

———, ed. *The Literary Structure of Scientific Argument: Historical Studies*. Philadelphia: University of Pennsylvania Press, 1991.

Deffis de Calvo, Emilia Inés. "El viaje como modelo narrativo en la novela española del siglo XVII." *Filología (Temas de literatura española: Homenaje a Marcos A. Morínigo)* 26 (1993): 89–106.

Del Río Parra, Elena. *Una era de monstruos. Representaciones de lo deforme en el Siglode Oro español*. Frankfurt, Madrid: Vervuert, Iberoamericana, 2003.

Devitt, Michael. *Realism and Truth*. Princeton: Princeton University Press, 1997.

Díaz de Urma, J. Bosco. *La tercera dimensión del espejo: Ensayo sobre la mirada renacentista*. Seville: Universidad de Sevilla, 2004.

Dick, Hugh G. "The Telescope and the Comic Imagination." *MLN* 58, no. 7 (1943): 544–48.

Díez Borque, José María. "Teatro de palacio: Excesos económicos y protesta pública." *Literatura, política y fiesta en el Madrid de los Siglos de Oro*. José María Díez Borque, Esther Borrego Gutiérrez, and Catalina Buezo Canalejo, eds. Madrid: Visor Libros, 2006, 43–78.

Diksterhuis, Eduard Jan. *The Mechanization of the World Picture*. Oxford: Clarendon Press, 1961.

Dille, Glenn F. *Antonio Enríquez Gómez*. Boston: Twayne and G. K. Hall, 1988.

Dixon, Thomas. *Science and Religion: A Very Short Introduction*. Oxford: Oxford University Press, 2008.

Dixon, Victor. "Lope's Knowledge." *A Companion to Lope de Vega*. Alexander Samson and Jonathan Thacker, eds. Woodbridge: Tamesis, 2008, 15–28.

Doble Gutiérrez, Samuel. "La consagración del telescopio como instrumento científico." *Laguna: Revista de filosofía* 14 (2004): 165–84.

Domínguez, Julia. "Coluros, líneas, paralelos y zodíacos: La cosmografía en el *Quijote*." *Cervantes* 29, no. 2 (2009): 139–57.

Doncel, Manuel. "Juan Pablo II y los *studii galileiani*." *Largo campo di filosofare*. José

Montesinos and Carlos Solís, eds. Santa Cruz de Tenerife: Fundación Canaria Orotava de Historia de la Ciencia, 2001, 733–52.

Drake, Stillman. "Galileo in English Literature of the Seventeenth Century." *Galileo, Man of Science*. Ernan McMullin, ed. New York: Basic Books, 1968, 415–31.

Durán López, Fernando. *Juicio y chirinola de los astros: Panorama literario de los almanaques y pronósticos astrológicos españoles (1700–1767)*. Gijón: Ediciones Trea, 2015.

Eamon, William. "Brecht and the Historical Galileo." *Gestus: Journal of Brechtian Studies* 3 (1989): 19–23.

———. *Science and the Secrets of Nature: Books of Secrets in Medieval and Early Modern Culture*. Princeton: Princeton University Press, 1996.

Egginton, William. *The Theater of Truth: The Ideology of (Neo)Baroque Aesthetics*. Stanford: Stanford University Press, 2010.

———. "The Baroque as a Problem of Thought." *PMLA* 124, no. 1 (2009): 143–49.

Egido, Aurora. *De la mano de Artemia: Literatura, emblemática, mnemotecnia y arte en el Siglo de Oro*. Barcelona: José J. de Olañeta and Universitat de les Illes Balears, 2004.

———. "La historia de Momo y la ventana en el pecho." *Las caras de la prudencia y Baltasar Gracián*. Madrid: Castalia, 2000, 49–90.

———. *Cervantes y las puertas del sueño: Estudios sobre "La Galatea," "El Quijote" y "El Persiles."* Barcelona: Promociones y Publicaciones Universitarias, 1994.

———. *Fronteras de la poesía en el Barroco*. Barcelona: Crítica, 1990.

Egido, Teófanes. *Sátiras políticas de la España moderna*. Madrid: Alianza, 1973.

Elena, Alberto. *Las quimeras de los cielos: Aspectos epistemológicos de la revolución copernicana*. Madrid: Siglo XXI, 1985.

Elías, Carlos. "En la gran ciencia también hay literatura: Análisis de elementos literarios en las obras científicas de Galileo y Darwin." *Espéculo: Revista de estudios literarios* 33 (2006): https://pendientedemigracion.ucm.es/info/especulo/numero33/grancien.html.

Elliott, John H. *Imperial Spain, 1469–1716*. London: Penguin, 2002.

———. *El Conde-Duque de Olivares: El político en una época de decadencia*. Barcelona: Crítica, 1990.

Enoch, Jay M. "Introducción a la historia de las lentes y correcciones visuales: Una referencia a España y a los territorios del Nuevo Mundo." *Óptica avanzada*. María Luisa Calvo Padilla, ed. Barcelona: Ariel, 2002.

Étienvre, Jean-Pierre. *Márgenes literarios del juego*. London: Tamesis, 1990.

———, ed. *Las utopías en el mundo hispánico*. Madrid: Universidad Complutense de Madrid and Casa de Velázquez, 1990.

Etreros, Mercedes. *La sátira política en el siglo XVII*. Madrid: Fundación Universitaria Española, 1983.

Fernández Álvarez, Manuel. *Copérnico y su huella en la Salamanca del Barroco*. Salamanca: Universidad de Salamanca, 1974.

Fernández de Navarrete, Manuel. *Biblioteca marítima española*. Madrid: Viuda de Calero, 1851, 585–90.

Fernández Luzón, Antonio. "Galileo: La ciencia contra la inquisición." *Clío: Revista de historia* 68 (2007): 74–81.

Fernández-Santamaría, José Antonio. *Razón de Estado y política en el pensamiento español del Barroco (1595–1640)*. Madrid: Centro de Estudios Constitucionales, 1986.

Ferrer, Teresa. "*El golfo de las sirenas* de Calderón: Égloga y mojiganga." *Giornate Calderoniane, Calderón 2000: Atti del Convegno Internazionale Palermo 14–17 de dicembre 2000*. Enrica Cancelliere, ed. Palermo: Flaccovio Editore, 2003, 293–308.

Findlen, Paula. *Possessing Nature*. Berkeley: University of California Press, 1994.

Finocchiaro, Maurice A. *Retrying Galileo, 1633–1992*. Berkeley: University of California Press, 2007.

Firpo, Luigi. *I "Ragguagli di Parnaso" di Traiano Boccalini: Bibliografia delle edizioni italiane*. Florence: Sansoni antiquariato, 1955.

Fletcher, Angus. *Time, Space, and Motion in the Age of Shakespeare*. Cambridge: Harvard University Press, 2007.

Floristán Imízcoz, José Manuel. "Informe de Juan Bautista Labaña, cosmógrafo real, sobre el sistema de cálculo de la longitud de Galileo Galilei." *Lógos hellenikós: Homenaje al profesor Gaspar Morocho Gayo*. Jesús-María Nieto Ibáñez, ed. León: Universidad de León, Secretariado de Publicaciones y Medios Audiovisuales, 2003, vol. 1, 817–36.

Frago Gracia Juan A., and José A. García-Diego. *Un autor aragonés para Los veintiún libros de los ingenios y de las máquinas*. Zaragoza: Diputación General de Aragón, 1988.

Frangerberg, Thomas. "A Private Homage to Galileo: Anton Domenico Gabbiani's Frescoes in the Pitti Palace." *Journal of the Warburg and Courtauld Institutes* 59 (1996): 245–73.

Freedberg, David. *The Eye of the Lynx: Galileo, His Friends, and the Beginnings of Modern Natural History*. Chicago: University of Chicago Press, 2002.

Frye, Northrop. *Anatomy of Criticism: Four Essays*. Princeton: Princeton University Press, 2000.

Gal, Ofer, and Raz D. Chen-Morris. *Baroque Science*. Chicago: University of Chicago Press, 2013.

Gállego, Julián. *Visión y símbolos en la pintura española del Siglo de Oro*. Madrid: Cátedra, 1987.

Gallison, Peter. *Image and Logic: A Material Culture of Microphysics*. Chicago: University of Chicago Press, 1997.

García Gavilán, Inmaculada. "La *Fábula de Prometeo y Pandora* de Miguel (Daniel Leví) de Barrios: Unas notas sobre la diégesis mítica." *Lectura y signo* 10 (2010): 211–40.

García Gibert, Javier. *Sobre el viejo humanismo: Exposición y defensa de una tradición*. Madrid: Marcial Pons Historia, 2010.

García López, Jorge. "Quevedo y Saavedra: Dos contornos del seiscientos." *La perinola: Revista de investigación quevediana* 2 (1998): 237–62.

García Melero, José Enrique. *Literatura española sobre las artes plásticas*. Vol. 1, *Bibliografía aparecida en España entre los siglos XVI y XVIII*. Madrid: Ediciones Encuentro, 2002.

García Santo-Tomás, Enrique. *La musa refractada: Literatura y óptica en la España del Barroco*. Frankfurt: Vervuert; Madrid: Iberoamericana, 2014.

———. "Visiting the Virtuoso in Early Modern Spain: the Case of Juan de Espina." *Journal of Spanish Cultural Studies* 13, no. 2 (2012): 129–47.

———. "Ruptured Narratives: Tracing Defeat in Diego Duque de Estrada's *Comentarios del desengañado de sí mismo* (1614–1645)." *eHumanista: Journal of Iberian Studies* 17 (2011): 78–98.

———. "Saavedra Fajardo en la encrucijada de la ciencia." *Crítica Hispánica* 32, no. 2 (2010): 83–102.

———. "Outside Bets: Disciplining Gamblers in Early Modern Spain." *Hispanic Review* 77, no. 1 (2009): 147–64.

———. "Fortunes of the *Occhiali Politici* in Early Modern Spain: Optics, Vision, Points of View." *PMLA* 124, no. 1 (2009): 59–75.

———. *Modernidad bajo sospecha: Salas Barbadillo y la cultura material del siglo XVII*. Madrid: Consejo Superior de Investigaciones Científicas, 2008.

———. "Fragmentos de un discurso doméstico (pensar desde los interiores masculinos)." *Ínsula* 714 (2006): 1–2.

———. *Espacio urbano y creación literaria en el Madrid de Felipe IV*. Frankfurt: Vervuert; Madrid: Iberoamericana, 2004.

———. "Artes de la ciudad, ciudad de las artes: La invención de Madrid en *El diablo Cojuelo*." *Revista canadiense de estudios hispánicos* 25, no. 1 (2000): 117–35.

García Soormally, Mina. *Magia, hechicería y brujería: entre La Celestina y Cervantes*. Seville: Renacimiento, 2011.

García Tapia, Nicolás. *Un inventor navarro: Jerónimo de Ayanz y Beaumont (1553–1613)*. Pamplona: Universidad de Navarra, 2001.

———. "Los códices de Leonardo en España." *Boletín del Seminario de Estudios de Arte y Arqueología* 63 (1997): 371–95.

———. *Ingeniería y arquitectura en el Renacimiento español*. Valladolid: Universidad de Valladolid, Secretariado de Publicaciones e Intercambio Editorial, 1990.

———. *Técnica y poder en Castilla durante los siglos XVI y XVII*. Salamanca: Universidad de Salamanca, 1989.

Garin, Eugenio. *El zodiaco de la vida*. Barcelona: Península, 1981.

Gasta, Chad M. "Cervantes's Theory of Relativity in *Don Quixote*." *Cervantes* 31, no. 1 (2011): 51–82.

Gaukroger, Stephen. *The Emergence of a Scientific Culture: Science and the Shaping of Modernity, 1210–1685*. Oxford: Oxford University Press, 2007.

Gelabert, Juan. *Castilla convulsa, 1631–1652*. Madrid: Marcial Pons Historia, 2001.

Ginzburg, Carlo. *Wooden Eyes: Nine Reflections on Distance*. Martin Ryle and Kate Soper, trans. New York: Columbia University Press, 2001.

Glenn, Richard F. "The Optics of Illusions: Considerations of Fernández de Ribera's *Los anteojos de mejor vista*." William C. McCrary and José A. Madrigal, eds. *Studies in Honor of Everett W. Hesse*. Lincoln: Society of Spanish and Spanish-American Studies, 1981, 123–33.

Gómez-Centurión Jiménez, Carlos María. "La sátira política durante el reinado de Carlos II." *Cuadernos de historia moderna y contemporánea* 4 (1983): 11–34.

Gómez Moreno, Ángel. *España y la Italia de los humanistas: Primeros ecos*. Madrid: Gredos, 1997.

Gómez Trueba, Teresa. *El sueño literario en España: Consolidación y desarrollo del género*. Madrid: Cátedra, 1999.

González, Francisco J. *Astronomía y navegación en España, siglos XVI y XVII*. Madrid: Mapfre, 1992.

González-Cano, Agustín. "Eye Gymnastics and a Negative Opinion on Eyeglasses in the *Libro del exercicio* by the Spanish Renaissance Physician Cristóbal Méndez." *Atti della Fondazione Giorgio Ronchi* 49 (2004): 559–63.

———. "Un poema del Siglo de Oro español sobre los anteojos." *Óptica pura y aplicada* 37 (2004): 33–43.

González Cañal, Rafael. "El conde de Rebolledo y los albores de la Ilustración." *Criticón* 103–4 (2008): 69–80.

González de Amezúa, Agustín. "Un escritor olvidado: El Dr. Juan Enríquez de Zúñiga." *Opúsculos histórico-literarios*. Madrid: Consejo Superior de Investigaciones Científicas, 1951, vol. 1, 280–306.

González Echevarría, Roberto. "Infinito e improvisación en *La vida es sueño*." *Bulletin of the Comediantes* 66, no. 2 (2014): 141–60.

González García, José M. "Saavedra Fajardo, en los múltiples espejos de la política barroca." *Res publica* 19 (2008): 13–39.

González Molina, Encarna. "Bertolt Brecht, la dialéctica y la ciencia." *Bajo palabra: Revista de filosofía* 3 (2008): 185–94.

Goodman, David C. *Power and Penury: Government, Technology and Science in Philip II's Spain*. Cambridge: Cambridge University Press, 1988.

Goulet, Andrea. *Optiques: The Science of the Eye and the Birth of Modern French Fiction*. Philadelphia: University of Pennsylvania Press, 2006.

Grande Yáñez, Miguel. "La relevancia de la disimulación en Saavedra Fajardo." *Res publica* 19 (2008): 189–99.

Green, Otis H. *Spain and the Western Tradition: The Castilian Mind in Literature from "El Cid" to Calderón*. Madison: University of Wisconsin Press, 1963–1966, 4 vols.

———. "Bartolomé Leonardo de Argensola, secretario del Conde de Lemos." *Bulletin hispanique* 53, no. 4 (1951): 375–92.

———. "*Ni es cielo ni es azul*: A note on the 'barroquismo' of B. L. de Argensola." *Revista de filología española* 34 (1950): 137–50.

———. "The Literary Court of the Conde de Lemos at Naples, 1610–1616." *Hispanic Review* 1, no. 4 (1933): 290–308.

Greer, Margaret R. *The Play of Power: Mythological Court Dramas of Calderón de la Barca*. Princeton: Princeton University Press, 1991.

Guillén, Claudio. *El primer Siglo de Oro: Estudios sobre géneros y modelos*. Barcelona: Crítica, 1988.

Hafter, Monroe Z. "Toward a History of Spanish Imaginary Voyages." *Eighteenth-Century Studies* 8, no. 3 (1975): 265–82.

———. "Saavedra Fajardo plagiado en *El no importa de España* de Francisco Santos." *Bulletin hispanique* 61 (1959): 5–11.

Halstead, Frank "The Optics of Love: Notes on a Concept of Atomistic Philosophy in the Theatre of Tirso de Molina." *PMLA* 58 (1943): 108–21.

———. "The Attitude of Tirso de Molina Towards Astrology." *Hispanic Review* 9, no. 4 (1941): 417–39.

———. "The Attitude of Lope de Vega toward Astrology and Astronomy." *Hispanic Review* 7, no. 3 (1939): 205–19.

Hammond, John H. "Francisco Santos and Zabaleta." *Modern Language Notes* 66, no. 3 (1951): 166.

———. *Francisco Santos' Indebtedness to Gracián*. Austin: University of Texas Press, 1950.

———. "References to Cervantes in the Works of Francisco Santos." *Cervantes Quadricentennial* (1949): 100–102.

———. "A Plagiarium from Quevedo's *Sueños*." *Modern Language Notes* 64 (1949): 329–31.

Hampton, Timothy. *Fictions of Embassy: Literature and Diplomacy in Early Modern Europe*. Ithaca: Cornell University Press, 2009.

Haney, Robert Warren. *The Prose Satires of Rodrigo Fernández de Ribera*. Doctoral dissertation, University of Kentucky, 1982.

Harbison, Robert. *Reflections on Baroque*. Chicago: University of Chicago Press, 2000.

Harkness, Deborah E. *The Jewel House: Elizabethan London and the Scientific Revolution*. New Haven: Yale University Press, 2008.

Harries, Karsten. *Infinity and Perspective*. Cambridge: MIT Press, 2001.

Haynes, Roslynn Doris. *From Faust to Strangelove: Representations of the Scientist in Western Literature*. Baltimore: Johns Hopkins University Press, 1994.

Hazañas de la Rúa, Joaquín. *Biografía del poeta sevillano Rodrigo Fernández de Ribera, y juicio de sus principales obras*. Seville: Torres y Daza, 1889.

Helmstutler di Dio, Kelley. "The Chief and Perhaps Only Antiquarian in Spain: Pompeo Leoni and His Collection in Madrid." *Journal of the History of Collections* 18, no. 2 (2006): 137–67.

Henchman, Anna. "The Telescope as Prosthesis." *Victorian Review: An Interdisciplinary Journal of Victorian Studies* 35, no. 2 (2009): 27–32.

Heninger, S. K., Jr. *Touches of Sweet Harmony: Pythagorean Cosmology and Renaissance Poetics*. San Marino: Huntington Library, 1974.

Herrero García, Miguel. *Ideas de los españoles del siglo XVII*. Madrid: Gredos, 1966.

———. *El alumbrado en la casa española en el tiempo de los Austrias*. Madrid: Consejo Superior de Investigaciones Científicas, Instituto Jerónimo Zurita, "Diana" Artes Gráficas, 1954.

Hersey, George L. *Architecture and Geometry in the Age of the Baroque*. Chicago: University of Chicago Press, 2002.

Hitchcock, A. G. "Entertaining Strangers: A Dialogue between Galileo and Descartes." *Comparative Criticism: An Annual Journal* 20 (1998): 63–85.

Hurtado Torres, Antonio. "La 'Esphera' de Sacrobosco en la España de los siglos XVI y XVII: Difusión bibliográfica." *Cuadernos bibliográficos* 44 (1982): 50–51.

Iffland, James. "'Apocalipsis más tarde': Ideología y *La Hora de todos*." *Co-Textes* 2 (1981): 27–94.

———. *Quevedo and the Grotesque*. London: Tamesis, 1978, 2 vols.

Ilardi, Vincent. *Renaissance Vision from Spectacles to Telescopes*. Philadelphia: American Philosophical Society, 2007.

———. "Eyeglasses and Concave Lenses in Fifteenth-Century Florence and Milan: New Documents." *Renaissance Quarterly* 29 (1976): 341–60.

Israel, Jonathan I. *Empires and Entrepots: The Dutch, the Spanish Monarchy and the Jews, 1585–1713*. London: Hambledon Press, 1990.

———. *The Dutch Republic and the Hispanic World, 1601–1661*. Oxford: Oxford University Press, 1982.

Jalón, Mauricio. "El 'orden de las ciencias' en el siglo XVI y la *Plaza universal*." *Península: Revista de Estudios Ibéricos* 5 (2008): 65–82.

———. "Empresas científicas: Sobre las políticas de la ciencia en el siglo XVII." *Madrid, ciencia y Corte*. Antonio Lafuente and Javier Moscoso, eds. Madrid: Comunidad de Madrid, 1999, 155–70.

Jauralde, Pablo. *Francisco de Quevedo (1580–1645)*. Madrid: Castalia, 1998.

Jay, Martin. *Downcast Eyes: The Denigration of Vision in Twentieth-Century French Thought*. Berkeley: University of California Press, 1994.

Johnson, Carroll. *Matías de los Reyes and the Craft of Fiction*. Berkeley: University of California Press, 1973.

Joucla-Ruau, André. *Le tacitisme de Saavedra Fajardo*. Paris: Éditions Hispaniques, 1977.

Juárez Almendros, Encarnación. *Italia en la vida y obra de Quevedo*. New York: Peter Lang, 1990.

Kamen, Henry. *The Spanish Inquisition: A Historical Revision*. New Haven: Yale University Press, 1997.

Kelter, Irving A. "The Refusal to Accommodate: Jesuit Exegetes and the Copernican System." *Sixteenth Century Journal* 26, no. 2 (1995): 273–83.

Kemp, Martin. *The Science of Art: Optical Themes in Western Art from Brunelleschi to Seurat*. New Haven: Yale University Press, 1990.

Kent, Conrad. "Política en *La Hora de todos*." *Journal of Hispanic Philology* 2, no. 1 (1977): 99–119.

Kerr, Richard A. "Finally, an End for Galileo." *Science* 301, no. 5641 (2003): 1831.

Kimmel, Seth. "Interpreting Accuracy: The Fiction of Longitude in Early Modern Spain." *Journal of Medieval and Early Modern Studies* 40, no. 2 (2010): 299–323.

Kollerstrom, Nicholas. "Galileo's Astrology." *Largo campo di filosofare*. José Montesinos and Carlos Solís, eds. Santa Cruz de Tenerife: Fundación Canaria Orotava de Historia de la Ciencia, 2001, 421–32.

Koslofsky, Craig M. *Evening's Empire: A History of the Night in Early Modern Europe*. Cambridge: Cambridge University Press, 2011.

Kramer, Kirsten. "Mitología y magia óptica: Sobre la relación entre retrato, espejo y escritura en la poesía de Góngora." *Olivar: Revista de Literatura y Cultura Españolas* 9, no. 11 (2008): 55–86.

Kramer-Hellinx, Nechama. *Antonio Enríquez Gómez: Literatura y sociedad en El siglo pitagórico y Vida de don Gregorio Guadaña*. New York: Peter Lang, 1992.

Kuhn, Thomas S. *The Structure of Scientific Revolutions*. Chicago: University of Chicago Press, 1962. Published in Spanish as *La estructura de las revoluciones científicas*. Mexico City: Fondo de Cultura Económica, 1971.

———. *The Copernican Revolution: Planetary Astronomy in the Development of the Western Thought*. Cambridge: Harvard University Press, 1957.

Lacadena y Calero, Esther. "El discurso oral en las academias del Siglo de Oro." *Criticón* 41 (1988): 87–102.

Ladra, David. "El debate sobre ciencia y progreso: Galileo Galilei." *Primer acto: Cuadernos de investigación teatral* 282 (2000): 16–20.

Lambert, Gregg. *On the (New) Baroque*. Aurora: Davies Group, 2009.

Lara Alberola, Eva. *Hechiceras y brujas en la literatura española de los Siglos de Oro*. Valencia: Universitat de València, 2010.

Latour, Bruno. *Petites leçons de sociologie des sciences*. Paris: La Découverte, 1993.

Layna Ranz, Francisco. "Dicterio, conceptismo y frase hecha: A vueltas con el vejamen." *Nueva revista de filología hispánica* 44, no. 1 (1996): 27–56.

Levin, Michael J. *Agents of Empire: Spanish Ambassadors in Sixteenth-Century Italy.* Ithaca: Cornell University Press, 2005.

Levisi, Margarita. "Los aspectos teatrales de *El diablo Cojuelo.*" *Antigüedad y actualidad de Luis Vélez de Guevara: Estudios críticos.* George C. Peale, ed. Amsterdam: Benjamins, 1983, 207–18.

———. "Hieronymus Bosch y los *Sueños* de Francisco de Quevedo." *Filología* 9 (1963): 163–200.

Lewis, John. *Galileo in France: French Reactions to the Theories and Trial of Galileo.* Bern: Peter Lang, 2006.

Lezama Lima, José. "La curiosidad barroca." *Ensayos—La expresión americana: Obras completas.* Mexico City: Aguilar, 1977, 302–25.

Lima, Robert. *Dark Prisms: Occultism in Hispanic Drama.* Lexington: University Press of Kentucky, 1995.

Lindberg, David C. *Theories of Vision from al-Kindi to Kepler.* Chicago: University of Chicago Press, 1976.

López Piñero, José María. *Historia de la ciencia y de la técnica en la Corona de Castilla.* Vol. 3, *Siglos XVI y XVII.* José María López Piñero, dir. Salamanca: Junta de Castilla y León, Consejería de Educación y Cultura, 2002.

———. *Ciencia y técnica en la sociedad española de los siglos XVI y XVII.* Barcelona: Labor Universitaria, 1979.

———. "Galileo en la España del siglo XVII." *Boletín de la Sociedad Española de Historia de la Medicina* 5 (1965): 51–58.

———, Víctor Navarro Brotons, and Eugenio Portela Marco, comps. *Materiales para la historia de las ciencias en España, s. XVI–s. XVII.* Valencia: Pre-Textos, 1976.

López Poza, Sagrario. "La *Política* de Lipsio y las *Empresas Políticas* de Saavedra Fajardo." *Res publica* 19 (2008): 209–34.

Luciani, Frederick. *Literary Self-Fashioning in Sor Juana Inés de la Cruz.* Lewisburg: Bucknell University Press, 2004.

Lynn, Kimberly. *Between Court and Confessional: The Politics of Spanish Inquisitors.* London: Cambridge University Press, 2013.

Madroñal Durán, Abraham. *De grado y gracias: Vejámenes universitarios en los Siglos de Oro.* Madrid: Consejo Superior de Investigaciones Científicas, Instituto de la Lengua Española, 2005.

Mancho Duque, María Jesús. "La divulgación científica y sus repercusiones léxicas en la época de *El Quijote.*" *La ciencia y El Quijote.* José Manuel Sánchez Ron, dir. Barcelona: Crítica, 2005, 257–78.

Maravall, José Antonio. *Antiguos y modernos.* Madrid: Alianza, 1998.

———. *Culture of the Baroque: Analysis of a Historical Structure.* Terry Cochran, trans. Minneapolis: University of Minnesota Press, 1986.

———. "Saavedra Fajardo: Moral de acomodación y carácter conflictivo de la libertad." *Estudios de historia del pensamiento español: Serie tercera: Siglo XVII.* Madrid: Ediciones Cultura Hispánica, 1975, 161–96.

Marcaida López, José Ramón. *Arte y ciencia en el Barroco español: Historia natural, coleccionismo y cultura visual.* Madrid: Marcial Pons Historia, 2014.

Marchitello, Howard. "Telescopic Voyages: Galileo and the Invention of Lunar Cartography." *Travel Narratives, the New Science, and Literary Discourse, 1569–1750*. Judy A. Hayden, ed. Aldershot: Ashgate, 2012, 161–77.
———. *The Machine in the Text: Science and Literature in the Age of Shakespeare and Galileo*. Oxford: Oxford University Press, 2011.
Marotti, Ferruccio. *Lo spazio scenico: Teorie e tecniche scenografiche in Italia dall'età Barocca al settecento*. Roma: Bulzoni, 1974.
Martin, Adrienne Laskier. *An Erotic Philology of Golden Age Spain*. Nashville: Vanderbilt University Press, 2008.
Martín Vega, Arturo. *La obra narrativa de Andrés Dávila y Heredia*. Madrid: Universidad Complutense, 1998.
Martinengo, Alessandro. *La astrología en la obra de Quevedo: una clave de lectura*. Pamplona: Universidad de Navarra, 1992.
Martínez, Miguel. "'Quien me entendiere me declare': España, Holanda y los indios de América en *La Hora de todos*." *Voz y letra: Revista de literatura* 17, no. 1 (2006): 93–119.
Martínez Hernández, Santiago. *Rodrigo Calderón—La sombra del valido: Privanza, favor y corrupción en la corte de Felipe III*. Madrid: Marcial Pons Historia, 2009.
Mas i Usó, Pascual. *Academias y justas literarias en la Valencia barroca*. Kassel: Reichenberger, 1996.
Mataix, Carmen. "Galileo: La actualidad de un renacentista." *Largo campo di filosofare*. José Montesinos and Carlos Solís, eds. Santa Cruz de Tenerife: Fundación Canaria Orotava de Historia de la Ciencia, 2001, 131–38.
Mayer, Thomas F. *The Trial of Galileo, 1612–1633*. Toronto: University of Toronto Press, 2012.
Mayr, Otto. *Authority, Liberty & Automatic Machinery in Early Modern Europe*. Baltimore: Johns Hopkins University Press, 1989.
McAdam, Ian. "Milton, Satan, Galileo, and Gunpowder." *Notes and Queries* 55, no. 3 (2008): 289–91.
McKendrick, Melveena. "'Retratos, vidrios y espejos': Images of Honour, Desire and the Captive Self in the Comedia." *Revista canadiense de estudios hispánicos* 20, no. 2 (1996): 267–83.
McLuhan, Marshall. *Understanding Media: The Extensions of Man*. Cambridge: MIT Press, 1994.
Mele, Eugenio. "Tra viceré, csienziati e poeti . . ." *Bulletin hispanique* 31, no. 3 (1929): 256–67.
Melero Jiménez, Elisa Isabel. *La tarasca de parto en el mesón del infierno, y días de fiesta por la noche: Su autor Francisco Santos, criado del Rey Nuestro Señor, y natural de Madrid. Dedicada a Juan Díaz Rodero. Edición y estudio de los preliminares y del libro primero*. Presentation, Department of Spanish Philology, School of Philosophy and Literature, Universidad de Extremadura, September 25, 2004.
Menéndez y Pelayo, Marcelino. *Historia de los heterodoxos españoles*. Madrid: V. Suárez, 1911–1932.
Metz, Christian. *The Imaginary Signifier: Psychoanalysis and the Cinema*. Bloomington: Indiana University Press, 1986.
Moll, Jaime. "Diez años sin licencias para imprimir comedias y novelas en los reinos de Castilla, 1625–1634." *BRAE* 54 (1974): 97–103.

Morán Turina, José Miguel, and Fernando Checa Cremades. *El coleccionismo en España: De la cámara de maravillas a la galería de pinturas*. Madrid: Cátedra, 1985.

Morby, Edwin S. "Two Notes on *La Arcadia*." *Hispanic Review* 36, no. 2 (1968): 110–23.

———. "Franz Titelmans in Lope's *Arcadia*." *MLN* 82, no. 2 (1967): 185–97.

Mosley, Adam. *Bearing the Heavens: Tycho Brahe and the Astronomical Community of the Late Sixteenth Century*. Cambridge: Cambridge University Press, 2007.

Muñoz Marquina, Francisco. *Bibliografía fundamental sobre la literatura española (Fuentes para su estudio)*. Madrid: Castalia, 2003.

Mussio, Thomas E. "Galileo, the New Endymion: Progress and Knowledge in G. B. Marino's *Adone*." *Italian Quarterly* 38, nos. 147–8 (2001): 15–26.

Natal Álvarez, Domingo. "Galileo y el copernicanismo español: El caso de Diego de Zúñiga." *La filosofía española en Castilla y León: De los orígenes al Siglo de Oro*. Maximiliano Fartos Martínez and Lorenzo Velázquez Campo, coords. Valladolid: Universidad de Valladolid, 1997, 413–20.

Navarro Brotons, Víctor. "Galileo y España." *Largo campo di filosofare*. José Montesinos and Carlos Solís, eds. Santa Cruz de Tenerife: Fundación Canaria Orotava de Historia de la Ciencia, 2001, 809–30.

———. "La ciencia en la España del siglo XVII: El cultivo de las disciplinas físico-matemáticas." *Arbor* 153, nos. 604–5 (1996): 197–252.

———. "The Reception of Copernicus's Work in Sixteenth-Century Spain: The Case of Diego de Zúñiga." *Isis* 86 (1995): 52–78.

———. "El copernicanismo en España." *Historia 16* 23 (1978): 61–66.

———. "La renovación de las ciencias físico-matemáticas en la Valencia preilustrada." *Asclepio* 24 (1972): 367–70.

———, and William Eamon, "Spain and the Scientific Revolution: Historiographical Questions and Conjectures." *Más allá de la leyenda negra: España y la Revolución científica/Beyond the Black Legend: Spain and the Scientific Revolution*. Víctor Navarro Brotons and William Eamon, coords. Valencia: Consejo Superior de Investigaciones Científicas and Instituto de Historia de la Ciencia y Documentación, 2007, 27–38.

Navarro Pérez, Milagros. *Francisco Santos, un costumbrista del siglo XVII*. Madrid: Consejo Superior de Investigaciones Científicas, 1975.

Negredo del Cerro, Fernando. "Las atalayas del mundo: Los púlpitos y la explicación eclesiástica de la decadencia de la monarquía." *La declinación de la monarquía hispánica en el siglo XVII: Actas de la VIIa reunión científica de la Fundación española de Historia moderna*. Francisco José Aranda Pérez, ed. Cuenca: Ediciones de la Universidad de Castilla-La Mancha, 2004, 87–102.

Norton, Marcy. *Sacred Gifts, Profane Pleasures: A History of Tobacco and Chocolate in the Atlantic World*. Ithaca: Cornell University Press, 2008.

Oelman, Timothy. "The Religious Views of Antonio Enríquez Gómez: A Profile of a Marrano." *Bulletin of Hispanic Studies* 60 (1983): 201–9.

Okasha, Samir. *Philosophy of Science: A Very Short Introduction*. Oxford: Oxford University Press, 2002.

Page, Sophie. *La astrología en los manuscritos medievales*. London: British Library, 2006.

Palley, Julian. *The Ambiguous Mirror: Dreams in Spanish Literature*. Valencia: Albatros; Chapel Hill: Hispanófila, 1983.

Pardo Tomás, José. *Un lugar para la ciencia: Escenarios de práctica científica en la sociedad hispana del siglo XVI*. La Orotava, Tenerife: Fundación Canaria Orotava de Historia de la Ciencia, 2006.

———. *Ciencia y censura: La Inquisición española y los libros científicos en los siglos XVI y XVII*. Madrid: Consejo Superior de Investigaciones Científicas, 1991.

Patterson, Alan. "Calderón's 'Deposición a favor de los profesores de la pintura': Comment and Text." *Art and Literature in Spain, 1600–1800: Studies in Honour of Nigel Glendinning*. Charles Davis and Paul Julian Smith, eds. London: Tamesis, 1993, 153–66.

Peale, C. George. *La anatomía de "El diablo Cojuelo": Deslindes de un género anatomístico*. Chapel Hill: Department of Romance Languages, University of North Carolina, 1977.

Pérez, Joseph. *The Spanish Inquisition*. Janet Lloyd, trans. New Haven: Yale University Press, 2005.

Pérez Lasheras, Antonio. *Fustigat mores: Hacia el concepto de la sátira en el siglo XVII*. Zaragoza: Universidad de Zaragoza, 1994.

Pérez Magallón, Jesús. *Construyendo la modernidad: La cultura española en el tiempo de los novatores (1675–1725)*. Madrid: Consejo Superior de Investigaciones Científicas, 2002.

Peterson, Mark. *Galileo's Muse: Renaissance Mathematics and the Arts*. Cambridge: Harvard University Press, 2011.

Picatoste, Felipe. *Apuntes para una biblioteca científica española del siglo XVI*. Madrid: Ollero & Ramos, 1999.

Piccus, Jules. "El memorial al rey Don Felipe IV de D. Juan de Espina." *Anuario Musical* 41 (1986): 199–228.

Pimentel, Juan. "La monarquía hispánica y la ciencia donde no se ponía el sol." *Madrid, ciencia y Corte*. Antonio Lafuente and Javier Moscoso, eds. Madrid: Comunidad de Madrid, 1999, 41–62.

———, and José Ramón Marcaida. "Dead Natures or Still Lifes? Science, Art, and Collecting in the Spanish Baroque." *Collecting Across Cultures: Material Exchanges in the Early Modern Atlantic World*. Daniela Bleichmar and Peter C. Mancall, eds. Philadelphia: University of Pennsylvania Press, 2011, 99–120.

Piñeiro, Manuel Esteban. "La ciencia de las estrellas." *La ciencia y El Quijote*. José Manuel Sánchez Ron, dir. Barcelona: Crítica, 2005, 25–35.

Piñeiro, Mariano Esteban. "Las academias técnicas en la España del siglo XVI." *Quaderns d'història de l'enginyeria* 5 (2002–3): 10–19.

———. "La Casa de la Contratación y la Academia Real Matemática." *Historia de la ciencia y de la técnica en la Corona de Castilla*. Luis García Ballester, ed. Valladolid: Junta de Castilla y León, 2002, vol. 3, 35–52.

———. "Los cosmógrafos del Rey." *Madrid, ciencia y Corte*. Antonio Lafuente and Javier Moscoso, eds. Madrid: Comunidad de Madrid, 1999, 121–34.

———, and Félix Gómez Crespo. "La Primera versión castellana de *De Revolutionibus Orbium Caelestium*: Juan Cedillo Díaz (1620–1625)." *Asclepio* 43, no. 1 (1991): 131–62.

Popper, Karl. *The Logic of Scientific Discovery*. London: Routledge, 2002.

Portuondo, María M. *Secret Science: Spanish Cosmography and the New World*. Chicago: University of Chicago Press, 2009.

Psillos, Stathis. *Scientific Realism: How Science Tracks Truth*. London: Routledge, 1999.

Raimondi, Ezio. *El museo del discreto: Ensayos sobre la curiosidad y la experiencia en la literatura*. Madrid: Akal, 2002.

R. de la Flor, Fernando. "El cetro con ojos: Representaciones del 'poder pastoral' y de la monarquía vigilante en el Barroco hispano." *Visiones de la monarquía hispánica*. Víctor Mínguez, ed. Castellón: Universitat Jaume I, 2007, 57–87.

———. *Pasiones frías: Secreto y disimulación en el Barroco hispano*. Madrid: Marcial Pons Historia, 2005.

———. "El *Quijote* espectral: Desarreglos visuales y óptica anamórfica a comienzos del Seiscientos en el ámbito hispano." *Topos y tropos* 6 (2005). http://www.toposytropos.com.ar/N6/dossier_el_quijote/flor.htm.

———. *Barroco: Representación e ideología en el mundo hispánico (1580–1680)*. Madrid: Cátedra, 2002.

———. "*La ciencia del cielo*: Representaciones del saber cosmológico en el ámbito de la Contrarreforma española." *Millars: Espai i historia* 99 (1996): 91–121.

Reed, Cory. "War Machines: Instrumentality and Empire in Early Modern Spanish Drama." *Laberinto* 6 (2012): 57–83.

Reeves, Eileen. *Galileo's Glassworks: The Telescope and the Mirror*. Cambridge: Harvard University Press, 2008.

———. *Painting the Heavens: Art and Science in the Age of Galileo*. Princeton: Princeton University Press, 1997.

Rehn, Jürgen, ed. *Galileo in Context*. Cambridge: Cambridge University Press, 2002.

Reiss, Timothy J. *The Discourse of Modernism*. Ithaca: Cornell University Press, 1982.

Reula Baquero, Pedro. "Casa encantada, tramoyas y tropelías: Don Juan de Espina y Segundo de Chomón." *Tropelías: Revista de teoría de la literatura y de literatura comparada* 23 (2015): 407–43.

Rey Bueno, Mar, and Miguel López Pérez, eds. *The Gentleman, the Virtuoso, the Inquirer: Vincencio Juan de Lastanosa and the Art of Collecting in Early Modern Spain*. Cambridge: Cambridge Scholars Publishing, 2008.

Rey Pastor, Julio. *La ciencia y la técnica en el descubrimiento de América*. Barcelona: Austral, 1942.

Riandière La Roche Saint-Hilaire, Josette. *Recherches sur la penseé politique de Francisco de Quevedo Villegas: L'homme, l'historien, le pamphlétaire*. Paris: Université de la Sorbonne Nouvelle, 1993, 4 vols.

Riccini, Raimonda, ed. *Taking Eyeglasses Seriously: Art, History, Science and Technologies of the Vision*. Milan: Silvana, 2002.

Rico, Francisco. *El pequeño mundo del hombre: Varia fortuna de una idea en la cultura española*. Barcelona: Destino, 2005.

Robbins, Jeremy. "The Spanish Literary Response to the Visit of Charles, Prince of Wales." *The Spanish Match: Prince Charles's Journey to Madrid, 1623*. Alexander Samson, ed. Aldershot: Ashgate, 2006, 107–22.

———. *Arts of Perception: The Epistemological Mentality of the Spanish Baroque, 1580–1720*. Abingdon: Routledge, 2006.

Robinson, Maureen "The History and Myths Surrounding Johannes Hispalensis." *Bulletin of Hispanic Studies* 80, no. 4 (2003): 443–70.

Rodríguez Mansilla, Fernando. "El romance 'A Don Juan de Espina' de Castillo Solórzano: maravilla y *self-fashioning*." *Calíope* 14, no. 2 (2008): 5–26.

Rodríguez-Puértolas, Julio. "Francisco Santos y los mitos del casticismo hispano." *Studia Hispanica in Honorem Rafael Lapesa*. Dámaso Alonso, ed. Madrid: Cátedra, 1975, vol. 3, 419–30.

Romanoski, Christian. *Tacitus Emblematicus: Diego Saavedra Fajardo und seine Empresas Políticas*. Berlin: Wiedler Buchverlag, 2006.

Romero-González, Antonio Félix. *La sátira menipea en España: 1600–1699*. Doctoral dissertation, SUNY-Stony Brook, 1991.

Roncero López, Victoriano. "Sátira contra los venecianos de Francisco de Quevedo." *El Crotalón* 1 (1984): 359–72.

Rose, Constance H. "Antonio Enríquez Gómez and the Literature of Exile." *Romanische Forschungen* 85 (1973): 63–77.

Rosen, Edward. "The Invention of Eyeglasses." *Journal for the History of Medicine and Allied Sciences* 11 (1956): 13–46, 183–218.

Rossi, Paolo. *The Birth of Modern Science*. Chyntia De Nardi Ipsen, trans. Oxford: Blackwell, 2001.

Roth, Cecil. *The Spanish Inquisition*. New York: W. W. Norton, 1964.

Rudavsky, T. M. "Galileo and Spinoza: Heroes, Heretics, and Hermeneutics." *Journal of the History of Ideas* 62, no. 4 (2001): 611–31.

Ruiz Morales, Mario. "Pragmática astronómica del rey Felipe II." *Mapping* 136 (2009): 47–51.

Salas, Xavier de. *El Bosco en la literatura española*. Barcelona: Imprenta de J. Sabater, 1943.

Salcedo y Ruiz, Ángel. *La literatura española: Resumen de historia crítica*. 2nd ed. Madrid: Calleja, 1915–1917, vol. 2.

Sánchez, José. *Academias literarias del Siglo de Oro español*. Madrid: Gredos, 1961.

———. "Nombres que reemplazan a capítulo en libros antiguos." *Hispanic Review* 11, no. 2 (1943): 143–61.

Sánchez Cantón, Francisco J. "Los manuscritos de Leonardo que poseía don Juan de Espina." *Archivo español de arte* 14 (1940): 39–42.

Sánchez Jiménez, Antonio. "Algunos chistes astrológicos de Lope de Vega." *Criticón* 122 (2014): 41–52.

Sánchez Navarro, Jesús. "El juego de la imaginación: Galileo y la longitud." *Largo campo di filosofare*. José Montesinos and Carlos Solís, eds. Santa Cruz de Tenerife: Fundación Canaria Orotava de Historia de la Ciencia, 2001, 61–84.

Sánchez Ron, José Manuel. "Ciencia, técnica, Cervantes y *El Quijote*." *La ciencia y El Quijote*. José Manuel Sánchez Ron, dir. Barcelona: Crítica, 2005, 7–12.

San Emeterio Cobo, Modesto. "La patria de Don Juan de Espina." *Boletín de la Biblioteca Menéndez Pelayo* 34 (1958): 355–57.

Schama, Simon. *The Embarrassment of Riches: An Interpretation of Dutch Culture in the Golden Age*. New York: Vintage, 1997.

Schmidt, Benjamin. "Exotic Allies: The Dutch-Chilean Encounter and the (Failed) Conquest of America." *Renaissance Quarterly* 52, no. 2 (1999): 440–73.

Schwartz-Lerner, Lía. "Golden Age Satire: Transformations of Genre." *MLN* 105, no. 2 (1990): 260–82.

———. *Metáfora y sátira en la obra de Quevedo*. Madrid: Taurus, 1983.

Schmelzer, Felix K. E. "La utopía centífica del Siglo de Oro: El Estado ideal como tópico de la prosa científica y técnica en castellano (1526–1613)." *RILCE: Revista de Filología Hispánica* 30, no. 1 (2014): 201–19.

Selcer, Daniel. "The Mask of Copernicus and the Mark of the Compass: Bruno, Galileo, and the Ontology of the Page." *Thinking Allegory Otherwise*. Brenda Machosky, ed. Stanford: Stanford University Press, 2010, 60–86.

Serres, Michel. *Feux et signaux de brume: Zola*. Paris: Grasset, 1975.

Shapin, Steven. *The Scientific Revolution*. Chicago: University of Chicago Press, 1996.

Shea, William. *Galileo in Rome: The Rise and Fall of a Troublesome Genius*. Oxford: Oxford University Press, 2003.

———. "Galileo the Copernican." *Largo campo di filosofare*. José Montesinos and Carlos Solís, eds. Santa Cruz de Tenerife: Fundación Canaria Orotava de Historia de la Ciencia, 2001, 41–60.

Signorotto, Gianvittorio. *Milán español: Guerra, instituciones y gobernantes durante el reinado de Felipe IV*. Félix Labrador Arroyo, trans. Madrid: La Esfera de los Libros, 2006.

Simón Díaz, José. *Historia del Colegio Imperial de Madrid*. Madrid: Consejo Superior de Investigaciones Científicas, 1952–1959, 2 vols.

Simón de Guilleuma, José María. "Juan Roget, óptico español inventor del telescopio." *Actes du XIe Congrés International d'Histoire des Sciences (Barcelona-Madrid, 1959)*. Barcelona: Asociación para la Historia de la Ciencia Española; Paris: Hermann, 1960, 708–12.

Simón Palmer, Carmen. "La Cuaresma en el Palacio Real de Madrid." *Revista de dialectología y tradiciones populares* 43 (1988): 579–84.

Skal, David J. *Screams of Reason: Mad Science in Modern Culture*. New York: W. W. Norton, 1998.

Slater, John D., and Andrés Prieto, "Was Spanish Science Imperial?" *Colorado Review of Hispanic Literatures* 7 (2009): 3–10.

Sliwa, Krzysztof. *Cartas, documentos, y escrituras del Dr. Frey Lope de Vega Carpio*. Newark: Juan de la Cuesta, 2007.

Smet, Ingrid A. R. de. *Menippean Satire and the Republic of Letters, 1581–1655*. Geneva: Droz, 1996.

Smith, Pamela H. "Science on the Move: Recent Trends in the History of Early Modern Science." *Renaissance Quarterly* 62, no. 2 (2009): 345–75.

Smith, Wendell P. "'Ver Mundo': Enchanted Boats, Magic, and Imperial Atlases in the Second Part of *Don Quijote*." *Cervantes* 32.2 (2012): 37–81.

Snyder, Jon R. *Dissimulation and the Culture of Secrecy in Early Modern Europe*. Berkeley: University of California Press, 2009.

Solé-Leris, Amadeu. *The Spanish Pastoral Novel*. Boston: Twayne, 1980.

Spiller, Elizabeth A. "Reading through Galileo's Telescope: Margaret Cavendish and the Experience of Reading." *Renaissance Quarterly* 53, no. 1 (2000): 192–221.

Stein, Louise. *Songs of Mortals, Dialogues of the Gods: Music and Theatre in Seventeenth-Century Spain*. Oxford: Clarendon Press, 1993.

Stewart, Susan. *On Longing: Narratives of the Miniature, the Gigantic, the Souvenir, the Collection*. Durham: Duke University Press, 1993.

Stoll, André. "El antojo del pirata o la corrupción de las Indias: Revolución científica y 'episteme' conceptista en *La Hora de todos y Fortuna con seso de Quevedo*." *Voz y letra: Revista de literatura* 12, no. 2 (2001): 35–61.

Suvin, Darko. "*Vida de Galileo* de Bertolt Brecht: el alimento celestial negado." *ADE teatro: Revista de la Asociación de Directores de Escena de España* 126 (2009): 65–77.

Tato Puigcerver, José Julio. "El léxico científico de Quevedo (IV)." *La perinola: Revista de investigación quevediana* 14 (2009): 375–80.

———. "El léxico científico de Quevedo (III)." *La perinola: Revista de investigación quevediana* 8 (2004): 545–58.

———. "El léxico científico de Quevedo (II)." *La perinola: Revista de investigación quevediana* 7 (2003): 447–58.

———. "El léxico científico de Quevedo (I)." *La perinola: Revista de investigación quevediana* 6 (2002): 371–86.

———. "Una nota sobre Quevedo, Copérnico y Galileo." *Espéculo: Revista de estudios literarios* 16 (2000–2001). http://www.ucm.es/info/especulo/numero16/galileo.html.

Thomas, Richard. *Trajano Boccalini's "Ragguagli di Parnaso" and Its Influence upon English Literature*. Aberystwyth: Aberystwyth Studies, 1922.

Thorndike, Lynn. *The Sphere of Sacrobosco and Its Commentators*. Chicago: University of Chicago Press, 1949.

Tobar Quintanar, María José. "El decoro cómico del *Buscón*: Parodia de la 'Atalaya' de Mateo Alemán." *La perinola: Revista de investigación quevediana* 16 (2012): 259–79.

Topper, David. "Galileo, Sunspots, and the Motions of the Earth." *Isis* 90 (1999): 757–67.

Trabulse, Elías. *Los orígenes de la ciencia moderna en México (1630–1680)*. Mexico City: Fondo de Cultura Económica, 1994.

———. "Un científico mexicano del siglo XVII: Fray Diego Rodríguez y su obra." *Historia mexicana* 24 (1974): 36–39.

Tuck, Richard. *Philosophy and Government, 1572–1651*. Cambridge: Cambridge University Press, 1993.

Ulreich, John C. "Two Great World Systems: Galileo, Milton, and the Problem of Truth." *Cithara: Essays in the Judaeo-Christian Tradition* 43, no. 1 (2003): 25–36.

Vaíllo, Carlos. "Afinidades selectivas: De *La torre de Babilonia* de Antonio Enríquez Gómez a la 'Babilonia común' de Gracián en *El Criticón*." *Los conceptos de Gracián: Tercer Coloquio Internacional sobre Baltasar Gracián en ocasión de los 350 años de su muerte (Berlín, 27–29 de noviembre de 2008)*. Sebastian Neumeister, ed. Berlin: Verlag Walter Frey, 2010, 193–216.

———. "Un libro híbrido y marginado de pastores del siglo XVII: *El amor con vista*, de Juan Enríquez de Zúñiga." *Arkadien in den romanischen Literaturen*. Roger Friedlein, Gerhard Poppenberg, and Annett Volmer, eds. Heidelberg: Universitätsverlag Winter, 2008, 387–94.

———, and Ramón Valdés, eds. *Estudios sobre la sátira española en el Siglo de Oro*. Madrid: Castalia, 2006.

Valbuena Briones, Ángel. "El juego de los espejos en 'El divino Narciso,' de Sor Juana Inés de la Cruz." *RILCE: Revista de filología hispánica* 6, no. 2 (1990): 337–48.

Valdés, Ramón. "Rasgos distintivos y *corpus* de la sátira menipea española en su Siglo de Oro." *Estudios sobre la sátira española en el Siglo de Oro*. Carlos Vaíllo and Ramón Valdés, eds. Madrid: Castalia, 2006, 179–208.

———. *Los "Sueños y discursos" de Quevedo: El modelo del sueño humanista y el género de la sátira menipea*. Barcelona: Universidad Autónoma de Barcelona, 1990.

Vallés Belenguer, José. *Miguel de Cervantes y la física*. Zaragoza: Mira, 2007.

Valverde, Nuria, and Mariano Esteban Piñeiro. "El Colegio Imperial." *Madrid, ciencia y Corte*. Antonio Lafuente and Javier Moscoso, eds. Madrid: Comunidad de Madrid, 1999, 187–94.

Van Fraassen, Bas C. *The Scientific Image*. Clarendon Library of Logic and Philosophy. London: Oxford University Press, 1980.

Vega, Jesusa. *Ciencia, arte e ilusión en la España ilustrada*. Madrid: Consejo Superior de Investigaciones Científicas, Ediciones Polifemo, 2010.

Vélez Sainz, Julio. "De las líneas espirales a la 'Piramidal, funesta': Sor Juana Inés de la Cruz y la matemática post-euclidiana." *Bulletin of Hispanic Studies* 92, no. 5 (2015): 519–30.

Verene, Donald Phillip. *Philosophy and the Return to Self-Knowledge*. New Haven: Yale University Press, 1997.

Vicente García, Luis Miguel. "Torres Villarroel: El canto del cisne de la Astrología culta." *Edad de Oro* 31 (2012): 369–96.

———. "Lope y la polémica sobre astrología en el Seiscientos." *Anuario Lope de Vega* 15 (2009): 219–43.

———. *Estrellas y astrólogos en la literatura medieval española*. Madrid: Ediciones del Laberinto, 2006.

Vicente Maroto, María Isabel, and Mariano Esteban Piñeiro. "La ciencia: Interés de un monarca, indiferencia de un pueblo." *Aspectos de la ciencia aplicada en la España del Siglo de Oro*. Valladolid: Junta de Castilla y León, 2006, 493–521.

Vilar Berrogain, Jean. *Literatura y economía: La figura satírica del arbitrista en el Siglo de Oro*. Francisco Bustelo García del Real, trans. Madrid: Revista de Occidente, 1973.

Vinge, Louise. *The Five Senses: Studies in a Literary Tradition*. Lund: Royal Society of Letters at Lund, 1975.

Vives Coll, Antonio. *La influencia de Luciano de Samosata en el Siglo de Oro*. San Cristóbal de La Laguna: Universidad de La Laguna, 1950.

Vosters, Simon A. "Lope de Vega y Titelmans: Cómo el Fénix se representaba el universo." *Revista de literatura* 21 (1962): 5–33.

———. "Dos adiciones a mi artículo 'Lope de Vega y Titelmans.'" *Revista de literatura* 22 (1962): 90.

———. "Levinus Lemnius and Leo Suabius in *La Dorotea*." *Hispanic Review* 20, no. 2 (1952): 108–22.

Wagman, Frederick. *Magic and Natural Science in German Baroque Literature: A Study in the Prose Forms of the Later Seventeenth Century*. New York: Columbia University Press, 1942.

Warshawsky, Matthew. "A Spanish Converso's Quest for Justice: The Life and Dream Fiction of Antonio Enríquez Gómez." *Shofar: An Interdisciplinary Journal of Jewish Studies* 23, no. 3 (2005): 1–24.

———. *Longing for Justice: The New Christian Desengaño and Diaspora Identities of Antonio Enríquez Gómez*. Doctoral dissertation, Ohio State University, 2002.

Wescott, Howard B. "From Garcilaso to Argensola: The Cosmos Reconsidered." *Calíope: Journal of the Society for Renaissance & Baroque Hispanic Poetry* 10, no. 1 (2004): 55–67.

Whitby, William M. "Pinturas, retratos y espejos en la obra dramática de Luis Vélez de Guevara." *Estudios sobre el Siglo de Oro en homenaje a Raymond R. MacCurdy*. Ángel González et al., eds. Albuquerque: University of New Mexico; Madrid: Editorial Cátedra, 1983, 241–51.

Wilkinson, Alexander. "Exploring the Print World of Early Modern Iberia." *Bulletin of Spanish Studies* 89, no. 4 (2012): 491–506.

Williams, Robert Haden. *Boccalini in Spain: A Study of his Influence on Prose Fiction of the Seventeenth Century*. Menasha: George Banta Publishing, 1946.

Wilson, Edward M. "El texto de la 'Deposición a favor de los profesores de la pintura,' de D. Pedro Calderón de la Barca." *Separata de la Revista de archivos, bibliotecas y museos* 77, no. 2 (1974): 709–27.

Wood, Derek N. C. "Milton and Galileo." *Milton Quarterly* 35, no. 1 (2001): 50–52.

Woolard, Kathryn A., and E. Nicholas Genovese. "Strategic Bivalency in Latin and Spanish in Early Modern Spain." *Language in Society* 36, no. 4 (2007): 487–509.

Zagorin, Perez. *Ways of Lying: Dissimulation, Persecution and Conformity in Early Modern Europe*. Cambridge: Harvard University Press, 1990.

Zappala, Michael O. *Lucian of Samosata in the Two Hesperias: An Essay in Literary and Cultural Translation*. Potomac: Scripta Humanistica, 1990.

Zuese, Alicia R. "Devil, *Converso*, *Duende*: Anamorphosis and the View of Spain in Luis Vélez de Guevara's *El diablo Cojuelo*." *Hispania* 93, no. 4 (2010): 563–74.

Index

Page numbers followed by *f* indicate figures. Page numbers followed by *n* plus a number indicate endnotes.